Solubility in Pharmaceutical Chemistry

Also of interest

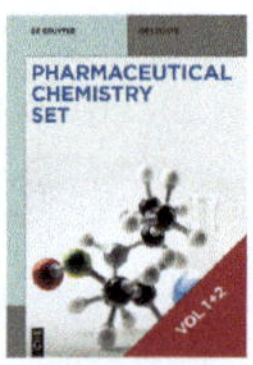

Pharmaceutical Chemistry.
Vol 1: Drug Design and Action
Vol 2: Drugs and Their Biological Targets
Campos Rosa, Camacho Quesada (Eds.), 2017
Set-ISBN 978-3-11-056680-2

Formulation Science and Technology.
Vol 1: Basic Theory of Interfacial Phenomena and Colloid Stability
Vol 2: Basic Principles of Formulation Types
Vol 3: Pharmaceutical, Cosmetic and Personal Care Formulations
Vol 4: Agrochemicals, Paints and Coatings and Food Colloids
Tadros, 2018
Set-ISBN 978-3-11-061201-1

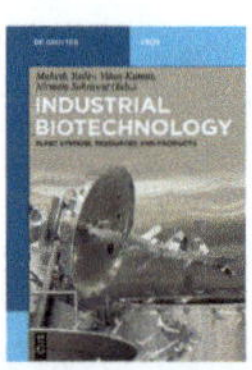

Industrial Biotechnology.
Plant Systems, Resources and Products
Yadav, Kumar, Sehrawat (Eds.), 2019
ISBN 978-3-11-056330-6, e-ISBN 978-3-11-056333-7

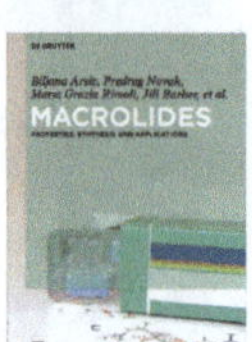

Macrolides.
Properties, Synthesis and Applications
Arsic, Novak, Kragol, Barber, Rimoli, Sodano, 2018
ISBN 978-3-11-051502-2, e-ISBN 978-3-11-051575-6

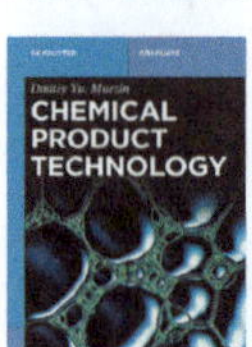

Chemical Product Technology.
Murzin, 2018
ISBN 978-3-11-047531-9, e-ISBN 978-3-11-047552-4

Solubility in Pharmaceutical Chemistry

Edited by
Christoph Saal, Anita Nair

DE GRUYTER

Editors
Dr. Christoph Saal
Merck KGaA
Frankfurter Str. 250
64293 Darmstadt
Germany
E-Mail: christoph.saal@merckgroup.com

Dr. Anita Nair
Merck KGaA
Frankfurter Str. 250
64293 Darmstadt
Germany
E-Mail: anita.nair@merckgroup.com

ISBN 978-3-11-054513-5
e-ISBN (PDF) 978-3-11-055983-5
e-ISBN (EPUB) 978-3-11-055888-3

Library of Congress Control Number: 2019947079

Bibliographic information published by the Deutsche Nationalbibliothek
The Deutsche Nationalbibliothek lists this publication in the Deutsche Nationalbibliografie; detailed bibliographic data are available on the Internet at http://dnb.dnb.de.

Cover image: SEAN GLADWELL <https://www.gettyimages.de/search/photographer?family=creative&photographer=SEAN+GLADWELL> / Moment / gettyimages.de
Typesetting: Integra Software Services Pvt. Ltd.
Printing and binding: CPI books GmbH, Leck

MIX
Papier aus verantwortungsvollen Quellen
FSC www.fsc.org FSC® C083411

www.degruyter.com

Contents

List of contributors

Annette Bauer-Brandl
Department of Physics, Chemistry and Pharmacy
University of Southern Denmark
Odense
Denmark
annette.bauer@sdu.dk

Wolfgang Beckmann
Bayer AG
Research & Development, Pharmaceuticals
Material Science
42096 Wuppertal
Germany
wolfgang.beckmann@googlemail.com

Joerg Berghausen
Fabrikstrasse 14.2.02-10
Novartis Campus
4002 Basel
Switzerland
joerg.berghausen@novartis.com

Martin Brandl
Department of Physics, Chemistry and Pharmacy
University of Southern Denmark
Odense
Denmark
mmb@sdu.dk

Sandrine Desrayaud
Fabrikstrasse.28.6.73.01
Novartis Campus
4002 Basel
Switzerland
sandrine.desrayaud@novartis.com

Stephanie Dodd
Novartis Institutes for BioMedical Research
250 Massachusetts Avenue
Cambridge, MA
United States of America
stephanie.dodd@novartis.com

Jennifer Dressman
Goethe University Frankfurt
Frankfurt
Germany
dressman@em.uni-frankfurt.de

Bernard Faller
Novartis
Institutes for BioMedical Research
Fabrikstrasse 14.2.02-01
Novartis Campus
4002 Basel
Switzerland
bernard.faller@novartis.com

David Elder
Hertford
Hertfordshire SG14 2DE
United Kingdom
davidelder2110@gmail.com

René Holm
Drug Product Development
Janssen Research and Development
Johnson & Johnson
Turnhoutseweg 30
2340 Beerse
Belgium
and
Department of Science and Environment
Roskilde University
4000 Roskilde
Denmark
rholm@ITS.JNJ.com

Marc Laisney
WSJ-386-10.48.58
Novartis Campus
4002 Basel
Switzerland
marc.laisney@novartis.com

https://doi.org/10.1515/9783110559835-201

James L. McDonagh
IBM Research UK
Hartree Centre
STFC Daresbury Laboratory
Sci-Tech Daresbury
Keckwick Lane
Daresbury Cheshire WA4 4AD
United Kingdom
james.mcdonagh@uk.ibm.com

John B. O. Mitchell
EaStCHEM School of Chemistry and
Biomedical Sciences Research Complex
University of St Andrews
North Haugh
St Andrews KY16 9ST
United Kingdom
jbom@st-andrews.ac.uk

David S. Palmer
Department of Pure and Applied Chemistry
University of Strathclyde
Thomas Graham Building
295 Cathedral Street
Glasgow
Scotland G1 1XL
United Kingdom
david.palmer@strath.ac.uk

Christos Reppas
Department of Pharmacy
National and Kapodistrian University of
Athens
157 84 Zografou
Athens
Greece
reppas@pharm.uoa.gr

Christoph Saal
Merck KGaA
Frankfurter Str. 250
64293 Darmstadt
Germany
christoph.saal@merckgroup.com

Rachael E. Skyner
Diamond Light Source Ltd.
Diamond House
Harwell Science and Innovation Campus
Fermi Ave
Didcot OX11 0DE
rachael.skyner@diamond.ac.uk

Klara Valko
School of Pharmacy
Bio-Mimetic Chromatography Ltd
Unit 5B BTC Bessemer Drive
Stevenage, Herts SG1 2DX
United Kingdom
and
UCL School of Pharmacy
29-34 Brunswick Square
London WC1N 1AX
United Kingdom
k.valko@ucl.ac.uk

Maria Vertzoni
Department of Pharmacy
National and Kapodistrian University of
Athens
157 84 Zografou
Athens
Greece
vertzoni@pharm.uoa.gr

David Elder

1 Solubility – definition and basic physicochemical considerations

1.1 Introduction

Solubility is often regarded as one of the most important attributes of a drug substance [1]. However, solubility of a solid active pharmaceutical ingredient (API) in a solvent (or mixed solvent) is a complicated phenomenon, and it is generally considered to be a dynamic equilibrium between the opposing forces of dissolution and reprecipitation. Under certain scenarios the equilibrium solubility may be exceeded to produce a supersaturated solution, which is metastable in nature [2]. The first stage in the process leading to a solution in an aqueous or organic solvent is disintegration of the crystal lattice and hydration or solvation of the API molecules. The thermodynamic driving force for this process is defined by the concentration gradient and resulting chemical–potential gradient between the solid (μ^s) and solid–liquid interface (μ^l). Then, the hydrated or solvated molecules diffuse from the "solid–liquid interface into the solution bulk phase" [3]. Similarly, the thermodynamic driving force for this latter process is defined by the concentration gradient and resulting chemical–potential gradient between the solid–liquid interface (μ^l) and the solution phase (μ^{sol}).

Solubility can be simplistically defined as the "amount of a substance that will dissolve in a given amount of another substance" [4]. This is often further refined as the amount of a solute that will dissolve in a given amount of solvent at a specified temperature and pressure. The latter caveats of temperature and pressure are important as most solutes become more soluble as the temperature increases, but the exact relationship is usually not simple [3].

However, these definitions omit an important factor, which is the nature of the solid-state form of the API. Dependent on the type of solubility measurement selected, this can change, as is typically seen with kinetic solubility or usually remain the same, that is, equilibrium solubility (see Table 1.1). IUPAC [5] tries to address this deficiency by defining solubility as "the analytical composition of a saturated solution expressed as a proportion of a designated solute in a designated solvent". The term "designated" implies no change in solid-state form, but this isn't implicitly stated. Solubility may be expressed in units of concentration, mole ratio, mole fraction, percentage, that is, 1% w/v, molality, or indeed other units [5].

Interestingly, changes in temperature play slightly different roles in the initial dissolution process depending on the intrinsic solubility of the API. For highly soluble compounds, it affects the diffusion rate constant and an increase in the intrinsic

David Elder, Hertford, Hertfordshire, SG14 2DE, UK

https://doi.org/10.1515/9783110559835-001

Table 1.1: Definitions of differing types of solubility.

Type of solubility measurement	Definition[1]
Kinetic	The concentration of a solute in solution when an induced precipitation first appears; this precipitate is often a thermodynamically metastable solid-state form.
Thermodynamic or equilibrium	A saturated solution in equilibrium with the thermodynamically stable solid-state form. No phase change occurs during the experiment if the thermodynamically stable solid-state form is introduced into the assay.
Intrinsic	The thermodynamic solubility at pH where API is in its neutral form (S_0).
Apparent	The solubility measured under given assay conditions.
Biorelevant (see Chapter 6)	The solubility measured using biorelevant media, for example, SGF, SIF, but more typically using FeSSGF, FaSSGF, FeSSIF, or FaSSIF media. Measurements are often performed at controlled body temperature, that is, 37 °C ± 1 °C

[1]Performed at controlled room temperature, that is, 25 °C ± 1 °C, unless specified otherwise.
SGF, simulated gastric fluid; SIF, simulated intestinal fluid; FeSSGF fed state simulated gastric fluid; FaSSGF, fasted state simulated gastric fluid; FaSSGF fed state simulated intestinal fluid; FaSSIF, fasted state simulated intestinal fluid.

thermodynamic driving force. In contrast, for poorly soluble APIs, it affects the surface reaction rate constant as well as the intrinsic thermodynamic driving force [3].

The pharmacopoeias such as the USP (United States Pharmacopoeia) [6] tend to describe solubility using much broader based terminology, for example, very soluble, freely soluble and soluble (see Table 1.2), which are based on the amount of solvent (in mL) needed to dissolve a specified amount of solute (1 g). The same

Table 1.2: USP definitions of solubility [6].

Descriptive term	Solubility (g/mL)
Very soluble	<1 part solvent needed to dissolve 1 part solute
Freely soluble	1–10 parts solvent needed to dissolve 1 part solute
Soluble	10–30 parts solvent needed to dissolve 1 part solute
Sparingly soluble	30–100 parts solvent needed to dissolve 1 part solute
Slightly soluble	100–1,000 parts solvent needed to dissolve 1 part solute
Very slightly soluble	1,000–10,000 parts solvent needed to dissolve 1 part solute
Practically insoluble	>10,000 parts solvent needed to dissolve 1 part solute

terminology and definitions that are used in the USP are equally applicable in other pharmacopoeias, for example, European Pharmacopoeia and British Pharmacopoeia. Although all the pharmacopoeias provide information on the solubility of majority of the test articles in specified solvents (typically water and certain stated organic solvents), the broad-based nature of these definitions renders this information to be less than useful for more than just a characterization of the respective substance.

1.2 Why is solubility important?

Generally, solubility plays a major role within pharmaceutical research and development with regard to different areas:
- Discovery, that is, utility in assay formats, for example, high-throughput screening (HTS)
- API manufacturing
- Formulation development for preclinical, clinical, and commercial formulations
- Drug bioavailability for per-oral drugs

1.2.1 Drug discovery

During "hit" identification and lead discovery phases of drug discovery, it is necessary to start to develop compound screening assays. This typically involves either (i) HTS of the company's entire compound library using biochemical or cell-based assays to screen for activity against the drug target and other proteins to get an understanding of selectivity of research compounds, (ii) fragment-based screening using small molecular weight (MW) compound libraries, or (iii) a tissue-based screening approach [7]. In all cases, compound solubility in DMSO (dimethyl sulfoxide) or, to a lesser extent, ethanol is required. These solvents are typically used because of their near universal solubilizing power and water miscibility [8]. Handling research compounds that are dissolved in such solvents facilitates compound handling to a large extent. Instead of handling and weighing of solid material, compounds can just be dosed by pipetting. This reduces time required for compound handling, allows for automation, and reduces consumption of research compounds.

The various compound libraries are typically stored as frozen DMSO solutions at storage conditions varying between −20 and 4 °C, at various concentrations (2–30 mM) [9]. These frozen solutions are then diluted further with buffers or water to perform the subsequent assays, which are typically performed at 1–10 μM concentrations. However, it is important to be aware of the final DMSO concentrations in these assays, as biochemical assays can be performed at DMSO concentrations of up to 10% v/v, whereas cell-based assays are much less tolerant and need DMSO concentrations of <1 % v/v.

However, sometimes the compound can show poor DMSO solubility [10]. Approximately, 10–20 % of compounds in compound libraries are not soluble in DMSO at the preferred concentrations [11]. In much the same way that the in silico prediction of aqueous solubility is useful in early-phase screening programmes, similar efforts to predict DMSO solubility have been undertaken [12]. In addition, DMSO solubility can change on storage. A combination of storage time, freeze–thaw cycling, DMSO hygroscopicity, and intrinsically low DMSO solubility can result in drug precipitation – often as a less soluble crystalline solid-state form [10]. Indeed, some researchers have advocated that concentrations of drugs in compound libraries should be reduced to 1 mM to address precipitation issues [13], whereas some researchers have also advocated automated storage in single-use mini-tubes [11].

Finally, DMSO has a non-linear effect on aqueous solubility of research compounds. Therefore, for a typical early-phase solubility assay utilizing 0.5–1.0 mL of aqueous buffer, only 10–50 μL of DMSO stock solution can be meaningfully added to the aqueous buffer component before the results become meaningless [14].

1.2.2 API manufacturing

Solubility in non-aqueous solvents at different temperatures is critical in selecting an appropriate solvent system for crystallization of the drug substance, which is a major factor in defining the purity and solid-state form, for example, polymorph, hydrate, solvate, co-crystal, or pharmaceutical salt of the drug substance [15]. These aspects will also be discussed in chapter 9 of this book. Also working with supersaturated solutions during API manufacturing without being aware of this can lead to uncontrolled precipitation of the API or precursors that can be difficult to control and lead to real manufacturing challenges. Nonetheless, the selection of the optimal solvent(s) and crystallization conditions for novel APIs is typically still mainly trial and error. However, in silico approaches aimed at optimizing solvent selection have seen greater utilization [16]. For example, a non-random two-liquid segment activity coefficient (NRTL-SAC) model was utilized for solvent selection as part of optimizing the crystallization process design. NRTL-SAC was used to screen crystallization solvents with the objective of optimizing API solubility and minimizing solvent usage. The NRTL-SAC model parameters for the candidate molecule are first identified from a small set of solubility experiments in selected solvents. The solubility behaviour of the API in other solvents and mixed solvents was then modelled. The optimal solvent systems were validated in the laboratory and utilized for process scale-up [16]. A more in-depth discussion of solubility in API manufacturing as well as in silico prediction of solubility will be given in chapter 10 of this book.

1.2.3 Formulation development for pre-clinical, clinical, and commercial formulations

Increased solubility can be achieved using several different formulation strategies. As high concentrations of the API are desirable in early animal experiments such as pharmacokinetic (PK) studies or pharmacodynamic (PD) studies, especially in toxicological studies that require administration of high doses and in human trials, realizing appropriate solubility of the API by formulations is key. As an example, high solubility of the API by a formulation can reduce the required administration volume and accordingly allow formulations that are more convenient to administer.

1.2.3.1 Using buffer systems to optimize solubility

pH also affects the solubility of ionizable drugs as it influences the degree of ionizability and the amount of drug present in the neutral and charged forms. The former is much less soluble than the latter based on Henderson–Hasselbalch equation [1]. Modification of the formulation pH is the simplest and most common approach to increasing the solubility of poorly soluble drugs [17]. Solubility enhancements of several orders of magnitude ($\geq 10^3$) can be readily achieved by modifying, then controlling the formulation pH (using buffer systems), at values of >3 pH units away from the respective pK_a [18]. Typically, strong acids or bases, for example, HCl or NaOH, will be used for making large changes in formulation pH, and buffer systems will be used to control the pH at the designated value. Citrates, acetates, phosphates, glycine, and TRIS (tris(hydroxymethyl)aminomethane) are commonly used buffer systems [17, 19]. The pH of maximal solubility isn't always the pH of optimal stability, and selection of the optimal formulation pH can involve “trade-offs” between solubility and stability.

1.2.3.2 Use of co-solvents to optimize solubility

Co-solvents are water-miscible solvents that enhance aqueous solubility. The most commonly used co-solvents for formulations are glycerine, propylene glycol, polyethylene glycol 400, DMSO, and ethanol. Typically, solubility increases in a logarithmic fashion with increasing fraction of the co-solvent. However, there may be physicochemical, regulatory, or safety considerations that constrain the absolute amount of the co-solvent within the formulation, particularly for paediatric use [17–19]. The EMA has recently published useful background information on propylene glycol and ethanol [20, 21].

Co-solvents are used in about one-sixth of all FDA-approved injectable products [22], and this figure is almost certainly higher now, given the increase in the numbers of poorly soluble APIs over the last two decades. Many of these injectable formulations are intended for infusion use and must be diluted with isotonic media, for example, saline and dextrose, prior to use. This significantly affects the ability of the co-solvent to maintain the drug in a solubilized form, with the inherent risk of precipitation.

1.2.3.3 Use of surfactants

Drugs with high lipophilicity can have poor wetting properties, and solubilization can be facilitated by surfactants. In addition, surfactants can solubilize poorly soluble drug molecules by micelle formation or by acting as co-solvents [23, 24]. Non-ionic surfactants are widely used, and some typical examples are polysorbate 20 and 80 (Tween 20 and 80), sorbitan monooleate 80 (Span 80), polyoxyl 40 stearate, solutol HS-15, polyoxyl 35 castor oil (Cremophor EL), polyoxyl 40 hydrogenated castor oil (Cremophor RH 40), D-α-tocopherol polyethylene glycol 1000 succinate (TPGS), and various polyglycol glycerides [18, 19]. The latter class of surfactants, for example, Softigen 767, Labrafil M-1944CS, Labrafil M-2125CS, Labrasol, and Gellucire 44/14, are useful in preparing lipid-based formulations that can significantly enhance solubility and thereby oral bioavailability using the various "self-emulsifying" systems, for example, self-emulsifying drug delivery systems [23–25]. Microemulsions, which are thermodynamically clear dispersions, can also be used to solubilize hydrophobic APIs [18, 19].

For drugs that are both hydrophobic and lipophilic, where a food effect may be encountered, a useful formulation strategy is to develop a softgel product [26, 27]. Here the drug is typically dissolved (although suspensions may be applicable if the dose is high) in a wide range of non-ionic surfactants, oils, and co-solvents. As such, solubility in these various lipidic vehicles will be important to ongoing development activities [26, 27]. However, accurate prediction of lipid solubility is complicated because interfacial effects can play a fundamental role in these formulations and the solubility can be affected by the lipid microstructure, that is, emulsions, oily solutions, micro-emulsions, nano-emulsions, and so on; as well as by the more fundamental physicochemical properties of the oil, surfactant, co-solvent, and the API [28].

1.2.3.4 Use of complexing agents

Complexation between a solute and a complexing agent can enhance the APIs aqueous solubility. The complexation reaction is dependent on relative size of the solute and the complexing agent, charge, and lipophilicity. Complexing agents form non-covalent inclusion complexes with the hydrophobic API or the most non-polar part of the API molecule within the complexation agent. In contrast to co-solvents, this has the advantage compared to other approaches that after dilution, a

1:1 complex will not precipitate. Complexation agents are typically pharmacologically inert and readily dissociate in the system or gastrointestinal tracts [29]. Cyclodextrins (CDs) are commonly used complexation agents [30]. They are α-(1–4) linked oligosaccharides comprising α-D-glucopyranose sub-units and they form a relatively hydrophilic outer surface (facilitating aqueous solubility), with a relatively hydrophobic inner surface that can accommodate the hydrophobic API. There are three types of CDs (α, β, and γ), which are comprised of 6, 7, or 8 subunits, and form cavities with diameters of 5.0 ± 0.3, 6.25 ± 0.25, and 7.9 ± 0.4 Å, respectively [28]. The β-form is the most commonly used, but covalent modifications (hydroxypropyl-β-CD or sulfobutylether-β-CD) can dramatically enhance the aqueous solubility [28]. For example, 400 mg/mL solubility with itraconazole (<5 μg/mL solubility in water) is achievable [29]. Common development themes for using CDs are low CD:drug ratios (<2:1), low dose (<100 mg), low drug solubility (<1 mg/mL), medium drug hydrophobicity (calculated log P, clog $P > 2.5$) and moderate binding constants (<5000 M^{-1}) [31].

1.2.4 Drug bioavailability for per-oral drugs

Aqueous solubility is also linked with the drugs' biopharmaceutical properties as discussed in Chapter 4 of this book. Thus, for oral drug products, solubility is required in biorelevant media, that is, gastric and intestinal fluids before a molecule can pass across a biological membrane of the intestine via either passive permeability or active transport. As such, without adequate biorelevant solubility, molecules can show solubility-limited absorption, with resultant non-linear kinetics [1] or insufficient bioavailability. The correlation of in vitro drug product solubility and in vivo bioavailability was first developed by Amidon et al. [32]. They developed a four-class system linking solubility and permeability properties to in vivo bioavailability. The four classes are shown in Table 1.3.

Table 1.3: 4-Box model for solubility and permeability: biopharmaceutical classification system (BCS) [32].

I	High solubility/high permeability	III	Low solubility/high permeability
II	Low solubility/high permeability	IV	Low solubility/low permeability

Compounds showing pK_a in the pH range of 1–8 tend to show pH-dependent solubility across the gastrointestinal (GI) tract. Tsume et al. [33] proposed a sub-classification of BCS II drugs into IIa and IIb. Both exhibit pH-dependant solubility; the former are weak acids, for example, naproxen and ibuprofen, that are poorly soluble at gastric pH but show good solubility at intestinal pH. In contrast, class IIb drugs are weak bases, for example, ketoconazole, that show the inverse solubility relationship. Interestingly, class IIb drugs are prone to supersaturation and/or precipitation as they move from the gastric into the intestinal compartments [34]. This will also be addressed in chapter 11 of this book.

1.3 In silico approaches

It is just over 20 years since the publication of Lipinski's seminal paper on experimental and computational, i.e., in silico approaches to estimate the solubility and permeability of drug candidates [35]. The iconic "Rule of 5" forecasts that absorption from the GI tract will be adversely impacted by several physicochemical parameters, including when the clog *P* is greater than 5, when MW is greater than 500 g/mol, when there are more than 5 H-bond donors or more than 10 H-bond acceptors. The related concept of "drug-likeness" importantly focused on both biological potency and physicochemical attributes, using tools such as lipophilic efficiency [36] or ligand efficiency [37]. This is important, as historically, biological potency was always seen as the most important parameter, and limited efforts were undertaken to try and simultaneously optimize the physicochemical attributes. Drug-likeness and related concepts are now widely used across the pharmaceutical industry to try and reduce the very high attrition rates currently seen with unprecedented pharmacological targets. Unfortunately, both combinatorial chemistry and HTS tend to favour leads with higher MW, higher clog *P*, and lower solubility [38].

As such, successful drug discovery strategies need to be a balance between optimizing both the "hydrophobicity-driven potency and hydrophilicity-driven biopharmaceutics properties" [38, 39]. Accordingly, an over-reliance on potency optimization resulting in non-optimal physicochemical properties will yield inferior ADMET (absorption, distribution, metabolism, excretion, and toxicity) properties and reduce the likelihood of clinical success [40]. Although the sub-optimal physicochemical characteristics can often be addressed using sophisticated formulation strategies [41, 42], and deficiencies in these properties can often still be rate limiting to the progression of drug candidates, particularly with respect to ADMET properties. Therefore, computational methodologies that can qualitatively predict certain physicochemical properties, for example, solubility, before a compound is even synthesized, based on molecular structural attributes are an essential requirement within drug discovery.

In silico approaches have also been applied to predicting solubility of the API in various organic solvents or mixtures of solvents [12, 43–45]. Computational models for predicting DMSO solubility showed a twofold decrease in the number of non-soluble compounds. However, a significant, that is, four- to ninefold increase was observed if only the most reliable predictions were considered. The structural features that influenced DMSO solubility were also assessed [43]. Models for predicting API solubility in various organic solvents (up to 85) have been described. The premise is that the relative partitioning of a solute between water and an immiscible organic solvent is given by the ratio of the solubilities in these solvents. Therefore, the solubility in an organic solvent can be predicted using the partition coefficient and solubility in water [44]. In addition, solubility prediction in mixed solvents, that is, water/co-solvent 1, water/co-solvent 1/co-solvent 2, using partial solubility parameters have been reported [45]. Within this book we have dedicated chapter 3 to the in silico prediction of solubility.

1.4 Relationships between solubility and physicochemical properties

There are significant numbers of in silico methods reported within the literature for predicting solubility from underlying molecular properties. However, these computational methodologies need to be able to cope with significant numbers of compounds and filter out “non-drug-like” compounds and/or attributes to focus chemistry initiatives on programmes with improved physicochemical attributes, thereby enhancing productivity. Importantly, it should be clearly appreciated that these early discovery methodologies will provide qualitative and not quantitative outcomes [46].

The intrinsic difficulties inherent in solubility prediction were graphically highlighted by the recent solubility challenge. An academic research group [47] measured the equilibrium solubility of 100 “drug-like” molecules under defined conditions, that is, fixed temperature(25 ± 2 °C), media (KCl buffer), and ionic strength (0.15 M). Utilizing this “training data set” they publicly requested other research groups to predict, using their own preferred computational approach(es), the intrinsic solubility of a further 32 “drug-like” compounds. The “training set” was selected to represent a broad chemical space with MW ranging from 115 (proline) to 645 (amiodarone), which had pK_a in the range of 1–12. The intrinsic solubility of the “training set” covered about seven orders of magnitude ranging from poorly soluble, that is, amiodarone to highly soluble compounds such as acetaminophen, with a relatively even distribution of intermediate values.

The authors received over 100 entries to the solubility challenge [48]. Participants used the full spectrum of available computational tools and approaches. Therefore, this solubility challenge provided an over-arching view of the industry’s ability to accurately predict aqueous solubility. However, the authors felt constrained in their ability to recommend an optimal approach. Rather they highlighted several methodologies that were equally successful at predicting aqueous solubility. Several participants in the solubility challenge were surprised that the simpler methodologies were better than the more complex approaches [49]. Some authors [50] went further claiming that any perceived advantages of these complex approaches were debatable, preferring instead a simple clog *P* correlation [51]. Hewitt et al. [49] highlighted that data quality was fundamental to the predictivity of any computational model. Indeed, even the "high-quality" data set provided by the organizers of the solubility challenge elicited concerns and questions concerning data quality. As such it is critical to recognize and understand the applicability domain, that is, the chemical space, where the model works best. Understandably, predictions made outside of this domain will be less reliable, but no "hard-and-fast" guidance can be provided.

Despite the impressive size of some of the participants' training sets, that is, in one case 46,000 compounds of known solubility, their methods still performed suboptimally for both soluble and insoluble compounds [52]. Interestingly, Kramer et al. [53] demonstrated enhanced solubility predictions with their meta-classifier approach, notwithstanding the fact that their "training set" was based on kinetic rather than equilibrium solubility. The authors showed a high prediction accuracy for the solubility of three quarters of these unknown compounds, but typically and perhaps unsurprisingly, they also showed a high bias, probably because their training set used small levels of DMSO as a co-solvent. However, despite this high level of predictivity, their model still only correctly predicted about one-third of the insoluble compounds in the data set. Finally, the accuracy of these in silico models needs to be further improved so that they mimic better the experimental determinations [40].

By far the biggest impediment to accurate solubility predictions is still the unpredictable nature of the solid-state forms, that is, presence of polymorphs, solvates and hydrates [54]. In other words, how to effectively model enthalpy and entropy within the system, that is, moving from an ordered, structured low entropy solid-state form to a disordered, unstructured high entropy solution state. Thus far, polymorphs still cannot be reliably predicted [55] and accordingly their effect on solubility cannot be accurately predicted by in silico tools. Case studies have shown that the reported solubility can be affected by a factor of 2 or more by factors such as temperature, differences in solid-state form, impurities, and water (in the case of solubility in anhydrous organic solvents) [56]. For more in-depth discussions, see chapters 9 and 10 of this book.

1.5 Solubility theory

Yalkowski and co-workers [57] derived the general solubility equation (GSE), to try and better model solubility:

$$\log S_o = -\log P - 0.01 * (\mathrm{MPt} - 25) + 0.5 \tag{1.1}$$

where S_o is the intrinsic solubility mg/mL, P is the octanol/water partition coefficient, and MPt is the melting point.

The GSE describes the influence of solvation energy, which the system gains after dissolution, arising from the log P term. Similarly, the crystal lattice energy that must be overcome prior to dissolution is addressed by the melting point term. Thus, the general concept as qualitatively introduced in Section 1.1. is defined quantitatively by the GSE.

However, the melting point term is only partially successful in helping to address solid-state complexity and its impact on aqueous solubility. It is also evident from the GSE that log P is the major variable in the GSE equation [57]. Indeed, medicinal chemists can usually modify log P far more easily than the melting point. This is because the melting point is more difficult to predict or indeed to control. The melting point today is not typically measured anymore during early discovery initiatives, as it was the case during the old days of medicinal chemistry. Consequently, optimizing log P tends to be the focus in many discovery organizations. Most marketed drugs have clog P of about 2.5 and it is probably no coincidence that this value also corresponds to the upper limit of "good solubility" predicted by the GSE [58]. Regrettably, poor aqueous solubility is therefore the logical outcome of introducing overly hydrophobic characteristics into potential new drug candidates.

The GSE limitation of clog P of >2.5 is probably the worst-case scenario as it does not accurately reflect the positive impact that ionization can have in improving aqueous solubility; therefore, replacing log P with log $D_{\mathrm{pH}\ 7.4}$ produces a more predictive GSE:

$$\log S_{\mathrm{pH}\,7.4} \quad = \ -\log D_{\mathrm{pH}\,7.4} - 0.01 * (\mathrm{MPt} - 25) + 0.5 \qquad (1.2)$$

Hill and Young [38] evaluated a large data set of ca. 20,000 compounds, utilizing measured log $D_{\mathrm{pH}\ 7.4}$, together with calculated values for hydrophobicity (i.e. clog P and clog $D_{7.4}$), accurate kinetic solubility measurements at pH 7.4, MW, and the number of aromatic rings in the molecules. The authors showed pronounced differences between the measured and calculated hydrophobicity with compounds of decreasing solubility. Indeed, poorly soluble compounds, that is, <30 µM showed a particularly bad correlation, that is R^2 = 0.11. This correlation improved slightly, that is, R^2 = 0.32, as the solubility increased from 30 to 200 µM, with the optimal correlation occurring with compounds exhibiting "good" solubility, that is, >200 µM, with R^2 = 0.462. Interestingly, these data supported the perspective that calculated log $D_{7.4}$ (or clog P) might be a better predictor of hydrophobicity rather than using the measured value [38].

Recently, the undesirable effects of aromaticity on aqueous solubility have been reported. These include the aromatic portion [59, 60], the number of aromatic rings [61, 62], and the percentage of sp^3 hybridized atoms [63] within the molecule. Molecules with limited lipophilicity are more likely to display poor aqueous solubility due to solid-state issues, i.e. 'brick dust molecules'; whereas highly lipophilic compounds are typically solubility limited due to inadequate solvation (poorly wetting), i.e. 'grease ball molecules' [60]. Numerous scenarios were modelled, and they showed that for compounds with a melting point of >250 °C and clog P of >2, the GSE establishes that solid-state considerations will prevail (over 50%); whereas, when the clog P is increased above 6, then the solid-state issues decrease markedly (about 25%). Thus, planar, flat and rigid molecules with extended ring systems have a high-likelihood (86%) of demonstrating reduced aqueous solubility [60]. How molecular planarity reduces aqueous solubility and how solubility can in turn be improved by modifying planarity has been evaluated by Ishikawa [64]. This is explainable by considering the increased lattice energy and consequently higher melting point that is arising from enhanced π- π stacking of the planar aromatic systems. Hill and Young [38] also demonstrated extended correlations between the number of aromatic ring systems and clog $D_{\mathrm{pH}\,7.4}$ (as opposed to log P) and ultimately aqueous solubility. Consequently, they proposed a solubility forecast index (SFI):

$$\mathrm{SFI} = \mathrm{clog}\, D_{\mathrm{pH}\,7.4} + \text{number of aromatic rings} \tag{1.3}$$

In those cases where SFI < 5, there is typically good aqueous solubility and the authors contended that each aromatic ring system was equivalent to one extra log unit of clog $D_{\mathrm{pH}\,7.4}$. They noted that the average number of aromatic ring systems in marketed oral products is 1.6 [38] and thus the average SFI would be 2.4.

Two key parameters that impact solubility but are not directly covered by these various "solubility" eqs. (1.1)–(1.3) are (i) purity and (ii) particle size. In the former case, impurities can affect the melting point term in eqs. (1.1) and (1.2), by introducing disorder into the crystal lattice and changing the chemical potential of the solid phase [65]. However, the nature of the impurities can also radically influence outcomes. Some impurities can increase solubility, whereas and perhaps counter-intuitively (given the above explanation), others can decrease solubility. Perhaps the best-known example of an impurity significantly decreasing aqueous solubility was that of ritonavir. The presence of a newly emerging, but poorly purging impurity was responsible for a four- to fivefold decrease in aqueous solubility. The impurity was a cis-geometrical isomer, whereas up to that point ritonavir in its known polymorph had exhibited trans-geometry. The impurity that was less soluble than the parent acted as a template during the crystallization process for the formation of a new conformational polymorph, which had cis-geometry [66]. Therefore, if the source, grade, or purity of either the solute or the solvent is modified in any way, then the solubility of the solute can be affected.

Particle size reduction is a well-known strategy for improving the bioavailability of poorly soluble compounds [26]. This approach increases the surface area that is available for dissolution and also increases the available surface energy. This in turn increases the dissolution rate, but typically not the solubility, at least not markedly unless the particle size is <1 μm [67].

The effect of particle size on solubility constant can be quantified as follows, using a modification of the Kelvin equation [68]:

$$\mathrm{Log}\,({}^{*}K_{A}) = \frac{\mathrm{Log}\,({}^{*}K_{A\to 0}) + \gamma A_{\mathrm{m}}}{3.454\,RT} \tag{1.3}$$

where $^{*}K_A$ is the solubility constant for the solute particles with the molar surface area A, $^{*}K_{A\to o}$ is the solubility constant for substance with molar surface area tending to zero (i.e. when the particles are large), γ is the surface tension of the solute particle in the solvent, A_{m} is the molar surface area of the solute (in $\mathrm{m}^2/\mathrm{mol}$), R is the universal gas constant, and T is the absolute temperature [69].

Nonetheless, particle size is rarely reported as being an important parameter in the determination of solubility. Indeed, accurate measurement of the equilibrium solubility of nano-sized drugs is often complicated by the inability to separate out a supernatant fraction, even after ultra-filtration or centrifugation, due to the presence of very small, suspended particles [26, 67]. Light scattering and turbidity measurements have been utilized to address this issue [67, 70].

1.6 Approaches to measuring solubility during different phases of research and development

Once the new chemical entity (NCE) has been initially synthesized in appropriate quantities, solubility can be measured the first time. Procedural approaches for solubility measurements at this early discovery stage vary from organization to organization. Solubility could be measured for every NCE developed by the organization or alternatively solubility could be measured upon request. However, whatever the process, solubility measurements will be required for a very large number of NCEs and therefore efficient procedures must be in place. There are two main purposes of measuring solubility at this stage [71]:

- The initial solubility measurement tries to answer the fundamental question: is the compound dissolved in the assay medium or has it precipitated out? This question is relevant for many types of assays, for example, biochemical and cellular assays that demonstrate the intrinsic activity of the compound. The same question also applies to assays that support non-clinical safety testing, which are now initiated at much earlier stages of research. In this case, low solubility of an NCE might result in a false negative and consequently hide safety-related risks of a compound or a whole series or scaffold.
- Second, solubility is an important parameter for compound optimization. The goal should be to deliver NCEs with appropriate solubility to ensure sufficient bioavailability and to simplify formulation development and clinical progression.

From a technical standpoint, delivering the required throughput to fulfil both objectives require a high degree of automation. The key to this – as for many other assay formats – is to use pre-dissolved compounds as described in Section 1.2.1. Typically, 10 mmol solutions in DMSO are utilized. This avoids handling of the solid material, which might be non-crystalline, oily, sticky, or highly electrostatic. This overcomes a potential tricky weighing stage and instead compound handling can be carried out by simple volumetric dispensing, that is, pipetting steps. Accordingly, it becomes feasible to implement solubility determinations on robotic systems that carry out manipulation such as volumetric dispensing, compound precipitation, and solid–liquid phase separation by filtration or centrifugation. Typically, these liquid handling systems can be combined with highly sensitive analytical systems, for example, high-performance liquid chromatography (HPLC) or ultra-HPLC (UPLC) utilizing generic methodologies and can be applied to automated solubility assessments with throughputs of 10–100 of compounds per day [65, 72–75]. See further discussions on the analytical approaches in Chapter 7.

However, one must bear in mind that this type of kinetic solubility does not answer the critical question "to what extent does my compound dissolve?" but instead provides the answer to the related question "to what extent does my compound precipitate?" As most drugs are intended for oral administration using solid dosage forms, the first question is more relevant during later research and development phases. The key differentiating point between kinetic solubility obtained using the pre-dissolved compound and thermodynamic solubility obtained using the solid compound is that metastable phases, that is, metastable polymorphs or amorphous phases, are often generated by the former technique. Solubility by the kinetic assay refers to these metastable forms, whereas the thermodynamically stable form will typically be used for further development. In a kinetic solubility assay, the compound will have only very limited time to precipitate out and accordingly will be mainly amorphous in nature. Consequently, solubility will be significantly higher compared to thermodynamic solubility, which typically utilizes the stable crystalline phase [76, 77].

Kinetic solubility is designed to facilitate high-throughput measurements, rather than necessarily providing accurate estimations of the true solubility. Consequently, turbidimetric or similar methods are often used, which allows the rapid determination of solubility using small amounts of compounds (5–50 μg) [78]. The main – and in many cases only – difference between kinetic and thermodynamic solubility assays is the use of DMSO stock solutions, rather than solid material. Handling steps for the solid materials can be difficult to automate and typically become more labour intensive and can constrain throughput of thermodynamic solubility assays. Assessment of solubility by the kinetic or thermodynamic solubility assays is typically limited to generic conditions such as one pre-defined buffer system, typically at neutral pH.

To get a physiologically more relevant understanding – especially for orally administered drugs, thermodynamic solubility can be measured using biorelevant conditions simulating the prevailing conditions within the GI tract. Initially, a pH-solubility profile is typically generated using different buffers simulating the different pH conditions encountered during the transit of the GI tract. However, a recent Innovative Medicines Strategy (IMI) Innovative tools for oral biopharmaceutics (OrBiTo) collaborative survey challenged the consensus that this is "typically" generated. Margolskee et al. [79] showed that in over one-fifth of cases, pH-solubility measurements are not performed. pH solubility evaluations use simple inorganic or organic buffer systems and they allow investigations of the pH-dependent solubility of the compound. A typical example was recently reported by Sieger et al. [80]. A robotic 96-well-plate automated method was utilized. A small quantity of accurately weighed solute (1–10 mg) was added to the appropriate well, and aqueous buffers (0.5–1.0 mL) of varying pH (typically in the physiological range, i.e. pH 1.2–6.8) were added. The wells were then shaken for 24 h, the contents filtered using 0.45 μm polytetrafluoroethylene (PTFE) filters and assayed using UV spectroscopy. Other standard approaches include the miniaturized shake-flask method [81], potentiometric titrations [82], and small-scale dissolution baths [83].

The standardized saturation shake flask (SSF) methodology for equilibrium solubility determinations was harmonized and validated by Baka et al. [84] and Völgyi et al. [85]. This approach involves accurately weighing the solute and adding it to an excess of buffered medium at controlled room temperature (25 ± 1 °C). The sample is vigorously stirred for 6 h and then left for a further 18 h to sediment, giving a total "incubation" time of 24 h, before centrifuging and sampling the supernatant and measuring the concentration. However, the total incubation time is often defined by the intrinsic dissolution rate of the solute, which in turn is dependent on morphology, crystallinity, particle size, wettability, quantity of solute added, and the intensity of the agitation [86]. In all cases, the equilibrium time must be shown to be appropriate and, sometimes, very long incubation times are needed. This can range from several days or longer. The CheqSol, that is, *Ch*asing *Eq*uilibrium *Sol*ubility approach represents a systematic method to assess the time required for equilibrium to be established [82, 87]. Long times that are required to reach equilibrium bring their own challenges, that is, analyte stability, pH stability, evaporation of solvent, even in some cases microbial contamination of the aqueous buffer. The reader is referred to the excellent review article of Brittain [87] for a more detailed overview of recommended approaches to improve data quality.

An additional approach that can be used to reduce long equilibration times for solutes with low dissolution rates is the facilitated dissolution method (FDM) [88]. This approach employs a small volume, that is, ≤1 % v/v of a second organic solvent that is totally immiscible in the aqueous phase, for example, *iso*-octane, octanol, and dichloromethane. The organic solvent partially solubilizes the solute thereby rapidly facilitating its equilibrium with the aqueous phase. As long as the system continues to contain three phases, aqueous, non-aqueous, and undissolved solid, the thermodynamic solubility is unaffected by the presence of the non-aqueous phase. The solubility of the solute in the water-immiscible organic phase used in the FDM approach should be at least two orders of magnitude greater than the corresponding solubility in the aqueous phase [86]. For lipophilic solutes, octanol is the best solvent; for less lipophilic solutes, 1,2-dichloroethane is the preferred solvent. Aliquots of the separated aqueous layer are removed, diluted as appropriate, and measured using UV-spectroscopic or HPLC methods. Takács-Novák et al. [86] demonstrated that the FDM method gave similar outcomes to the standard SFF approach. They also commented that the FDM approach can identify those scenarios where the inadequate equilibrium has been attained using the classical SSF method, that is, SSF > FDM, and where the quality of the data from the latter approach should be questioned.

1.7 Application of biopharmaceutical solubility approaches

The GI tract is a complicated biological system with discreet compartments, pronounced changes in pH, ionic strength, and the presence of naturally occurring surfactants, that is, bile acids [89, 90]. As such, even the aforementioned approaches to determine pH-dependent solubility do not provide the complete picture of solubility behaviour of the drug substance that will underpin drug absorption considerations.

Consequently, biorelevant solubility and phase stability, that is, inter-conversion between different salt forms (and polymorphs) in the GI tract should also be assessed. For example, conversion of the free base form of a weak base to the corresponding hydrochloride salt may take place within the acidic conditions found within the stomach. Alternatively, conversion of the designated salt into the less or more soluble hydrochloride salt can occur. Hydrolysis of the designated salt into the less soluble free base (or free acid) form can also occur in these biorelevant media. Although most of the transitions are reported in the literature within the context of dissolution assessments, they are equally or more germane due to the extended duration of the equilibrium solubility experiment, that is, ≥24 h (see Table 1.4 for overview). See chapter 11 on the relationship between solubility and dissolution rate.

Accordingly, there must be a sound knowledge of the phase-stability or conversion of the designated solid-state form during the equilibrium solubility exercise [91]. However, evidence of phase-conversions is still useful information as it almost certainly has biorelevant implications (see Chapter 9). In addition, the solubility of hydrochloride salts can be less than the corresponding free base [92], and other salts of that NCE due to the common ion effect in gastric media [93, 94]. Some examples of such phase transitions are provided in Table 1.4.

Indeed, the challenge of accurately measuring solubility of pharmaceutical salts has necessitated the development of novel computational approaches not dependent on explicit solubility equations, such as p-DISOL-X™ [91, 95, 96]. Salt solubility can also be dependent on experimental design [95]. The reader is referred to

Table 1.4: Overview of typical physical transitions that can be observed in biorelevant solubility determinations.

Typical physical transitions	Overview	Media utilized	Reference
Conversion of free form to salt	Haloperidol[1] free base converted to hydrochloride salt	pH 1.2	[93, 94]
	CI-1041 free base converted to less soluble hydrochloride salt	pH 1.2 (with tween 80)	[97]
Conversion of salt to hydrochloride salt	Haloperidol mesylate salt converted to hydrochloride salt	pH 1.2	[93, 94]
	E2050 dihydrochloride converted to mono-HCl salt	Water (various pH)	[98]
Disproportionation	Haloperidol mesylate salt converted to free base	pH > ca. 5	[93, 94]
	Compound A converted to parent over a 10-h period	SGF (pH 1.2)	[91]
	Compound A converted to metastable form of parent over a 30-min period	FaSSIF (pH 5.0)	[91]
	Bromocriptine mesylate partly converted to amorphous free base	Phosphate buffer (pH 6.5), FaSSIF (pH 6.5)	[99]
	Flurbiprofen tromethamine converted to flurbiprofen once the concentration of salt exceeded 4.3 mM	Water (pH 6.15)	[100]

[1]pH max of haloperidol system was ca. 5. pH max is the pH of maximum solubility for that salt system. At pH max, both salt and free form (base or acid) can coexist in the solid state.

the excellent review article of Brittain [87] for a more detailed overview of this issue; in particular, the case study covering the various haloperidol salts: hydrochloride, phosphate, and mesylate.

Equilibrium solubility is more appropriate than kinetic solubility for later stage development, where API supply is not constrained. However, from a biorelevant solubility perspective, kinetic solubility or indeed dissolution testing (see Chapter 11) can have some advantages. Or rather the time-based, that is, temporal nature of thermodynamic solubility should also be studied. This is because residence times in the stomach and small intestine are typically significantly less than 24 h (see Table 1.5). If the solubility is time dependant or changes with time, this will be biorelevant but will be not be captured using the classical 24-h-based equilibrium solubility methodologies.

Table 1.5: Residence time in stomach and small intestine compartments[1].

GI compartment	Residence time (fasted state) (h)	Residence time (fed state) (h)
Stomach (fundus) [101][1]	0.4	1.04
Stomach (antrum) [101]	0.32	1.58
Small intestine (proximal) [102]	2.00	–
Small intestine (distal) [102]	1.50	–
Small intestine	3.2 [103]	4.76 [103]

[1]Only the stomach and small intestine have been reported. The colon is significantly less important from an absorption perspective, having reduced the surface area and blood volumes.

There are two scenarios where the equilibrium solubility could change significantly over the designated 24-h time period. The first scenario is where there is a change in the solid-state form, typically from a more soluble (metastable) to less soluble (stable) form, that is, polymorph, salt, or co-crystal. Changes to more soluble forms have also been reported. He et al. [91] described µDISS solubility assessments of a zwittterionic NCE (compound A) over a 10-h period in SGF and 1.5 h in FaSSIF. The NCE was a sulfate salt with pK_a of 3.9 (basic) and 7.1 (acidic). In SGF, the authors observed 10-fold higher solubility over a 5-h time course compared to the parent. Thereafter, there was a gradual decrease in aqueous solubility until it equalled that of the parent. The authors showed that after 10 h, the salt had converted to the parent NCE. In contrast, in FaSSIF the initial solubility of the sulfate was 10-fold higher compared to the parent and after 30 min it decreased to fivefold higher and retained this value for the rest of the experiment. Interestingly, in FaSSIF media the residual solid was found to be a higher solubility, metastable polymorph of the parent. The relative bioavailability of the salt was assessed in fasted dogs and the exposure was fivefold greater than the parent. The authors indicated that if the equilibrium solubility had been measured at completion, that is, 24 h, these transitions would have missed. A second example of in situ conversion was recently reported for bromocriptine mesylate in pH 6.5 phosphate buffer and pH 6.5 FaSSIF using µDISS approach [99]. In this case, the undissolved solid was partially amorphous free base.

The second scenario is where the API is poorly wetting over GI-relevant timescales (<4 h), but it will adequately wet and thereby solubilize over an extended 24-h time period.

This is a relatively common phenomenon for "grease ball molecules", but it is typically under reported [104]. The use of surfactants to facilitate wetting of these hydrophobic APIs is well established, both in dosage form development and dissolution testing [19]. Certain researchers have used wetting kinetics as an alternative approach towards understanding the enhanced dissolution rate of poorly soluble drugs [105]. However, this issue rarely receives any focus within the scope of solubility testing. An additional reason for using biorelevant media, for example,

FaSSIF and FeSSIF, is that they contain naturally occurring surfactants, for example, bile acid salts, which facilitate wetting of hydrophobic APIs. However, even here the time course of the biorelevant solubility determination should be assessed. For a more in-depth discussion on biorelevant media, see Chapter 6. Interestingly, in a recent IMI OrBiTo collaborative initiative, biorelevant solubility information was missing in nearly three quarters of cases [79].

Finally, methods to predict drug absorption and bioavailability have been improved significantly over the last decade. This includes approaches that go beyond allometric scaling [106] and use Biopharmaceutical Classification System (BCS) approaches [32]. During the last few years, the BCS has been refined into the Development Classification System (DCS) [107]. These systems address solubility, permeability, and the dose of the drug. The DCS approach additionally considers dissolution rate and distinguishes between solubility-limited absorption and dissolution-limited absorption, thus providing insights in formulation strategies for poorly soluble compounds. The use of PK simulation software has also become widespread [108] and allows a more detailed understanding of the behaviour of the research compound in humans and animals, including dissolution, solubility, and an understanding of how a drug might precipitate in the GI tract.

1.8 Conclusion

Solubility is one of the most important physicochemical parameters that is used across pharmaceutical research and development. Although solubility is relatively easy to define, it can be difficult to predict using computational methods primarily due to solid-state constraints, for example, enthalpy and entropy considerations. In parallel, experimental methods to assess solubility have been improved and automated during recent years and currently allow measurement of solubility for large numbers of compounds. This holds especially true for measurement of kinetic solubility. The use of kinetic and thermodynamic solubility should be clearly distinguished as the erroneous use of kinetic solubility for compound optimization can be misleading.

However, in later stage of development when API availability is increased, thermodynamic (or equilibrium) solubility, typically over 24 h, becomes significantly more relevant. Additionally, to gain a deeper understanding of the role of solubility in drug absorption in animals and humans, methods and media to mimic in vivo behaviour, that is, behaviour in biorelevant media, have become more and more widespread and easier to use over the recent years. Similarly, just as the transit times in the gastric and small intestinal compartments influence absorption, the temporal aspects of thermodynamic solubility should also be assessed. For example, the solubility should be measured after times such as 1, 4, and 24 h. There is also a significant overlap between solubility and dissolution. This is particularly important for poorly wetting hydrophobic APIs, where the time course of solubility, which is often dictated by the wetting of the API surface, can significantly influence drug absorption.

Acknowledgements: This chapter is loosely based on an earlier article in *American Pharmaceutical Review*; Elder DP, C. Saal C. 2012. Solubility in pharmaceutical R&D: predictions and reality. *Am. Pharm. Rev.*, 2014, 17(1), 1–6.

References

[1] Bergström, CAS., Holm, R., Jørgensen, SA., Andersson, SB., Artursson, P., Beato, S., Borde, A., Box, K., Brewster, M., Dressman, J., Feng, KI., Halbert, G., Kostewicz, E., McAllister, M., Muenster, U., Thinnes, J., Taylor, R., & Mullertz, A. Early pharmaceutical profiling to predict oral drug absorption: Current status and unmet needs. Eur. J. Pharm. Sci. 2014, 57, 173–199.
[2] Myrdal, PB., & Yalkowsky, SH. 2007, Solubilization of drugs in aqueous media, Swarbick J, ed., Encyclopedia of Pharmaceutical Technology, 3rd Edition, Informa Healthcare, New York, USA, 2007, 3311.
[3] Paus, R., & Ji, Y. Modeling and predicting the influence of variable factors on dissolution of crystalline pharmaceuticals. Chem. Eng. Sci. 2016, 145, 10–20.
[4] Solubility, Merriam-Webster on-line dictionary. https://www.merriam-webster.com/dictionary/solubility (Accessed on 28 December 2017).
[5] IUPAC gold book, 2017. Solubility. https://goldbook.iupac.org/html/S/S05740.html. (Accessed on 28 December 2017).
[6] Description and solubility. USP 31/NF 26, 2008, 840.
[7] Hughes, JP., Rees, S., Kalindijian, SB, & Philpott, KL. Principles of early discovery. Brit. J. Pharmacol. 2011, 162(6), 1239–1249.
[8] Balakin, KV. DMSO solubility and bioscreening. Curr. Drug. Discovery 2003, 8, 27–30.
[9] Di, L., & Kerns, EH. Biological assay challenges from compound solubility: Strategies for bioassay optimization. Drug Discov. Today 2006, 11(9/10), 446–451.
[10] Lipinski, CA. Solubility in water and DMSO: Issues and potential solutions, Biotechnology: Pharmaceutical Aspects (Pharmaceutical Profiling in Drug Discovery for Lead Selection, Volume 1, Borchardt RT, et al, Eds, AAPS Press, 2004, 93–125.
[11] Oldenburg, K, Poooler, D, Scudder, D, Lipinski, CA, & Kelly, M. High throughput sonication: Evaluation for compound solubilization. Com. Chem. High Throughput Screen 2005, 8, 499–512.
[12] Balakin, KV., Savchuk, NP., & Tetko, IV. In silico approaches to prediction of aqueous and DMSO solubility of drug-like compounds: Trends, problems and solutions. Curr. Med. Chem. 2006, 13(2), 223–241.
[13] Waybright, TJ., Britt, JR., & McCloud, TG. Overcoming problems of compound storage in DMSO: Solvent and process alternatives. J. Biomol. Screen 2009, 14(6), 708–715.
[14] Ruell, J., & Avdeef, A. A measured solution. Mod. Drug Discovery 2003, June, 47–49.
[15] Miller, JM., Rodriguez- Hornedo, N., Blackburn, AC., Macikenas, D., & Collman, BM. Solvent Systems for Crystallization and Polymorph Selection, Augustijns P, Brewster ME, eds. Solvent Systems and their Selection in Pharmaceutics and Biopharmaceutics, Biotechnology: Pharmaceutical Aspects, Vol. VI, Springer, New York, NY, 2007, 53–19.

[16] Kokitkar, PB., Plocharczyk, E., & Che, C_C. Modeling drug molecule solubility to identify optimal solvent systems for crystallization. Org. Proc. Res. Dev. 2008, 12(2), 249–256.
[17] Rowe R., Sheskey PJ, Weller PJ. Editors. Handbook of pharmaceutical excipients, fifth edition, Pharmaceutical Press, London, United Kingdom, 2006.
[18] Strickley, RG. Solubilizing excipients in oral and injectable formulations. Pharm. Res. Pharm. Res. 2004, 21, 201–230.
[19] Ackers MJ Editor. Sterile drug products, Formulation, packaging, manufacturing and quality, Informa, New York, London, 2010.
[20] EMA. Questions and answers on propylene glycol and esters in the context of the revision of the guideline on 'Excipients in the label and package leaflet of medicinal products for human use' (CPMP/463/00 Rev.1). EMA/CHMP/704195/2013, 20th November 2014.
[21] EMA. Questions and answers on ethanol in the context of the revision of the guideline on 'Excipients in the label and package leaflet of medicinal products for human use' (CPMP/463/00). EMA/CHMP/507988/2013, 23rd January 2014.
[22] Nerma, S., & Brendl, RJ. Excipients and their use in injectable products. PDA J. Pharm. Sci. Technol. 2011, 51, 166–171.
[23] Hauss, DJ. Oral lipid based formulations. Adv. Drug Deliv. Rev. 2007, 59, 667–676.
[24] Poulton, CW., & Porter, CJH. Formulation of lipid-based delivery systems for oral administration: Materials, methods and strategies. Adv. Drug Deliv. Rev. 2008, 60, 625–637.
[25] Shah, NH., Carvajal, MT., Patel, CI., Infeld, MH., & Malick, AW. Self-emulsifying drug delivery system (SEDDS) with polyglycolyzed glycerides for improving in vitro dissolution and oral absorption of lipophilic drugs. Int. J. Pharm. 1884, 106, 15–23.
[26] Williams, HD., Trevaskis, NL., Charman, SA., Shanker, RM., Charman, WM., Poulton, CW., & Porter, CJH. Strategies to address low drug solubility in discovery and development. Pharmacol. Rev. 2013, 65, 315–499.
[27] Gullapalli, RP. Gelatin capsules (softgels). J. Pharm. Sci. 2010, 99(10), 4107–4148.
[28] Rane, SS., & Anderson, BD. What determines drug solubility in lipid vehicles: Is it predictable?. Adv. Drug Deliv. Rev. 2008, 60(6), 638–656.
[29] Panakanti, R., & Narang, AS. Impact of excipient interactions on drug bioavailability from solid dosage forms. Pharm. Res. 2012, 29, 2639–2659.
[30] Brewster, ME., & Loftsson, T. Cyclodextrins as pharmaceutical solubilizers. Adv. Drug Deliv. Rev. 2007, 59, 645–666.
[31] Carrier, RL., Miller, LA., & Ahmed, I. The utility of cyclodextrins for enhancing oral bioavailability. J. Contr. Rel. 2007, 123, 78–99.
[32] Amidon, GL., Lennernas, H., Shah, VP., & Crison, JR. A theoretical basis for a biopharmaceutical drug classification: The correlation of in-vitro drug product dissolution and in-vivo bioavailability. Pharm. Res. 1995, 12, 413–420.
[33] Tsume, Y., Mudie, DM., Langguth, P., Amidon, GE., & Amidon, GL. The biopharmaceutics classification system: Sub-classes for in vivo predictive dissolution (IPD) and IVIVC. Eur. J. Pharm. Sci. 2014, 2014, 57, 152–163.
[34] Dressman, JB., Thelen, K., & Jantratid, E. Towards quantitative prediction of oral drug absorption. Clin Pharmacokinet 2008, 47, 655–667.
[35] Lipinski, CA., Lombardo, F., Dominy, BW., & Feeney, PJ. Experimental and computational approaches to estimate solubility and permeability in drug discovery and development settings. Adv. Drug Deliv. Rev. 1997, 23, 3–25.
[36] Edwards, MP, & Price, DA. Role of physicochemical properties and ligand lipophilicity efficiency in addressing drug safety risk. Ann Rep Med Chem 2010, 45, 381–391.
[37] Hopkins, AL., Groom, CR., & Alex, A. Ligand efficiency: A useful metric for lead selection. Drug Discov. Today 2004, 9(10), 430–431.

[38] Hill, AP., & Young, RJ. Getting physical in drug discovery: A contemporary perspective on solubility and hydrophobicity. Drug Discov. Today 2010, 15, 648–655.
[39] Elder, DP., & Holm, R. Aqueous solubility: Simple predictive methods (in silico, in vitro and bio-relevant approaches). Int. J. Pharm. 2013, 453, 3–11.
[40] Gleeson, MP., Hersey, A., Montanari, D., & Overington, J. Probing the links between in vitro potency, ADMET and physicochemical parameters. Nat. Rev. Drug Discov. 2011, 10, 197–208.
[41] Elder, DP. Effective formulation development strategies for poorly soluble active pharmaceutical ingredients (APIs). Am. Pharm. Rev. 2010, 13, 6(1), 28–34.
[42] Liu, R. Water-Insoluble Drug Formulation, Second Edition, CRC Press, US, 2008.
[43] Tetko, IV, Novotarskyi, S, Sushko, I, Ivanov, V, Petrenko, AE, Dieden, R, Lebon, F, & Mathieu, B. Development of dimethyl sulfoxide solubility models using 163 000 molecules: Using a domain applicability metric to select more reliable predictions. J. Chem. Inf. Model 2013, 53(8), 1990–2000.
[44] Abraham, MH., Smith, RE., Lutchefeld, R., Boorem, AJ., Luo, R., & Acree, WE. Prediction of solubility of drugs and other compounds in organic solvents. J. Pharm. Sci. 2010, 99(3), 1500–1515.
[45] Jouyban, A., Shayanfar, A., Panahi-Azar, V., Yousefi, BH., Acree, WE., & York, P. Solubility prediction of drugs in mixed solvents using partial solubility parameters. J Pharm Sci 2011, 100(10), 4368–4382.
[46] Lipinski, CA., Lombardo, F., Dominy, BW., & Feeney, PJ. Experimental and computational approaches to estimate solubility and permeability in drug discovery and development settings. Adv. Drug Deliv. Rev. 2012, 64, 4–17.
[47] Llinàs, A., Glen, RC., & Goodman, JM. Solubility challenge: Can you predict solubilities of 32 molecules using a database of 100 reliable measurements?. J. Chem. Inf. Model 2008, 48, 1289–1303.
[48] Hopfinger, AJ., Esposito, EX., Llinàs, A., Glen, RC., & Goodman, JM. Findings of the challenge to predict aqueous solubility. J. Chem. Inf. Model 2009, 49, 1–5.
[49] Hewitt, M., Cronin, MTD., Enoch, SJ., Madden, JC., Roberts, DW., & Dearden, JC. In silico prediction of aqueous solubility: The solubility challenge. J. Chem. Inf. Model 2009, 49, 2572–2587.
[50] Faller, B., & Ertl, P. Computational approaches to determining drug solubility. Adv. Drug Deliv. Rev. 2007, 59, 533–545.
[51] Hansch, C., Quinlan, JE., & Lawrence, GL. Linear free energy relationship between partition coefficients and the aqueous solubility of organic liquids. J. Org. Chem. 1968, 33, 347–350.
[52] Cheng, T., Li, Q., Wang, Y., & Bryant, SH. Binary classification of aqueous solubility using support vector machines with reduction and recombinant feature selection. J. Chem. Inf. Model 2011, 51, 229–236.
[53] Kramer, C., Beck, B., & Clark, T. Insolubility classification with accurate prediction probabilities using a metaClassifier. J. Chem. Inf. Model 2010, 50, 401–414.
[54] Hilfiker R. Ed. Polymorphism in the pharmaceutical industry, Wiley-VCH, Weinheim. 2006.
[55] Chan, HCS., Kendrick, J., & Leusen, FJJ. Predictability of the polymorphs of small organic compounds: Crystal structure predictions of four benchmark blind test molecules. Phys. Chem Chem Phys. 2011, 13, 20361–20370.
[56] Black, S., Dang, L., Liu, C., & Wei, H. On the measurement of solubility. Org. Proc. Res. Dev. 2013, 17, 486–492.
[57] Sanghvi, T., Jain, N., Yang, G., & Yalkowski, SH. Estimation of aqueous solubility by the General Solubility Equation (GSE) the easy way. QSAR Comb. Sci. 2003, 22, 258–262.

[58] Leeson, PD., & Springthorpe, B. The influence of drug-like concepts on decision making in medicinal chemistry. Nat. Rev. Drug Discov. 2007, 6, 881–890.
[59] Lamanna, C., Bellini, M., Padova, A., Westerberg, G., & Maccari, L. Straightforward recursive partitioning model for discarding insoluble compounds in the drug discovery process. J. Med. Chem. 2008, 51, 2891–2897.
[60] Wassvik, CM., Holmén, AG., Draheim, R., Artursson, P., & Bergström, CAS. Molecular characteristics for solid-state limited solubility. J. Med. Chem. 2008, 51, 3035–3039.
[61] Ritchie, TJ., & MacDonald, SJF. The impact of aromatic ring count on compound developability – Are too many aromatic rings a liability in drug design?. Drug Discov. Today 2009, 14, 1011–1020.
[62] Ritchie, TJ., MacDonald, SJF., Young, RJ., & Pickett, SD. The impact of aromatic ring count on compound developability: Further insights by examining carbo- and hetero-aromatic and -aliphatic ring types. Drug Discov. Today 2011, 16, 164–171.
[63] Lovering, F., Bikker, J., & Humblet, C. Escape from the flatland: Increasing saturation as an approach to improving clinical success. J. Med. Chem. 2009, 52, 6752–6756.
[64] Ishikawa, M., & Hashimoto, Y. Improvement in aqueous solubility in small molecule drug discovery programs by disruption of molecular planarity and symmetry. J. Med. Chem. 2011, 54, 1539–1554.
[65] Alsenz, J., & Kansy, M. High throughput solubility measurement in drug discovery and development. Adv. Drug Deliv. Rev. 2007, 59(7), 546–567.
[66] Chemburkar, SR., Bauer, J., Deming, K., Spiwek, H., Patel, K., Morris, J., Henry, R., Spanton, S., Dziki, W., Porter, W., Quick, J., Bauer, P., Donaubauer, J., Narayanan, BA., Soldani, M., Riley, D., & McFarland, K. Dealing with the impact of ritonavir polymorphs on the late stages of bulk drug process development. Org. Proc. Res. Dev. 2000, 4(5), 413–417.
[67] Van Eerdenburgh, B., Vermant, J., Martens, JA., Froyen, L., Humbeeck, JV., Van Den Mooter, G., & Augustijns, P. Solubility increases associated with crystalline drug nanoparticles: Methodologies and significance. Mol. Pharm. 2010, 7, 1858–1870.
[68] Nelson, KG. The Kelvin equation and solubility of small particles. J. Pharm. Sci. 1972, 61(3), 479–480.
[69] Hefter GT, Tompkins RPT Eds.The experimental determination of Solubilities, Wiley-Blackwell, Chichester, 2003.
[70] Lindfors, L., Forrsen, S., Skantze, P., Zackrisson, A., & Olsson, U. Amorphous drug nanosuspensions: 2 Experimental determination of bulk monomer concentration. Langmuir 2006, 22, 911–916.
[71] Petereit, A., & Saal, C. What is the solubility of my compound? Assessing solubility for pharmaceutical research and development compounds. Am. Pharm. Rev. 2011, 14, 68–73.
[72] Hoellke, B., Arlt, M., & Saal, C. Comparison of nephelometric, UV-spectroscopic and HPLC methods for high-throughput determination of aqueous drug solubility in microtiter plates. Anal. Chem. 2009, 81, 3165–3172.
[73] Wan, H., & Holmen, AG. High-throughput screening of physicochemical properties and in-vitro ADME profiling in drug discovery. Comb. Chem. High Through Screen 2009, 12, 315–329.
[74] Pitt, A. High-throughput screening to determine aqueous drug solubility. Pharm. Disc. 2005, 5, 46–49.
[75] Kerns, EH. High-throughput physicochemical profiling for drug discovery. J. Pharm. Sci. 2001, 90, 1838–1858.
[76] Petereit, A., & Saal, C. Optimizing solubility: Kinetic versus thermodynamic solubility – Temptations and risks. Eur. J. Pharm. Sci. 2012, 47, 589–595.

[77] Sugano, K., Kato, T., Suzuki, K., Keiko, K., Sujaku, T., & Mano, T. High-throughput solubility measurement with automated polarized light microscopy analysis. J. Pharm. Sci. 2006, 95, 2114–2122.
[78] Blasko, A., Leahy-Dios, WO., Nelson, SA., Austin, RB., Killion, GC., Visor, IJ., & Massey, B. Revisiting the solubility concept of pharmaceutical compounds. Monatshefte für Chemie / Chemical Monthly 2001, 132(7), 789–798.
[79] Margolskee, A., Darwich, AS., Pepin, X., Aarons, L., Galetin, A., Rostami-Hodjegan, A., Carlert, S., Hammarberg, M., Hilgendorf, C., Johansson, P., Karlsson, E., Murphy, D., Tannergren, C., Thörn, H., Yasin, M., Mazuir, F., Nicolas, O., Ramusovic, S., Xu, C., Pathak, SM., Korjamo, T., Laru, J., Malkki, J., Pappinen, S., Tuunainen, J., Dressman, J., Hansmann, S., Kostewicz, E., He, H., Heimbach, T., Wu, F., Hoft, C., Laplanche, L., Pang, Y., Bolger, MB., Huehn, E., Lukacova, V., Mullin, JM., Szeto, KX., Costales, C., Lin, J., McAllister, M., Modi, S., Rotter, C., Varma, M., Wong, M., Mitra, A., Bevernage, J., Biewenga, J., Van Peer, A., Lloyd, R., Shardlow, C., Langguth, P., Mishenzon, I., Nguyen, MA., Brown, J., Lennernäs, H., & Abrahamsson, B. IMI-oral biophasrmaceutics tools project – Evaluation of bottom-up PBPK prediction success part 1: Characterisation of the OrBiTo database of compounds. Eur. J. Pharm. Sci. 2017, 96, 598–609.
[80] Sieger, P., Cui, Y., & Scheuerer, S. pH-dependent solubility and permeability profiles: A useful tool for prediction of oral bioavailability. Eur. J. Pharm. Sci. 2007, 105, 82–90.
[81] Wyttenbach, N., Alsenz, J., & Grassmann, O. Minutarized assay for solubility and residual solid screening (SORESOS) in early drug development. Pharm. Res. 2007, 24, 888–858.
[82] Stuart, M., & Box, K. Chasing equilibrium: Measuring the intrinsic solubility of weak acids and bases. Anal. Chem. 2005, 77, 983–990.
[83] Avdeef, A., & Tsinman, O. Minuaturized rotating disk intrinsic dissolution rate measurements: Effects of buffer capacity in comparisons to traditional Wood's apparatus. Pharm. Res. 2008, 25, 2613–2627.
[84] Baka, E., Comer, J., & Takács-Novák, K. Study of the equilibrium solubility measurement by saturation shake-flask method using hydrochlorothiazide as model compound. J. Pharm. Biomed. Anal. 2008, 46, 335–341.
[85] Völgyli, G., Baka, E., Box, K., Comer, J., & Takács-Novák, K. Study of pH-dependent solubility of organic bases. Revisit of Henderson-Hasselbalch relationship. Anal Chim Acta 2010, 673, 40–46.
[86] Takács-Novák, K., Urac, M., Hovárth, P., Völgyli, G., Anderson, BD., & Avdeef, A. Equilibrium solubility measurement of compounds with low dissolution rate by Higuchi's facilitated dissolution method. A validation study. Eur J Pharm Sci 2017, 106, 133–141.
[87] Brittain, H. Thermodynamic vs. kinetic solubility: Knowing which is which. Amer. Pharm. Rev. 2014. http://www.americanpharmaceuticalreview.com/Featured-Articles/160452-Thermodynamic-vs-Kinetic-Solubility-Knowing-Which-is-Which/ Accessed on 24 December 2017.
[88] Higuchi, T., Shih F-M, L., Kimura, T., & Rytting, JH. Solubility determination of barely aqueous-soluble organic solids. J. Pharm. Sci. 1979, 83, 423–428.
[89] Diakidou, M., Vertzoni, M., Goumas, K., Soederlind, E., Abrahamsson, B., Dressman, JB., & Reppas, C. Characterization of the contents of ascending colon to which drugs are exposed after oral administration to healthy adults. Pharm. Res. 2009, 26, 2141–2151.
[90] Kalantzi, L., Goumas, K., Kalioras, V., Abrahamsson, B., Dressman, JB., & Reppas, C. Characterization of the human upper gastrointestinal contents under conditions simulating bioavailability/bioequivalence studies. Pharm. Res. 2006, 23, 165–176.
[91] He, Y., Ho, C., Yang, D., Chen, J., & Orton, E. Measurement and accurate interpretation of the solubility of pharmaceutical salts. J Pharm Sci 2017, 106, 1190–1196.

[92] Miyazaki, S., Oshiba, M., & Nadai, T. Unusual solubility and dissolution behavior of pharmaceutical hydrochloride salts in chloride containing media. Int J Pharm 1980, 6(1), 77–85.
[93] Li, S., Wong, S., Sethia, S., Almoazen, H., Joshi, YT., & Serajuddin, ATM. Investigation of solubility and dissolution of a free base and two different salt forms as a function of pH. Pharm. Res. 2005, 22(4), 628–635.
[94] Li, S., Doyle, P., Metz, S., Royce, AE., & Serajuddin, ATM. Effect of chloride ion on dissolution of different salt forms of haloperidol, a model basic drug. J Pharm Sci 2005, 94(10), 2224–2231.
[95] Völgyi, G, Marosi, A, Takács-Novák, K, & Avdeef, A. Salt solubility products of diprenorphine hydrochloride, codeine and lidocaine hydrochlorides and phosphates – Novel method of data analysis not dependent on explicit solubility equations. ADMET DMPK 2013, 1(4), 48–62.
[96] Avdeef, A. Suggested improvements for measurement of equilibrium solubility-pH of ionizable drugs. ADMET DMPK 2013, 3(2), 84–109.
[97] Chen, LR., Wesley, JA., Bhattachar, S., Ruiz, B., Bahash, K., & Babu, SR. Dissolution behavior of a poorly water soluble compound in the presence of tween 80. Pharm. Res. 2003, 20(5), 797–801.
[98] Wang, Z., Burrell, LS., & Lambert, WJ. Solubility of E2-50 at various pH: A case study in which the apparent solubility is affected by the amount of excess solid. J. Pharm. Sci. 2002, 91(6), 1445–1455.
[99] Andersson, SBE., Alvebratt, C., Bevernage, J., Bonneau, D., Da Costa Mathews, C., Dattani, R., Edueng, K., He, Y., Holm, R., Madsen, C., Müller, T., Muenster, U., Müllertz, A., Ojala, K., Rades, T., Sieger, P., & Bergström, CA. Interlaboratory validation of small-scale solubility and dissolution measurements of poorly water soluble drugs. J. Pharm. Sci. 2016, 105(9), 2864–2872.
[100] Anderson, BD., & Conradi, RA. Predictive relationships in the water solubility of salts of a nonsteroidal anti-inflammatory drug. J. Pharm. Sci. 1985, 74(8), 815–820.
[101] Bergstrand, M, Soderlind, E, Weitschies, W, & Karlsson, MO. Mechanistic modeling of a magnetic marker monitoring study linking gastrointestinal tablet transit, in vivo drug release, and pharmacokinetics. Clin. Pharmacol. Ther. 2009, 86(1), 77–83.
[102] Worsøe, J., Fynne, L., Gregersen, T., Schlageter, V., Christensen, LA., Dahlerup, JF., Rijkhoff, NJM., Laurberg, S., & Krogh, K. Gastric transit and small intestinal transit time and motility assessed by magnetic tracking system. BMC Gastroenterol 2011, 11, https://doi.org/10.1186/1471-230X-11-145, Accessed on 24 January 2018.
[103] Mudie, DM., Amidon, GL., & Amidon, GE. Physiological parameters for oral delivery and in vitro testing. Mol. Pharm. 2010, 7(5), 1388–1405.
[104] Kuentz, M., Egloff, P., & Rothlisberger, D. A technical feasibility study of surfactant-free drug suspensions using octenyl succinate-modified starches. Eur. J. Pharm. Biopharm. 2005, 63(1), 37–43.
[105] Verma, S., & Rudraradju, VS. Wetting kinetics: An alternative approach towards understanding the enhanced dissolution rate for amorphous solid dispersion of a poorly soluble drug. AAPS Pharm. Sci.Tech. 2015, 16(5), 1079–1090.
[106] West, GB., Brown, JH., & Enquist, BJ. A general model for the origin of allometric scaling laws in biology. Science 1997, 276, 122–126.
[107] Buttler, JM., & Dressman, JB. The developability classification system: Application of biopharmaceutical concepts to formulation development. J. Pharm. Sci. 2010, 99, 4940–4954.

[108] Parrott, N., & Lave, T. Computer models for predicting drug absorption. Drugs Pharm. Sci. 2010, 193, 338–355.
[109] Charkoftaki, G., Dokoumetzidis, A., Valsami, G., & Macheras, P. Elucidating the role of dose in the biopharmaceutics classification of drugs: The concept of critical dose, effective in vivo solubility, and dose dependent BCS. Pharm. Res. 2012, 29(11), 3188–3198.
[110] Avdeef, A., Fuguet, E., Llinàs, A., Ràfols, C., Bosch, E., Völgyi, G., Verbić, T., Boldyreva, E., & Takács-Novák, K. Equilibrium solubility measurement of ionizable drugs – Consensus recommendations for improving data quality. ADMET & DMPK 2016, 4(2), 117–178.

Annette Bauer-Brandl and Martin Brandl

2 Solubility and supersaturation

2.1 Introduction and fundamental considerations

Solubility and supersaturation are classical issues – for example in chemical engineering – regarding crystallization processes of chemical compounds. This aspect is addressed in detail in Chapter 10 of this book. Solubility and supersaturation are also relevant for pharmaceutical drug delivery: Supersaturable drug delivery systems are often used to increase the bioavailability of poorly soluble drugs that are administered as oral solid dosage forms (see Chapter 8). In order to understand their properties in a biopharmaceutical context, transient states of supersaturation need to be considered in settings, where dissolution media gradually change their compositions as is the case during passage through the digestive system. Under these circumstances, a clear distinction between solubility and supersaturation can become difficult, and the impact on dissolution rate, drug absorption rates, and bioavailability can be confusing. This chapter is intended to promote the understanding of these questions.

2.1.1 Definitions and significance of solubility and supersaturation

2.1.1.1 Classical definitions of solubility and supersaturation

2.1.1.1.1 *Solubility*

Solubility of a solute (e.g., a drug substance) refers to the qualitative and quantitative composition of its saturated solution. This means that the solution is in a dynamic equilibrium between the solid particles of the solute in the form of a suspension in a given solvent that is kept under constant conditions (with respect to temperature, solvent composition, etc.). The solubility is expressed as the proportion of the designated solute in the designated solvent system.

The equilibrium refers to a certain solid-state form of the solute. If the suspensions contained different crystalline polymorphs of the compound or the same compound in its amorphous state, each of them would strive towards its individual equilibrium according to different solubilities for different solid-state forms. For the same reason that there is an equilibrium between the solid state of the solute and the solvent, the solubility depends widely on solvent composition. If a pure solvent contains any compounds in

Annette Bauer-Brandl, Martin Brandl, Department of Physics, Chemistry and Pharmacy University of Southern Denmark, Campusvej 55, DK-5230 Odense M.

https://doi.org/10.1515/9783110559835-002

addition to the solute, again the equilibrium will be different. Examples of such additives that are frequently used in a pharmaceutical context are buffer salts, cosolvents, polymers, and surfactants.

2.1.1.1.2 Supersaturation

A supersaturated solution has a higher concentration of a given solute in a given solution as compared to the equilibrium state. A supersaturated solution is thus necessarily unstable.

Such instability of supersaturated states is frequently found in the literature designated as "metastability" (which translates to "beyond stability"). This term is designated to those cases where the "metastable" conditions can be preserved for some time. This time is not specified and may range between seconds and hundreds of years and beyond. Thus, IUPAC [1] in general suggests avoiding the term "metastable", because it relates a thermodynamic term (stability) with a kinetic property.

With respect to the solubility of drugs, it is worth to not only consult literature in chemistry but also in the pharmaceutical field, for example, FDA Guidelines for ANDAs [2]. The term "true thermodynamic solubility" is used for the condition "which is reached after infinite time". This definition is perfectly in line with the classical (IUPAC) definition of solubility, namely, the state that is reached at infinite equilibrium time. The term "solubility" is clearly distinguished from the term "apparent solubility", which literally means the solubility that is observed. However, when it comes to apparent solubility and supersaturation in a pharmaceutical context, FDA defines it as follows:

"Apparent solubility refers to the concentration of material at apparent equilibrium (supersaturation)."

This use of the terms is difficult to bring in line with the aforesaid. The difficulty arises from the IUPAC definition of supersaturation that is an instable state. Such instable states may be observed (i.e., they are apparent) for quite long-time periods, but still they are not at equilibrium.

The following text refers strictly to the IUPAC definition. However, when discussing apparent solubility and its impact on biopharmaceutics, the intention of the FDA wording will become more obvious.

2.1.1.2 The concept of differential solubility

Consider an apparatus containing a solution in a flowing system, for example, along a process line of a crystallization plant in chemical industry. For practical reasons, different parts of this system may be exposed to slightly altered conditions. Examples thereof are temperature gradients or concentration gradients that are generated by physico-chemical conversions or the gradual introduction of additives. Depending on stirring effectiveness, local differences in additive or temperature distributions induce local (small) differences in solubility. Such conditions may be denoted as "differential solubility" because here the infinite solubilities are connected to the global solubility value in the same system. It is acknowledged that differential solubility addresses the borderline between thermodynamic solubility and kinetic effects (gradually changing conditions) if the differences are very small. Therefore, differential solubility may have very limited effect in terms of solubility enhancement factors. However, if these small gradients persist over a long-time period, they will lead to considerable solute transport by dissolution and diffusion processes. In such a dynamic system where the composition of solutions changes with both time and location, possibly supersaturated transitional states may also be generated. They are much more difficult to describe quantitatively as compared to homogeneous systems. In most cases the existence of differential solubility is therefore disregarded, unless it leads to supersaturation followed by precipitation of the solute and thus its significance becomes obvious.

2.1.1.3 Significance of solubility, differential solubility and supersaturation for oral drug delivery systems

Solubility restricts the amount of a drug substance that will be dissolved at equilibrium, for example, from an oral dosage form, in a closed system. Typical closed systems in this context are shaked flasks in which solubility is measured experimentally, and for dissolution studies in single-vessel set-ups (including certain transfer models).

After the oral intake of a dose of a drug, high drug concentrations occur in the gastrointestinal fluids as compared to the concentration of the drug in the rest of the body, e.g. in the blood circulation or in tissues. High concentration gradients between the gastro-intestinal tract and the blood lead to high transfer rates for the passive transport processes of drug molecules from the inner lumen of the gastrointestinal tract into the blood stream. This process is called (passive) *drug absorption*. The higher the concentration gradient, the faster is the absorption. During the time period over which the concentration gradient is maintained, the amount of drug that is transported over the intestinal wall (but not metabolized) can be accumulated to be defined as the drug fraction absorbed. It is often denoted as "bioavailability" and expressed in percent of the full dose given reaching systemic circulation in unchanged form. In this view on transport processes, to enhance oral bioavailability, increasing drug concentrations in the gastro-intestinal tract as much as possible appears as a straightforward formulation approach for poorly soluble drugs, if possible even above the solubility limit. In other words, it is advantageous to choose those formulations that may induce supersaturation. Such formulations are designated as "supersaturable" or "supersaturating" formulations. Experimental data of such systems is accessible through dissolution studies.

However, for several reasons this perception may fall short because it is too simple as compared to the real biopharmaceutical situation:

First, classical solubility studies take (many) hours to reach equilibrium. In most cases 24 h is regarded appropriate. Typical oral drug formulations have a much shorter transit time along the sites of absorption in our gastro-intestinal tract. This is one reason why in many cases the solubility of the drug may not be reached in vivo, and supersaturation thus will not be reached either.

Second, solubility limits as well as supersaturation may not be indicative in vivo because absorption processes occur simultaneously with dissolution. Absorption continuously decreases the dissolved drug concentration. The more drug is dissolved, the higher is the concentration of the drug; thus, the higher is the concentration gradient and the higher is the absorption rate, which in turn lowers the concentration of the drug in the gastro-intestinal tract. Solubility limits (i.e. supersaturation) may thus not be reached and thus precipitation will not occur. In such dynamic biopharmaceutical system, it is decisive if dissolution or absorption has the higher rate to possibly reach or surpass solubility limits after a certain period of time.

Another effect on solubility that needs to be regarded in an in vivo situation is the highly dynamic properties in the gastro-intestinal tract in terms of media compositions: Media widely change composition along the gastro-intestinal tract. There is a pH difference between the acidic stomach and intestines where the pH increases gradually to neutral and slightly basic pH values. The composition also gradually changes with respect to bile, and thus bile salt and phospholipid composition, not even mentioning food components that may also be present. This scenario results in differential solubility, which in turn induces gradually changing concentration gradients. Furthermore, possibly supersaturation may be induced, especially for weak bases being well dissolved in the stomach, which may become supersaturated during the transit to the intestines due to the related pH shift. Depending on the dose of the drug and the volumes and compositions of the intestinal fluids, supersaturation of the drug may be kept over a considerable time period and thus contribute to better drug absorption.

The in vivo situation can be summarized analogously to the aforesaid a differential solubility. Bioavailability is governed by concentration gradients in a highly dynamic system. In general, better soluble and supersaturating systems are advantageous because they dissolve faster as compared to other systems, yielding higher concentration gradients and thus faster drug uptake and absorption.

It may be concluded that for drug development it is productive to first look at solubility, but not to forget dissolution processes in the dynamic system. In other words, drug development takes place at the intersection between thermodynamics and kinetics.

2.1.2 Factors affecting solubility

The solubility of a solute of a certain solid-state form in a given solvent is a single value under the given conditions. It depends on the interactions between drug molecules in the solid-state form and the intermolecular interactions between drug and solvent molecules in solution. In simple molecular solutions, a solvation shell of solvent molecules is formed around the solute. Such interactions comprise both enthalpic and entropic terms. Therefore, the solubility of solutes in solvents changes with temperature and typically increases with temperature. Increased temperatures are relevant in synthesis, chemical engineering, particle engineering, and processing. For physiological considerations, however, body temperature (and possibly room temperature) is most relevant.

The effect of the pressure on solubility of drugs is also of limited significance in the in vivo situation. It may play a role during manufacturing, for example, by supercritical gas methods.

A general perception of the energy relationships regarding dissolution of a solid and solvation of the molecules is given in Figure 2.1 where the thermodynamic cycle of sublimation, solvation, and dissolution is depicted. Energetic relationships for Gibbs free energy, enthalpy, and entropy are connected to each of these processes. The entropy has a considerable contribution to the dissolution process, in particular, for solvation. For several drugs, solvation enthalpies and entropies have been experimentally determined as the differences between the respective energy terms for sublimation and dissolution. Even structurally closely related drugs may differ widely in the mechanism of solvation and dissolution in terms of the contributions of entropy and enthalpy [3].

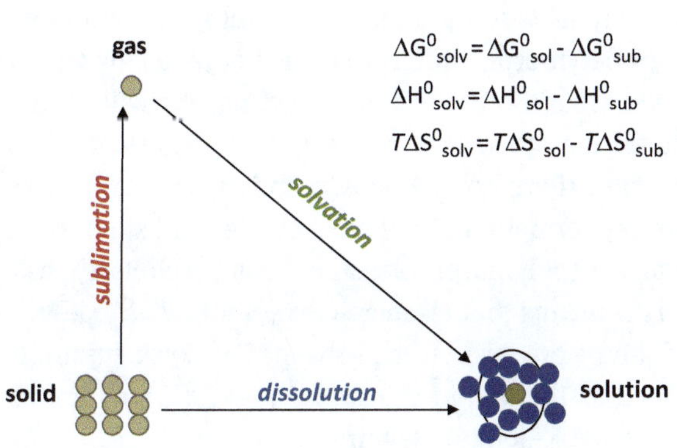

Figure 2.1: Thermodynamic cycle of sublimation, solvation, and dissolution; figure adapted from [3]. Grey dots represent API; blue dots represent solvent.

There are two principle causes that can make substances poorly soluble: either the solid-state form or the solved state is non-favourable to yield a high solubility.

Accordingly, the following distinction between the two properties has been made: solid-state-limited and solvation-limited solubility [4, 5]. In terms of poorly soluble drugs, they have also been denoted more figuratively as "brick dust" and "grease ball" type substances, respectively, as is outlined as follows:

The solid-state-limited type of poorly soluble drugs is restricted by the crystal lattice energy. The interactions between drug molecules in the solid-state form are very high and typically, melting temperatures are also high. Furthermore, it needs, high activation energy to remove a solute molecule from such a crystal. Thus, solubility is compromised by both thermodynamic and kinetic effects. This type of solutes is therefore referred to as "brick stone" type materials, referring to very low solubility, although the material is reasonably well wetted. The high energy needed to remove a molecule from the crystal is compensated for by the energy of solvation when this molecule is dissolved in the aqueous medium (enthalpy-entropy compensation).

The other group of poorly soluble drugs are not solid-state limited but they are difficult to wet like "grease balls". The removal of a molecule from the solid-state form requires comparably less energy; these substances typically have low melting points. However, in this case, the interaction between the drug and the solvent is less advantageous. The energy gained by building a solvation or more specifically hydration shell is limited because of largely lipophilic properties of the drug. This type of drugs is said to have solvation-limited solubility.

2.1.2.1 Impact of solid-state on solubility and supersaturation

2.1.2.1.1 Solid-state forms

Equilibrium solubility is defined for each solid-state form, for example, for each polymorph. Yet, the solubilities of different polymorphs differ from each other as discussed in chapter 9 of this book. At any given condition, one of the polymorphs will be the thermodynamically most stable, and all the others are thus less stable. Therefore, this most stable phase will be present when thermodynamic equilibrium is established. However, kinetic factors may prevent the formation of the most stable solid-state form, and a thermodynamically instable phase (polymorph or amorphous precipitate) may be preserved for considerable time. This means that an increased solubility is observed as compared to the stable system of solid-state form in the solvent. The solubility measured in such a case can be designated as the *apparent solubility*, and the solution as *supersaturated* with respect to the most stable solid-state form.

As to the relationship between different polymorphs, it is useful to consider the crystallization process. Starting with a concentrated solution of the solute, the Ostwald rule of stepwise crystallization suggests that typically the most unstable solid-state form crystallizes first, followed gradually by the more stable solid-state forms. The energy relationships of stepwise crystallization are schematically depicted in Figure 2.2.

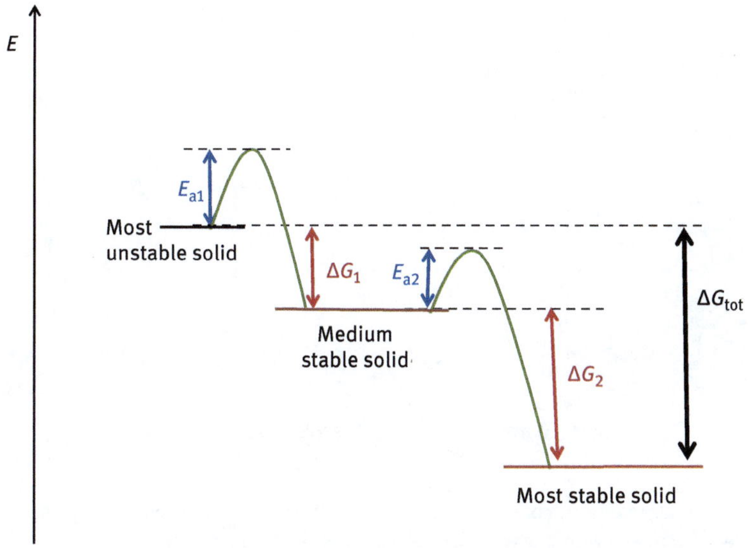

Figure 2.2: Schematic illustration of Ostwald rule of stepwise crystallization.

From the energy relationships it can be concluded that the solubility of the most stable solid-state form is lowest, while the solubility of the least stable solid-state form (of highest energy content) is highest. In other words, increased free energy of the solid-state form leads to instability, but less activation energy for the dissolution process is needed. Thereby a higher dissolution rate is achieved. Furthermore, due to a larger free energy difference between the solid state and the dissolved state the solubility is higher. This relationship is schematically shown in Figure 2.3.

In brief, the advantage of thermodynamic stability is a trade-off of decreased solubility and low dissolution rate. An advantage in solubility and dissolution rate comes with decreased stability.

2.1.2.1.2 Particle size effects

In a dynamic equilibrium of solubility under constant conditions, small crystals would tend to gradually disappear, while the large crystals grow even larger. The reason is a decrease in total surface energy for the system with the larger particles; it is due to the reduction of the overall surface area when the larger particles are present. The growth of large particles to the disadvantage of the small particles over time is called "Ostwald ripening" [6]. This process is technically utilized for the preparation of crystals of certain particle size-distributions. The extent to which the crystal enlargement occurs depends on both solvent and solute. It is typically induced by transient temperature variations, which lead to solubility

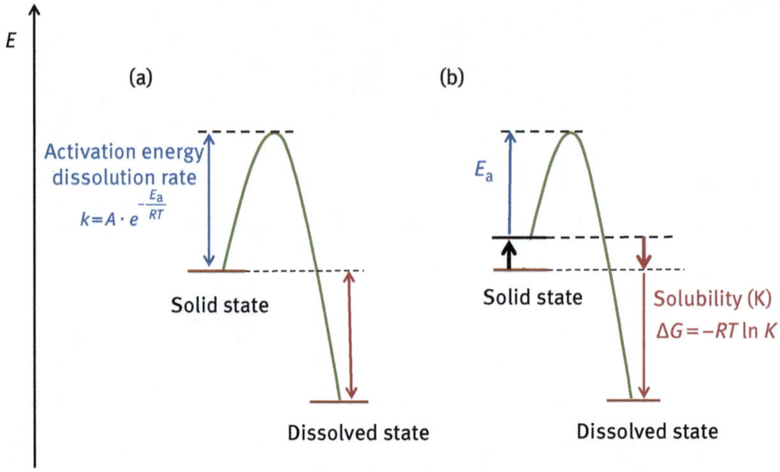

Figure 2.3: Relationship between energy contents of solid-state forms. (a) Stable solid phase as a reference; (b) unstable solid phase with increased energy content leads to decreased activation energy E_a; larger Gibbs free energy $|\Delta G|$ for dissolution, thus increased solubility and increased dissolution rate.

differences, whereby the equilibrium is disturbed on purpose to increase both the rate of dissolution and the rate of recrystallization.

Solubility constants are defined for comparably large particles of the solute. For very small particles (i.e., well below 1 μm in diameter) of materials of poor solubility, the *solubility* is postulated to increase with decreasing particle size. The reason for this phenomenon is the additional surface energy of such small (nano-) particles due to higher "curvature" of the particles. Based on this assumption, the solubility enhancement effect of small particle size on solubility can be derived from the Ostwald–Freundlich relationship (eq. (2.1)):

$$\ln\frac{C_s}{C_{\text{inf}}} = \frac{2\,\sigma\,V_M}{R\,T\,\rho\,r} \tag{2.1}$$

C_s is the saturation solubility of the small (nano-) particles, C_{inf} is the solubility of the solid with infinitely large particles, σ is the interfacial tension (surface tension) of the solute particle in the solvent, V_M is the molar volume, R is the universal gas constant, T is the absolute temperature, ρ is the density of the solid, and r is the particle radius.

For drug crystals, one may argue for an increased solubility of very small particles as follows: The enlarged surface-to-volume ratio of small particles provokes that a larger fraction of molecules is positioned on the surfaces, as well as a larger fraction at the edges and the corners of the crystals. At these positions there is less interaction with neighbouring molecules. Molecules at the surface, edges, and corners of a crystal exhibit less lattice interactions, as these are not present in all three dimensions around the molecule, but only to the direction towards the inner part of the crystals. Thus, larger fractions of less tightly bound molecules are exposed to the solvent. This may lead to a higher dissolution "pressure" as compared to more spacious particles. Furthermore, it has been argued that the diffusional layer thickness above the surface of small particles is also reduced. From theoretical considerations, it has been concluded that the discussed effects become practically significant below 10 nm of particle diameter [7]. Theoretical dependence of solubility on particle size is depicted in Figure 2.4.

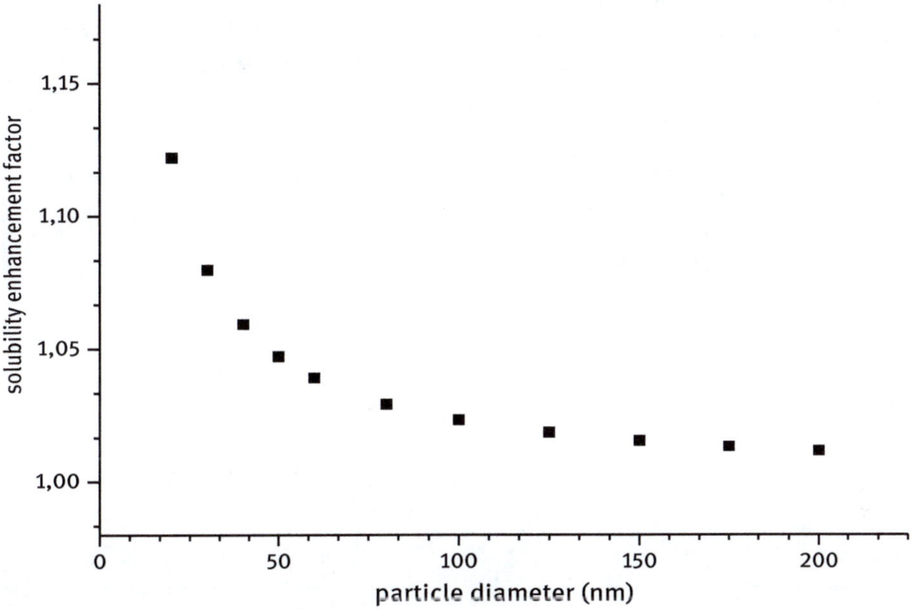

Figure 2.4: Dependence of solubility enhancement factor derived from Ostwald–Freundlich relationship (Equation 2.1), for a model system.

There are alternative scientific opinions about the particle size limits below which the solubility increases; however, all of them would agree that the particles must be at least below 1 μm in diameter.

For the estimation of the practical significance of the theoretical considerations regarding the particle size effect, it needs to be kept in mind that particle sizes of small drug particles are typically in the higher 100 nm range. The solubility of such particles is difficult to measure. For classical experimental methods (filtration and centrifugation), the results largely depend on the respective separation conditions used and are prone to artefacts (filter pore sizes, centrifugation conditions, caking of the centrifugation pellet and adsorption of drug on device surfaces). Experimental results also tend to be rather poorly reproducible. For example, solubilities have been reported to be increased by factors of 10- to 100-fold for nanocrystal suspensions treated by centrifugation if the separation between supernatant and pellet is difficult. Variation is expected to be less for samples treated with ultracentrifugation, but the absolute values will be different due to different cut-offs of particle size for different methods. However, much better reproducible results can be expected from bulk methods, for example, for turbidimetry or for light scattering [8]. In this case, the relative solubility for the same systems was found to be increased by max. 15 %. This value was in good agreement with the Ostwald–Freundlich relationship keeping the expected experimental error in mind.

Figure 2.5 depicts an example of experimental data for a drug nanocrystal suspension of particle sizes from 150 to 350 nm and their respective solubilities measured by bulk methods. The figure also shows calculated data.

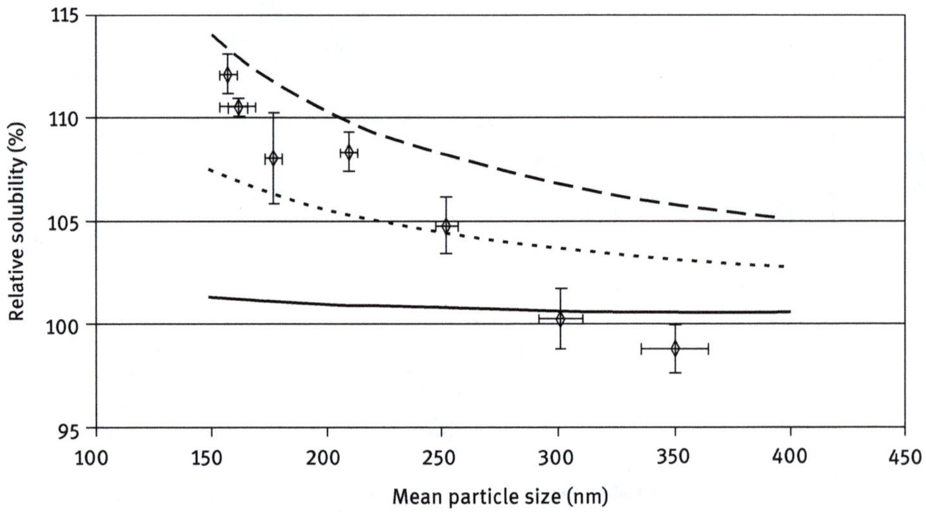

Figure 2.5: Experimental dependence of solubility on mean particle size, for loviride nanosuspensions as an example, analysed by bulk methods, and theoretical factors for the same system taken from Ref. [8]. Key: (◊) experimental values measured by scattering monitoring ($n = 3$); (____): calculated data using interfacial tension 5 mN/m; (-) (. . . .) calculated data using the estimated interfacial tension of 27.5 mN/m; (- - -) calculated data using an interfacial tension of 50 mN/m.

In contrast to the solubility, it is generally accepted that smaller particle size – given the same solid-state form and crystal habit – leads to faster dissolution. It is possible to take advantage of (possibly) increased solubility and even more of the higher dissolution rate for drug nanocrystals. Nanocrystals are often prepared from larger particles by intensive milling and particle comminution. These processes bring in a lot of energy into the bulk of the powder and into individual particles: they are mechanically activated. The milling is typically done in wet state in the form of suspensions to distribute the energy, to lower surface (interfacial) energies, and to prevent agglomeration. Nevertheless, the processes activate the material and may induce solid-state form changes such as amorphization – at least at the particle surfaces. After longer milling times, the material may entirely have transformed to the amorphous state. Furthermore, if surfactants are used as additives to reduce interfacial tension and to prevent agglomeration tendencies, the surface properties and surface energies of the particles are also largely affected. It is very difficult to distinguish and quantify these effects and to quantify their respective contributions to solubility. The solubility of nanocrystal dispersions may be increased not only by the sole effect of the small particle size but also by changes in solid-state form and by the presence of the surfactants.

For crystal particles larger than approx. 1 µm, particle size effects on solubility are generally negligible for crystals of the same quality. However, the effect of particle size on dissolution rate is highly significant. For highly soluble substances, the effect is less pronounced compared to poorly soluble substances.

2.1.2.1.3 Degree of crystallinity

Ideal crystals have no defects. Amorphous materials have no long-range order and no crystallinity. This leads to the definition of the degree of crystallinity [9]: Degree of crystallinity is a crystal's position on a scale between the ideal crystal (with no defects) and the amorphous state (containing the maximum number of defects). A totally amorphous particle corresponds to zero crystallinity. There is only a short-range order (interaction with the nearest neighbouring molecules) while the long-range order of a crystal structure, which continues indefinitely through the entire particle, is absent. Amorphous particles may contain ordered domains that under certain conditions can act as nuclei for crystallization. Decreased degrees of crystallinity for real crystals, for example the presence of lattice defects and amorphous domains, increase both the enthalpy and the entropy of the material. As the increase in enthalpy upon lowering the degree of crystallinity is not fully compensated by the increase in entropy, Gibbs free energy typically increases for such systems. Therefore, the lower the degree of crystallinity of a solid, the higher is its solubility and dissolution rate, but again the lower the thermodynamic stability as well.

Powders that contain particles with different degrees of crystallinity can be the reason for physical instability during storage, for example, if recrystallization is induced by moisture uptake leading to higher degree of crystallinity and to crystal growth.

In other cases, the crystallization of particles with a low degree of crystallinity is kinetically hindered even in the presence of crystalline domains. In these cases, the solubility of each sample depends on its specific degree of crystallinity and is observed as unchanged during considerable experimental time. The solubility of such a sample will also depend on the amount of suspended material that is not yet dissolved. Such observation should lead the analyst to the conclusion that the crystallization of this sample is hindered; the analyst may want to induce crystallization by other means in order to confirm this hypothesis.

This effect can be of practical consequence; an example of which has been described in detail with drug substance griseofulvin [10]. Thus, the detection and quantification of the degree of crystallinity by a suitable method is important for development, manufacture, and quality assurance of pharmaceutical preparations.

2.1.2.2 Impact of the solvent on solubility and supersaturation

2.1.2.2.1 Solvents and co-solvents

It is common knowledge that the solubility of a compound will largely vary in different solvents. Solubility may also strongly depend on the composition of solvent mixtures, that is, the presence of additional compounds dissolved in the same solvent, for example, co-solvent systems. The relationships are non-linear except for ideally miscible mixtures. For water–solvent mixtures, the solvent might influence the water structure (hydrotropic effect) and thus affect both solvation and solubility.

2.1.2.2.2 Common ions and common co-formers

The solubility of a pharmaceutical salt "AB" depends on the respective solubility of the ions A^+ and B^-, which it is composed of. The solubility of salts can be estimated by calculation of the solubility product. Increased concentration of one of the ions will affect the equilibrium and decrease the solubility of the counter ion. Solubility will therefore depend on the presence of common ions that increase the concentration of this ion in the solution (for $[A^+]$ or $[B^-]$, respectively). Typical examples are ions from buffer salts or from other additives. A similar effect is described for the common co-former effect that decreases the solubility of co-crystals.

Solubility also depends on the ionic strength of the solution, but typically to a minor degree unless high concentrations are used. However, in cases where salts change the cluster structure of the aqueous media with a significant entropic effect, solubility of a compound can change considerably ("salting in"; "salting out"; hydrotropic effect).

2.1.2.2.3 pH values of aqueous systems

Many drugs are weak acids or weak bases. In such cases, the pH value of an aqueous solvent has a huge impact on solubility. Depending on the difference between pK_a-value of the drug and pH value of the solution, solubility enhancement factors of 1,000 and above are common.

The effect of pH on solubility is depicted in classical (schematic) pH solubility plots as shown in Figure 2.6.

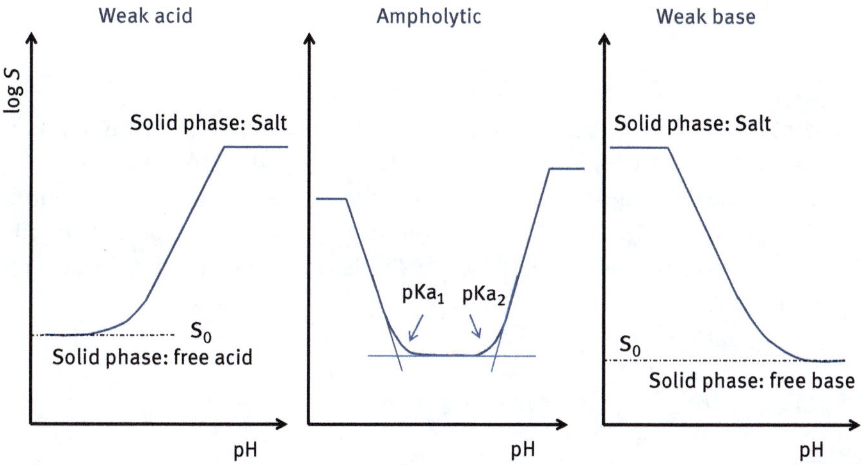

Figure 2.6: Schematic representation of the pH-dependence of solubility S for acidic drugs (left), ampholytic drugs (middle), and basic drugs (right). S_0: intrinsic solubility of free acid or base, respectively.

It should be noted that the difference in solubility of a compound at different pH values – even within the physiological range – can be huge. This fact leads to high degrees of transient supersaturation when solutions are exposed to a pH shift in the gastrointestinal tract, in particular, for bases at transition from the stomach to the intestines.

2.1.2.2.4 Solubilization/supramolecular assemblies

The use of surface-active ingredients – also termed *surfactants* – above their CMC (critical micelle concentration) is a common means to increase the solubility of poorly water-soluble drugs. Micelles are supramolecular assemblies of such amphiphilic molecules that can associate with drug molecules. The drug molecules are typically incorporated into the hydrophobic interior of the micelles that shield the drug molecules from the aqueous bulk. Thus, surfactants increase the saturation concentration of the drug. Complexes such as cyclodextrin (CD) inclusion complexes are very similar in this respect. As such micellar solutions and CD complex solutions are in equilibrium and thermodynamically stable. Sometimes they are wrongly denoted "supersaturated" just because the solubility of the drug is higher when compared to the pure solvent without the additives such as surfactant or CD.

The term "supersaturated" should be clearly defined with respect to solvent composition. If a drug is solubilized in a micellar solution, its saturation concentration is higher than that in the surfactant-free solvent, where the drug is dissolved (and not solubilized). The micellar solution is thermodynamically stable and thus not supersaturated. It may only become supersaturated if the concentration of surfactant decreases over time (e.g., due to digestion or absorption).

2.1.2.2.5 Local solubility differences (differential solubility)

Changes in pH values along the gastro-intestinal tract can determine the dissolution–precipitation-driven drug behaviour (see above). In addition, changes of the composition of intestinal fluids, most prominently regarding bile salt concentrations along the transit, affect the degree of solubilization of drug molecules in bile salt micelles and mixed micelles with phospholipids. This again implies local solubility differences and affects drug absorption.

2.1.3 Factors affecting supersaturation

The term "supersaturation" implies that the actual observed concentration is higher compared to equilibrium solubility in the same solvent under the same conditions. In a pharmaceutical context, typically mixtures consisting of physiological media or artificial biomimetic media and the respective drug together with the formulation additives are studied. Therefore, it is necessary to explicitly define the term supersaturation and its degree with respect to the reference solubility regarding the medium, for example, in water, buffer, biomimetic media, and/or in the presence of formulation additives.

Let us consider solubility in simple aqueous solvent without solubility enhancing additives: A commonly used means to express the degree of supersaturation (DS) is by the ratio of the observed dissolved solute concentration (Cb) and the equilibrium saturation concentration (Cs) in the solvent (eq. (2.2)):

$$\mathrm{DS} = \mathrm{Cb}/\mathrm{Cs} \tag{2.2}$$

The tendency to spontaneously recrystallize or precipitate depends on the degree of supersaturation in a solution: The higher the degree of supersaturation, the less activation energy needs to be overcome, and the more probable is spontaneous recrystallization or precipitation.

This relationship is illustrated by the classical Ostwald–Miers diagram as a function of temperature as shown in Figure 2.7. For any concentrations above the solubility line precipitation may occur, depending on the conditions such as cooling rate, stirring conditions, presence of foreign particles, and so on. Above the precipitation line, the degree of supersaturation is so high that precipitation necessarily occurs. The Ostwald–Miers region is between the lines, where precipitation of supersaturated systems may occur or delayed or fail to appear.

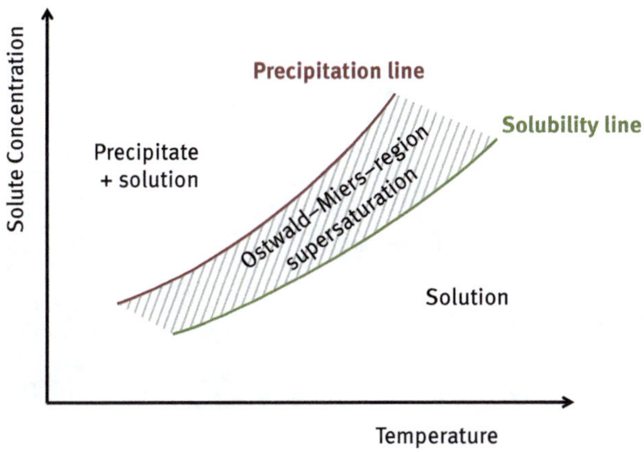

Figure 2.7: Classical schematic Ostwald–Miers diagram.

However, for drug delivery systems and in vivo situations, temperature changes can be largely disregarded. Therefore, Figure 2.8 illustrates the Ostwald–Miers region or metastable zone width introduced by supersaturation at constant temperature. Again there is a limit for the degree of supersaturation above which precipitation (phase separation) must occur; it is highlighted in Figure 2.8 with an asterisk. Phase separation may be in the form of immiscible fluids (saturated drug solutions) or amorphous precipitates; nucleation/growth mechanisms lead to crystalline precipitates.

Nucleation is either spontaneous (primary nucleation), or it may continue to occur in the presence of existing seed crystals (secondary nucleation). Primary nucleation refers to the first nuclei to form and those that form independently from other pre-existing nuclei of the new phase, while secondary nucleation depends on pre-existing crystals that cause the formation of secondary nuclei, for example, by shearing forces while stirring. Crystal growth attributes to solute that is deposited on the surfaces of the nuclei. These competing mechanisms determine the number of crystals that are formed in a system and via crystal growth thus also their size distribution.

The relationship between degree of supersaturation and nucleation/growth is defined by the following simplified equations (eqs. (2.3) and (2.4)):

$$\frac{dG}{dt} = k_g \cdot \mathrm{DS}^g \tag{2.3}$$

$$\frac{dN}{dt} = k_n \cdot \mathrm{DS}^n \tag{2.4}$$

dG/dt, growth rate; k_g , growth constant; g, growth order;
dN/dt, nucleation rate; k_n, nucleation constant; n, nucleation order; and DS, degree of supersaturation.

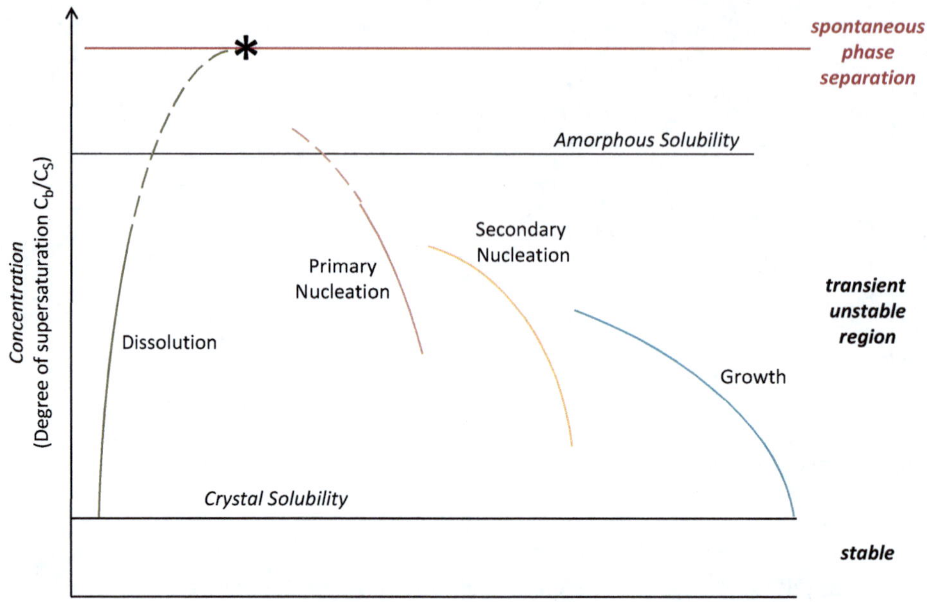

Figure 2.8: Concentration-dependent transient region (Ostwald–Miers region) at constant temperature for drug delivery systems. Dissolution beyond a certain concentration necessarily leads to spontaneous phase separation ("precipitation line") as marked by the asterisk. This may be followed by nucleation and growth. Under other circumstances, nucleation and growth may occur in the unstable region already at lower concentrations; solubility of an amorphous phase ("amorphous solubility") may not even be reached.

For small organic molecules such as drugs with a molar mass below 500 g/mol, the value for crystal growth order g is typically between 1 and 2. The value for nucleation order n is typically between 5 and 10. From the difference of the kinetics of nucleation and crystal growth it follows that at low degree of supersaturation, crystals grow rather than they nucleate resulting in fewer but larger crystals. In contrast, at higher degrees of supersaturation nucleation rate overpasses crystal growth, resulting in many small crystalls which occur.

Figure 2.9 illustrates how the degree of supersaturation affects particle size distribution of the precipitated crystals.

Depending on precipitation conditions, the solid phases differ not only in particle size distribution but also in the quality of the crystals. The probability of defects in the crystal lattice is also affected. The faster the crystallization, the higher the proportion of defects in the crystals. Similar relationships apply to the formation of amorphous domains, with an effect on the degree of crystallinity. The higher the crystallization rates, the lower the degree of crystallinity of the product.

Even at a low degree of supersaturation, a supersaturated solution would readily crystallize when seed crystals of a more stable phase are introduced. By the exposure

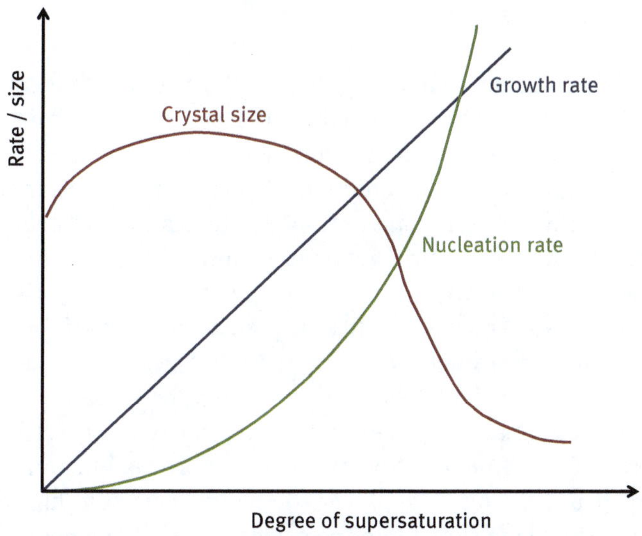

Figure 2.9: Schematic representation of the relationship between degree of supersaturation and the particle size distribution of the precipitated crystalline material.

to the crystal surface architecture of the stable structure, molecules would readily interact and the activation energy is thus lowered by the seed crystal. Crystal growth is induced and rapid equilibration from supersaturated state to solubility level will occur. There are also other conditions that can lower the activation energy for precipitation such as local concentration differences, temperature gradients, foreign nuclei, agitation, and so on. These may induce rapid precipitation and crystallization.

As nucleation rates and crystal growth depend on the degree of supersaturation and the composition of the medium, degree of mixing, viscosity, etc., different types, shapes and solid-state structures may be formed. Therefore, not only the composition but also the conditions under which a precipitate is formed determine its properties and thus solubility and dissolution rate.

2.1.4 Solubility and supersaturation in composed solvents

Pharmaceutical scenarios both in vivo and in vitro typically comprise aqueous media composed of water, buffer, and solubility-enhancing additives. These additives may be excipients used in drug formulations that can form micelles or complexes, or physiological gastric and intestinal fluids or biomimetic fluids containing bile salts and phospholipids.

2.1.4.1 Apparent solubility: alternative terms and definitions

Apparent solubility literally is the observed solubility. Only in the rare case of ideal experimental results it is identical with "true" solubility. The observed result depends on the method of solubility determination including several parameters such as suitable observation time given to reach equilibrium. It also depends on the method used for phase separation: in experimental solubility studies it is common to separate the solution from the solid by centrifugation or by filtration. Depending on centrifugation conditions or, respectively, the filter type, the limit to classify dispersed small particles as "dissolved" is set at a different particle size cut-off. The observed solubility result will vary accordingly.

Yet another situation applies to colloidal states of the drug, for example, in composed solvents comprising colloidal complexes and in micellar solutions: It is impossible to separate these colloids from the true molecularly dissolved material by classical methods. In this case, the term "apparent solubility" is used to highlight that the effect of the additives is included into the perception of the increased concentration of the solutions as compared to the pure compound in the plain solvent. The same term also covers dissolution media that contain multiple-component colloidal additives, such as biomimetic media containing bile salts, phospholipids, and so on. The use of the term "apparent solubility" stresses in this context the fact that there is a difference between the freely dissolved (solvated) molecules and those apparently dissolved, that is, supramolecularly assembled molecules. Both states of the solute molecules are embraced by the term "apparent solubility". Such solutions are stable and thus not supersaturated.

However, some scientific publications including the FDA guideline [2] use the term "apparent solubility" as a synonym for "supersaturation". The FDA guideline states as follows: "Apparent solubility refers to the concentration of material at apparent equilibrium (supersaturation). Apparent solubility is distinct from true thermodynamic solubility, which is reached at infinite equilibrium time." The reasoning behind this wording may be that it is an experimentally found value, and in that sense, it is apparent. It is distinct from the thermodynamic solubility. However, supersaturated states are not at apparent equilibrium, which would mean that they do not change concentration over time. Apparent solubility may, on the other hand, be a stable state, for example, in the case of micellar solutions, and thus not be supersaturated at all.

On the other hand, the term "apparent solubility" has been used for the solubility of partly amorphous materials in cases, where no recrystallization was observed during experiment times. The solubilities of the samples followed their degree of crystallinity for the observation periods [10].

2.1.4.2 Apparent supersaturation

The definition of supersaturation in general and thus the degree of "supersaturation" in a broad sense depends on the respective reference value, namely, if the true or the apparent solubility is used.

If the undissolved material is exclusively amorphous, the dissolved concentration of a solute is higher as compared to the solution above the crystalline form of the same material, as long as the unstable ("metastable") system persists and the amorphous material does not crystallize. This is a case of "true" supersaturation. It is an unstable state.

However, the same term "supersaturation" is frequently but less correctly found to be used in the literature for any increased drug concentration as compared to the solubility referring to the pure crystalline solid residue in simple aqueous solutions including approaches that are not only based on solid-state forms of the same material, but also include the use of additives to the solvents (composed solvents) or to the solid material (e.g., co-amorphous material, polymer matrix, surfactants). However, if additives are present, the system can very well be in thermodynamic equilibrium, presuming the additives in the composition increase equilibrium solubility. A good example would be the introduction of surfactants above their CMC, where the drug molecules are solubilized in the micelles to form stable micellar solutions. Therefore, in this case the use of the term "supersaturation" should be avoided and can be better denoted as "apparent supersaturation" because it describes a stable state [11].

These relationships are depicted in Figure 2.10.

Increasing the solubility of a drug substance by special additives in a formulation leading to apparent supersaturation can be a valid approach to reach a better absorption and bioavailability. Whether and how much this approach can increase the bioavailability in a specific case depends on the kinetics of the release of the drug molecules from the supramolecular assemblies by the disappearance of the solubility-enhancing agents from the solution via dilution, absorption, digestion, and other effects. In contrast, enhanced absorption is always expected for truly supersaturated solutions.

2.1.4.3 Kinetic solubility; dynamic solubility

The term "kinetic solubility" is sometimes used for transient states of improved solubility. The term is a contradiction in itself. In the current context, it would be a synonym for the well-known term "supersaturation". As the degree of supersaturation is variable, supersaturation is not a single value but depends on experimental conditions and may change over time. In general, it is unfortunate to connect kinetic issues and with thermodynamic nomenclature. "Dynamic solubility" has also been suggested as an alternative term for supersaturation which reflects the dependence of the found values on the actual experimental conditions in a better way [10].

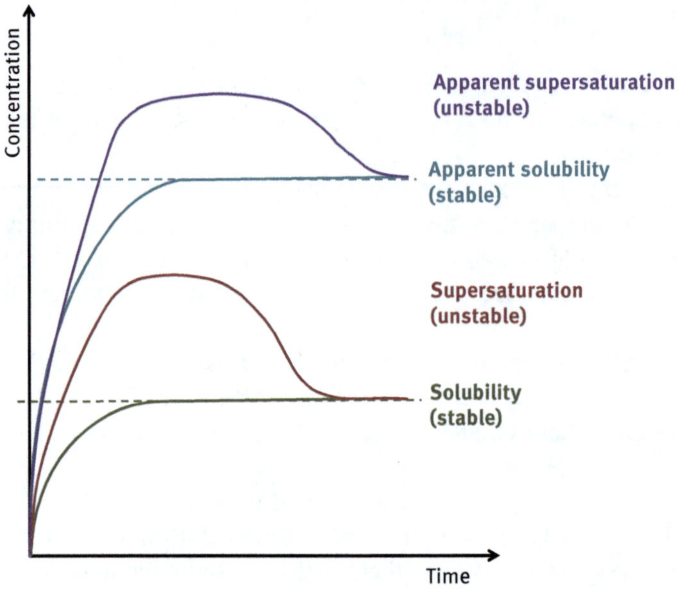

Figure 2.10: Relationships and terminology of solubility, supersaturation, and their respective apparent counterparts.

2.2 Distinguishing truly, molecular dissolved from apparently dissolved and supersaturated states

It is useful to distinguish solutes that are molecularly dissolved by solvation in a solution from those solubilized in micelles and complexes: In supramolecular assemblies, the drug molecules are not "free", not truly dissolved, but "apparently" dissolved. The difference of the solubilized state when compared to free, truly dissolved drug molecules is reflected in their diffusion rates. Even more importantly, colloidal states such as complexes and micelles cannot permeate through biological barriers. They need to first release the drug in its free form, the rate of which will be dependent on solution composition. These processes affect drug transport kinetics and oral absorption. This fact is relevant for the understanding of the performance of drug delivery systems including bioavailability.

2.2.1 Solubility/dissolution

The simplest mechanism of dissolution processes is described for crystals composed of neutral molecules that dissolve from the crystal surfaces by the formation of a saturated solution at the very surface of the crystal, and diffusion processes of the molecules into the bulk solution.

However, for weak acids and weak bases, solubility is widely pH dependent. The drug compound will act as a buffer in the aqueous solution. If such solutions are gradually changing their pH by titration with HCl and NaOH, respectively, supersaturated solutions can be formed for transient periods that may last for hours. Therefore, pH/solubility profiles measured by titration can be inaccurate.

Pharmaceutical salts are much more soluble than their corresponding free acids or bases; in many cases, the solubility is enhanced a thousand-fold or more. The salts would also dissolve much faster, and this is widely independent of the pH of the solution. An example of such relationship for salicylic acid and sodium salicylate this has been studied in detail [12]: even at low pH (pH 1) where the free acid is being formed in the bulk of the solution, the salt dissolves much faster than the acid (as has been measured by intrinsic dissolution rate to normalize for specific surface area). At pH 2 the substance is more than 300 times more soluble as compared to pH 7. However, at pH 2, the intrinsic dissolution rates of salicylic acid and sodium salicylate, respectively, are equal to those at pH 7. In both cases the salt dissolves approximately 100 times faster than the acid. The dissolution process appears to depend on the pH in the diffusion layer, which is determined by the solid rather than the pH of the bulk of the aqueous medium where the protonation/deprotonation reaction takes place. This is an example of reaction-limited dissolution processes. Such models are not frequently used. However, they may be a very useful alternative to mathematically describe dissolution processes of formulations [13].

The outlined processes, their respective kinetics, and related modelling are relevant for in vivo drug dissolution, for example, regarding pH changes during passage, presence of digestion fluids, food effect, changes of viscosities, and mechanical agitation. All these effects will affect solubility, supersaturation, and precipitation kinetics.

2.2.2 True, molecular supersaturation

For technical applications in the field of crystallization and particle engineering, supersaturation is used to induce precipitation under controlled conditions according to the desired precipitate properties. Supersaturation is often approached starting from a well-defined concentrated solution. Then a change of conditions, for example, solvent composition, pH, or temperature is introduced to enter the Ostwald–Miers region of supersaturation, and precipitation started by seeding.

Depending on the solvent properties, different crystal faces may be developed according to different kinetics and in turn in different specific surface area fractions. Therefore, the habit of the prepared crystals may vary with crystallization conditions. This will, in turn, have an impact on dissolution rate in addition to the crystal size (total specific surface area). Furthermore, transient supersaturation may occur during dissolution from the better soluble crystal faces.

If the crystallization process is very fast, partly amorphous material may be obtained. Amorphous systems can also be prepared by other means, for example, extensive milling, quench cooling, freeze drying or spray drying of solutions. They will exhibit true supersaturation during dissolution; however, being unstable, they may recrystallize at any time.

2.2.2.1 Examples of truly supersaturated systems

Truly supersaturated situations are unstable. In a strict sense, they apply to molecularly dissolved, free molecules in a certain solvent. The classical examples are amorphous and partly amorphous systems. Also different polymorphs will have different solubilities. Typically, their solubilities would not differ by more than approximately a factor of 2, in many cases much less. An unstable solution is being formed by the better soluble polymorph because the most stable polymorph may precipitate at any time. Similar relationships would apply to co-crystals. Hydrates, however, are typically less soluble than their anhydrates. A solution of a stable anhydrate polymorph may in contact with aqueous medium, spontaneously transform to the hydrate as discussed in chapter 9 of this book.

In dynamically changing solvent compositions, for example, pH shift, transient supersaturated states may occur.

For any other examples involving additives, it is necessary to study their effect on solvent, enhanced solubility, and enhanced apparent solubility, because enhanced concentrations may not be supersaturated.

2.2.3 Apparently supersaturated systems

Apparent supersaturation can be induced by using co-solvents or alternative solvents and other types of additives. The term is often used in connection with supramolecular assemblies, such as micellar solubilization. The solutions have a high concentration. However, they are not necessarily supersaturated in a strict sense, and not necessarily unstable, as is depicted in Figure 2.10.

By using additives, the solvation shell of a drug molecule is altered when compared to the pure solvent due to the interaction between drug molecule and the additives. There is a dynamic equilibrium between the truly (freely) dissolved and solubilized molecules. In this sense, compositions with several ingredients may be truly and apparently supersaturated at the same time.

2.2.3.1 Examples of apparently supersaturated systems

In many cases the formulation of poorly soluble drugs is based on the principle of apparent supersaturation as discussed in chapter 8 of this book. The poor solubility is enhanced by additives in the formulations, such as polymers, micelles, mixed micelles, CDs, and so on. The most common formulation principles of such formulations are discussed in Section 2.4.

2.2.4 Experimental considerations for the determination of solubility

2.2.4.1 Solubility by classical shake flask methods

Saturated solutions are made by exposing an excess of the solid material to the respective solvent in a closed system and waiting until the equilibrium has been reached. It is a dynamically equilibrated state where the fraction of material dissolving per time and the fraction that precipitates per time are equal, leading to a constant solute concentration. The samples for solubility measurements are kept at a certain temperature for a certain time, while they are shaken to ensure homogeneity and to decrease waiting time. Typically, this time is between several hours and days. This approach is too time consuming to be suitable for high-throughput studies, but it leads to valid data if it is carried out carefully. It is also good practice to prepare control samples at another, slightly higher temperature to supersaturate the solution with respect to the target temperature. These samples are then cooled down to the same target temperature as the standard samples. If the two experimental solubilities for the original samples and the control samples coincide, the solubility value found is probably close to the "true" value. The optimum experimental outcome of such approach is depicted in Figure 2.11.

In any case, it is useful to confirm that the solid-state form of the precipitate has remained unchanged after the incubation time in order to exclude any transformation to more stable polymorphs or hydrates/solvates. This is an important control experiment to confirm which solid-state form does the measured solubility value refers to.

2.2.4.2 Methods for phase separation

For solubility studies, the separation of solid from liquid is needed. Any phase separation method collects a different sub-set of the particle size distribution of the solid residue and defines it as undissolved material while the remaining solid particles are defined as dissolved.

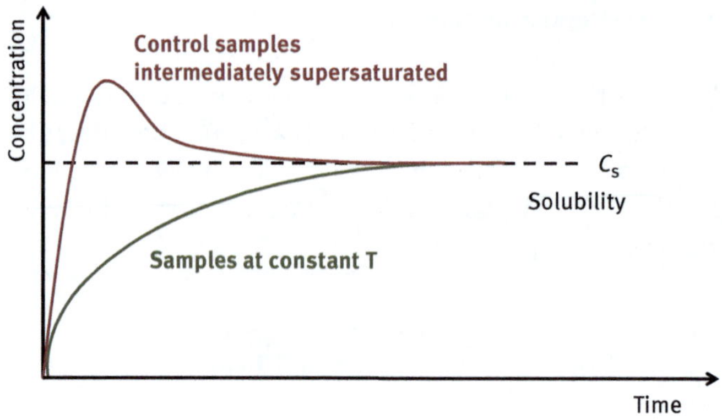

Figure 2.11: Idealized outcome of solubility measurement by shake flask approach including control experiment starting from conditions of higher solubility. Convergence of the two curves of both experiments confirms the "true" result.

The most common methods to separate solutions from the undissolved particles are filtration and centrifugation.

It is crucial not to change temperature of the equilibrated samples during the separation steps, for example, in centrifuges or during filtration, because the equilibrium will be lost instantly and additional dissolution or precipitation will occur, leading to wrong results.

For filtration, the filter material as well as pore size are the most obvious parameters that will significantly affect the separation. It is good practice to discard the first filtrate to remove manufacture additives that may have remained in the filter material. In addition, it needs to be checked whether the filter material may retain considerable amounts of solute by adsorption and whether adsorption occurs anywhere else during sample preparation, for example, to plastic disposable equipment. Non-specific adsorption is frequently observed for poorly soluble and lipophilic solutes. If this is the case, one may pre-saturate the adsorption capacity or coat surfaces prior to exposing the actual sample.

For centrifugation, the gravitational forces (e.g. rotation speed) and time period affect the particle cut-off of the separation into pellet and supernatant. These experimental parameters usually need to be optimized for each single case according to the stability of the pellet cake to avoid particles of a fluffy solid residue contaminate the supernatant by accident.

Both for filtration and for centrifugation, the definition of the fractions of the dissolved and the undissolved material is a matter of the chosen experimental details.

In the case of clear solutions, optical bulk methods can also be used to determine solubility (e.g., turbidimetry and diffraction). However, these methods may not be useful in the case of colloidal solutions.

If a distinction should be made between colloidal associates that contain both apparently dissolved drug and freely dissolved drug, other experimental methods than centrifugation/filtration or optical methods need to be used to separate the supramolecular assemblies from free molecules. The separation of truly dissolved drug and the colloids can be done by equilibrium dialysis [14]. These studies take several hours/days to reach equilibrium. The results depend on the molecular weight cut-off of the dialysis membrane.

2.2.4.3 Kinetic solubility

The term "kinetic solubility" in this context addresses an experimental approach to solubility that can be carried out in high throughput. It starts from a concentrated solution of the studied compound in a water-miscible solvent (usually DMSO) in which the solute is well soluble. This concentrated solution is added to the aqueous medium of interest until precipitation occurs. The mixture is left to reach equilibrium. Precipitation is a kinetic process and may be hindered, that is, supersaturated states may be present during the experiment. Due to the kinetic approach of the method, the solubility is designated "kinetic solubility". Different solid-state forms may be obtained where according to Ostwald rule of stepwise crystallization, the most unstable phase would precipitate first. Therefore, the kinetic solubility approach may lead in the first place to comparably high solubility values due to the equilibrium with an unstable phase. There is yet another effect that may lead to increased solubility values: The presence of even a small fraction of the "good" solvent in the aqueous medium may alter the solubility of the compound in comparison to the pure aqueous medium.

Therefore, kinetic solubility is generally higher or equal compared to thermodynamic solubility. It is very rarely lower.

Figure 2.12 depicts the time course of experiments illustrating the increased value of kinetic solubility as compared to thermodynamic solubility. Additional discussion of this topic is provided in chapter 7 of this book.

2.2.4.4 Experimental considerations for supersaturation assays

Transient instable states are difficult to detect experimentally because, depending on the conditions of preparation and sampling, they may reach different degrees of supersaturation for different time periods before they collapse. For example, amorphous material may recrystallize at any time if the activation energy in a single spot is reached. Special care must be taken to avoid sudden precipitation that may occur at any time during sample preparation and handling. The closer the

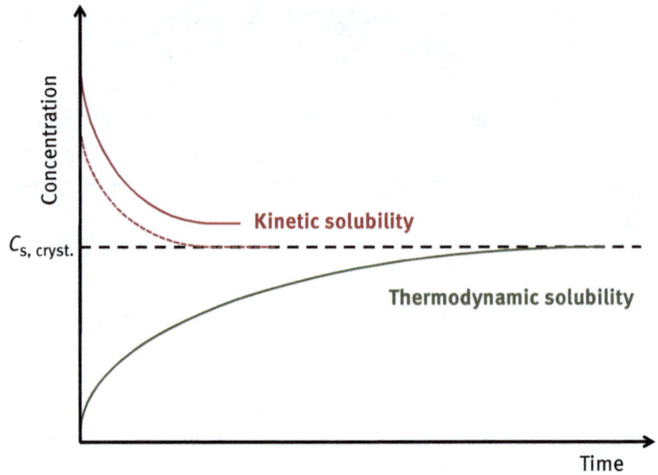

Figure 2.12: Schematic representation of the concentration time course for kinetic solubility in comparison to classical thermodynamic solubility.

concentration is to the maximum degree of supersaturation, the higher is the probability of premature precipitation. The experimental techniques used to determine concentrations of such unstable systems, therefore, need to be carefully chosen. Special care is also needed in terms of temperature control. Samples are diluted with a suitable solvent as soon as possible before the assay to make sure that they stay dissolved under any conditions they may meet. If this is not possible, an aliquot of the solution may be allowed to precipitate, and the precipitate is re-dissolved in a good solvent prior to quantification.

Furthermore, if the recrystallization just concerns the very surface of the amorphous particles, solubility will decrease. However, due to the fact that only a small fraction of recrystallized material is deposited on the surface , the bulk material will stay unchanged and the recrystallized phase will be difficult to detect. Similar considerations apply to hydrates being formed on surfaces of anhydrates.

Even for crystals of one single phase, there is a theoretical expectation of true supersaturation if there is a large difference of the dissolution rates above the different crystal faces. In this case alternative crystal habits can provoke a dissolution rate advantage of a certain habit. In such case, if the differences are large, transitionally enhanced concentrations may occur. The supersaturation along with habit differences will vanish with equilibration time. The dynamic equilibrium in the respective solvent will lead to the most advantageous habit and to large crystals for this current solvent (Ostwald ripening).

2.3 Supersaturation: biopharmaceutical aspects

The most obvious use of the supersaturation principle in the field of pharmaceuticals is the oral administration of poorly soluble drugs. Supersaturation may enable to enhance bioavailability.

Orally administered drugs experience LADME:
- Liberation
- Absorption
- Distribution
- Metabolism
- Elimination

To increase oral bioavailability, the first two, namely liberation, i.e. the release of the drug from the drug delivery system, and absorption are the most important steps that can be addressed by drug formulations. As far as liberation is concerned, enhanced solubility and thus true supersaturation is the goal. Concentration gradients drive the absorption process. For optimum bioavailability, they should be as large as possible and persist for a long time.

To predict oral bioavailability, it is beneficial to study the respective solubilities in fluids mimicking the content in the gastro-intestine ("biomimetic media"). Furthermore, the pH changes and the effect of food on both pH and the composition of the fluids should to be taken into account as well. The simulated biomimetic media (e.g., FaSSIF and FeSSIF) are complex blends revealing a set of different micelles and mixed micelles that interfere with both solubility and supersaturation. However, it has been shown that aspirated human intestinal fluids can show quite different degree of stabilization [15].

A simplified representation of absorption processes from solutions showing supersaturation by amorphization (enabling formulations) of the drug and/or micellization is given in Figure 2.13. It is restricted to passive transport. In all cases micelles are contained in intestinal fluids (bile salt micelles and mixed micelles) to a different extent depending on food intake. In addition, formulation additives may also form micelles and other supra-molecular structures. Thus, Figure 2.13 represents also the possible effect of food on permeation and absorption.

In vivo, dissolution and absorption are parallel and simultaneous processes. Fast dissolution rates will lead to a higher absorption rates and to increased bioavailability. The critical degree of supersaturation may thus not be reached and precipitation may not occur.

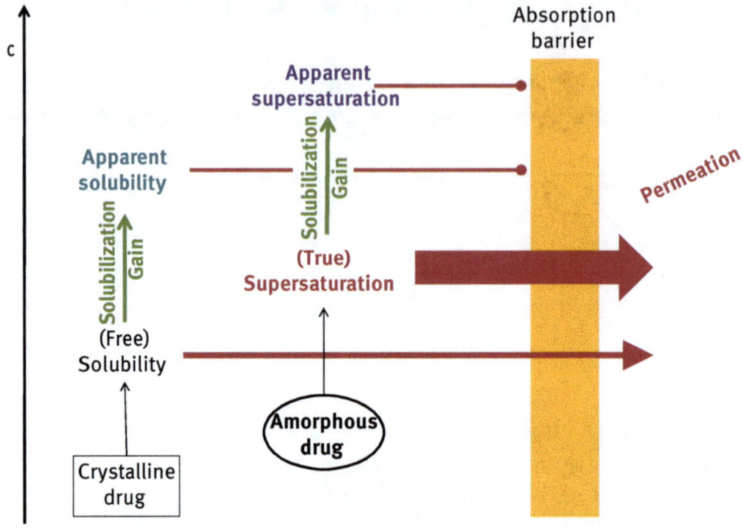

Figure 2.13: Schematic representation of the contribution of freely dissolved drug, true supersaturation, and micellar solubilized drug (i.e., apparent supersaturation) to the permeation and absorption.

2.3.1 Extent and persistence of supersaturation

To increase the rate and extent of absorption, fast dissolution to a high, preferably even supersaturated concentration is desired. Moreover, the high concentration should be kept for a considerable time at the site of absorption. Both the extent and the duration of supersaturation will increase the absorbed amount of drug and thus increase bioavailability.

This perception has been vividly illustrated by the "spring and parachute" principle as depicted in Figure 2.14. The figurative terms "spring" and "parachute" are used to illustrate a high concentration of the drug in gastrointestinal fluids maintained for a prolonged time period.

It is desirable to dissolve as much drug as possible and as fast as possible, while avoiding precipitation, because precipitates cannot be absorbed as such. They first need to re-dissolve when the solution is deprived of drug and its concentration decreased by the absorption process.

Keeping the spring and parachute model in mind, the dissolved drug over time that is prone to absorption correlates with the area under the curve (AUC) in concentration vs. time profile as depicted in Figure 2.14. Highly supersaturated systems that precipitate faster lead to less absorption leading to a smaller AUC and to a lower bioavailability. Thus, it appears beneficial to restrict the degree of supersaturation that is reached by a formulation to a lower level that in turn can last for a longer time.

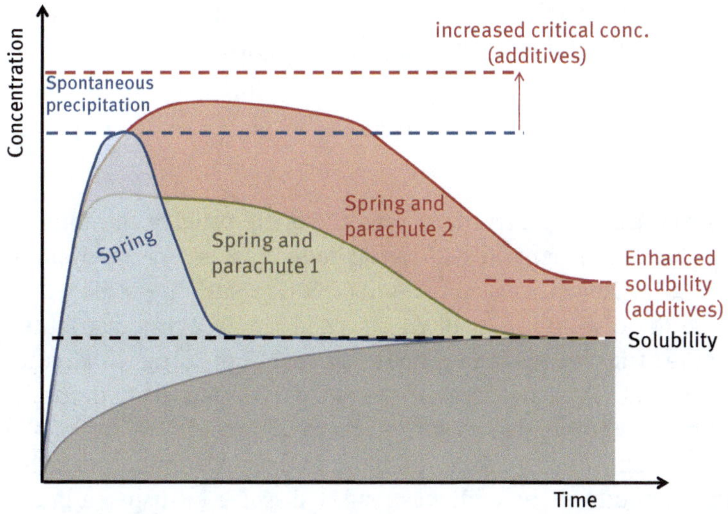

Figure 2.14: "Spring and parachute" principles affecting the available dissolved drug over time. The figure distinguishes parachutes that increase the (apparent) solubility of the drug as well as increase the concentration of spontaneous precipitation (red colour) and those that do not show this effect.

When comparing literature data, there are many cases of large discrepancies between reported kinetic solubilities and the degrees of supersaturation. This fact may be due to different rates at which the supersaturation has been generated: If a supersaturating system is produced at a faster rate, the degree of supersaturation will reach a higher value. Consequently, the real solubility advantage of supersaturating drug delivery systems cannot be determined accurately.

Precipitation inhibitors allow to keep a higher degree of supersaturation before precipitation occurs. They increase the metastable zone width: the critical concentration for the precipitation is increased, and higher degrees of apparent supersaturation can be reached without the risk of precipitation. In such case the AUC can be even further increased and bioavailability can be enhanced.

However, precipitation inhibitors – as well as any "parachutes" – are additives in the solutions that may increase the equilibrium solubility of the drug compound, and thus the degree of supersaturation may be less than estimated by comparison to the pure solvent. It can be disputed about the effect of "parachutes" on the molecular level: Would they let the drug remain as free, truly dissolved molecules, or are the drug molecules possibly associated with larger assemblies thus being apparently dissolved? If the latter is the case it will influence diffusion kinetics, drug transport, and bioavailability. By which mechanism a "parachute" works – may it leave the molecules dispersed, dissolved, associated, re-dispersed or re-dissolved, replenishing the solution from associates or from precipitate may be disputed. However, the common aim is to avoid the crystallization of large stable crystals, because in any case these hamper better bioavailability.

The "spring and parachute" principle is a useful illustration for the increased bioavailability of poorly soluble drugs by enabling formulations.

2.3.1.1 Springs

Fastest dissolution rates are not achieved by presenting the drug in the form of large crystals. The reduction of particle size can already increase the dissolution rate to a significant degree. However, even better dissolution rates are achieved by presenting the drug to the dissolution medium in already molecularly distributed form rather than as pure drug crystals. Amorphous "blends" of the drug with hydrophilic excipients are useful. Typical examples are solid solutions in hydrophilic polymers, co-amorphous systems or co-solvent solutions, and others; for a brief discussion of these, see Section 2.4. and Chapter 8. In all of those, the drug molecules are "separated" from each other. The additives readily dissolve or disperse in the respective dissolution medium, and "free" drug molecules are left alone in the solution. Supersaturation is generated. It is a matter of the kinetics of nucleation and growth how long this state can be preserved.

2.3.1.2 Parachutes

Avoiding precipitation and crystallization bears thermodynamic and kinetic aspects.

From a thermodynamic perspective, "parachute" additives may strictly speaking not only act by precipitation inhibition, but also by increasing thermodynamic solubility. Such solutions are by definition not supersaturated. Known examples for substances increasing solubility are co-solvents, surfactants above their CMC and CDs. The same types of excipients are frequently found in the literature denoted as "parachutes" and precipitation inhibitors although they have solubility-enhancing effects by solubilization or complexation [16, 17].

Kinetic parachutes delay drug precipitation from supersaturated states by delaying nucleation or growth of particles. Increased viscosity of the solutions affecting diffusion rates plays a role. Furthermore, hydrogen bonding between drug molecules and precipitation inhibitor molecules increase the activation energy for the nucleation, leading to a nucleation delay. The interaction between precipitation inhibitor and drug may also inhibit the growth by interaction with the surface of small nuclei.

Drug molecules rich in hydrogen bond donors typically interact best with polymers that act as precipitation inhibitor with hydrogen bond acceptors (e.g. polyvinylpyrrolidone, PVP), while drugs with H-bond acceptors interact with polymers of H-bond donor type (e.g. hydroxypropylmethylcellulose, HPMC). However, this is only a rule of thumb, as steric hindrance, intramolecular forces, and the flexibility of the molecules are also involved. In the ideal case, hydrogen-bonding patterns between the drug and the polymer with regard to numbers of hydrogen donors and acceptors as well as to their conformations fit together. However, simulations highlight that hydrophobic forces and Van der Walls forces may also play a prominent role, in particular, in the case of poorly soluble drugs [18]. It should also be mentioned that approaches to synthesize "designer excipients" have been reported where polymers were modelled by high-throughput controlled polymerization and screening. Such excipients have been found to effectively inhibit precipitation by molecular interaction [19]; however, their effect on solubility and drug transport properties has not been studied. In summary, the usefulness of parachutes is plausible. However, to find the optimum parachute for a given problem relies on screening approaches and the mechanism of action on a molecular level remains blurred.

2.3.2 Supersaturation for improved bioavailability

2.3.2.1 Truly supersaturated systems: impact on PK

It is the basis of the spring and parachute approach to favour systems that can keep the supersaturated state for a long period: Even if the concentration of the molecularly dissolved drug is kept at a relatively low degree of supersaturation, there is a significant impact and good advantage in terms of the total dose fraction that can be absorbed.

In a kinetic setting, the actual (free) drug concentration over time is decisive for bioavailability (AUC). Absorption processes continuously decrease the drug concentration while simultaneously the solution may be replenished from a reservoir. Thus, even if a precipitate has been formed, it is a matter of the rate of re-dissolving, in how far the replenishing process can balance the concentration decrease due to absorption processes. In some cases, it can be advantageous if the drug precipitates in small particles and/or amorphous form, both of which will re-dissolve fast, and above which a supersaturated state can be kept for a long time.

The higher the degree of supersaturation, the higher is the risk of spontaneous precipitation. Therefore, solubility enhancement leading only to restricted degree of supersaturation may be beneficial for bioavailability because such supersaturation may last for longer time. However, if the precipitate is of high energy, which is in many cases likely according to Ostwald rule of stepwise crystallization (see Figure 2.2), it will serve as a reservoir for further dissolution with again improved dissolution rate.

Precipitation inhibitors may or may not be needed in such systems. Other additives may even be more useful if they help to assure the generation of amorphous and other drug-rich particles. The use of some of the precipitation inhibitors can even be counter-productive with regards to enhancing bioavailability, especially if they interact closely with the free drug molecules, generate supramolecular assemblies, and hinder absorption.

In vivo, the effect of temperature on solubility and supersaturation may appear very limited at first glance. However, cooling of a supersaturated system can readily induce precipitation. This effect may need to be taken into account in some cases: a glass of cold water decreases the temperature in the stomach dramatically and it may take quite a while to warm up (approx. 20 min from room temperature). Temperature in the intestines is not as constant as expected swinging in the range of approx. 2 °C [20].

2.3.2.2 Apparently supersaturated systems: impact on PK

It has been widely acknowledged that the associated state of drug molecules has an impact on their transport properties. This perception dates back to the middle of the last century, where absorption from micellar solutions has been studied in animal experiments.

One of the first mechanistic studies of solubilization of drugs (weak acids and weak bases) provided the background to predict the total solubility of poorly soluble drugs (both freely dissolved and solubilized) according to pH based on equilibrium constants [21]. A similar system has been studied in terms of in vivo absorption, where predictions according to binding constants suggested that absorption is not connected to the apparently dissolved drug but the free drug [22].

The discussion of the impact of micellization/solubilization has become more significant since the number of poorly soluble drugs have increased during the past decades, and the use of enabling formulations as well. Mechanistic studies have demonstrated that there is a direct correlation between the passive (transcellular) diffusion and the molecularly (freely) dissolved drug concentration. There is a trade-off for enabling formulations; reducing the free drug concentration due to partitioning into micelles or other solubilizing carriers goes along with the fact that the free drug concentration may be enhanced in supersaturating formulations due to overall increased concentrations [23].

2.3.3 Redistribution from supramolecular assemblies

The replenishing of the solutions for absorbed drug may be decisive for bioavailability. Micellar solutions, including the dispersions in gastric fluids comprising bile salt micelles and mixed micelles that are generated during digestion, may serve as drug reservoirs and continuously replenish the solutions with free drug. It solely depends on the interaction between drug molecules and micelles determining the kinetics of exchange with the "true" solution, which of the processes is rate dependent.

In steady state, and in the absence of a solubilizer, the absorption is described as a first-order process:

$$\frac{dA}{dt} = -k \cdot A \tag{2.5}$$

with A being the amount of drug remaining in the digestive system at any time point t, and k being the absorption constant.

In the presence of solubilizer or complexing agent, a constant fraction of the drug is retained in supramolecular associates and if it is assumed that only the free drug is absorbed; eq. (2.5) becomes eq. (2.6):

$$\frac{dA}{dt} = -k \cdot f \cdot A \tag{2.6}$$

The absorption rate would decrease by a factor f, which is directly connected to the energy of the molecular interaction, and thus stability of the micelles/complexes/supramolecular aggregates. On the other hand, the re-distribution from the colloids for replenishing the solutions upon concentration decrease by absorption is faster from surfaces that have less energy of interaction. One can denote the supramolecular assemblies as compartments of different depth according to interaction energy.

It is difficult to clearly distinguish the different types of interactions of drug molecules with excipients such as solubilizers and precipitation inhibitors and the depth of these compartments. For the oral bioavailability, it is decisive to which extent the interactions affect the availability of drug molecules for transport processes and absorption.

2.3.4 Precipitates as a reservoir to maintain high, supersaturated concentrations

The occurrence of precipitates in the gastrointestinal tract is not always a dramatically severe issue for a formulation in terms of bioavailability. It needs to be considered that precipitation may not in all cases lead to the thermodynamically stable solid-state forms in the form of large, well-defined particles. Rather the opposite is the case: According to the Ostwald rule, it would be the most unstable and consequently best soluble solid-state form, which might be a metastable crystalline form or even an amorphous form of the drug. Depending on the nucleation rate, the particles may also be quite small. Such small and instable particles may even improve bioavailability as compared to fully solubilized systems: The particles will in the first place induce true supersaturation [24]. They will dissolve fast when conditions change, and they can act as a reservoir, providing more drug to be dissolved fast to replenish the medium with drug. An example of such behaviour is discussed regarding micro- and nanoparticles of fenofibrate in marketed formulations. It was observed that biomimetic media enhanced the apparent solubility of fenofibrate via micellar solubilization, but they did not increase transport rates in in vitro permeation studies. However, if nano- or microparticles were present, they served as a reservoir and maintained high levels of molecularly dissolved drug, which in turn caused high and constant permeation rates at different levels according to particle size as expected from bioequivalent doses (200 mg microparticle = 145 mg nanoparticles) [25].

2.4 Drug delivery approaches: supersaturable formulations and supersaturating formulations

2.4.1 Enabling formulations for poorly soluble drugs

Poorly soluble drug candidates are presented in so-called candidate enabling formulations, for which the terms "supersaturable formulation" and "supersaturating" formulations have also been used interchangeably.

Bioavailability enhancing formulation approaches for poorly soluble drugs have traditionally been focused on improving drug solubility. Thus, high-concentration gradients are achieved, and drug precipitation along the gastrointestinal pathway is avoided. The solubility and solubilization of drugs and formulations in settings mimicking physiologic conditions has widely been used to optimize such formulations. This includes the use of solubilizers, precipitation inhibitors, and other additives.

An overview over formulation principles is found in [26] and more specific discussion of the principle of supersaturation in [27]. A collection of preparation methods including high-throughput set-ups and decision guidance on which principle to prefer in [28].

In the following section very short representations of the most common formulation approaches are provided.

2.4.2 Formulation approaches

2.4.2.1 Solutions

Solutions are non-sophisticated formulations that might be used in commercial formulations. Due to their simplicity and comparably owing fast development, they are widely used for toxicological studies, pharmacological or pharmacokinetic studies in animals, which require prompt availability of formulations (also see Chapter 4). However, solutions are in general more prone to stability issues by chemical degradation as compared to solids [29]. This fact may be the reason for the limited use of solutions in drug formulations. In early animal and human studies the storage stability (shelf life) of the formulation is not a critical issue.

For poorly water-soluble drugs, the use of alternative solvents or the use of water–co-solvent mixtures is a straightforward approach. In many cases solvents such as PEG400 and DMSO are used because of their good dissolution ability. However, especially in the case of DMSO, the tolerability is limited [30]. Upon oral administration and thus "dilution" with gastrointestinal fluids the systems may become transitionally supersaturated. Excess drug may be solubilized in micelles and mixed micelles in the media, or precipitation may occur. The precipitate is expected to be either crystalline or amorphous as discussed above. Amorphous drug might be solid particles or droplets in a miscibility gap that can be composed of different components from the gastrointestinal content together with the drug.

If solutions are formulated using surfactants above their CMC, the solubility is increased by micellar solubilization leading to a high apparent solubility and stable solutions are achieved. It is difficult to confirm experimentally whether a real supersaturation may occur upon administration depending on the composition of the gastrointestinal media. The supersaturation effect is also highly dependent on food intake and transition times in the gastrointestinal tract. A re-distribution of the drug into further micelles may occur, hindering precipitation. The drug-filled micelles can replenish the solution with molecularly truly dissolved drug and thus increase bioavailability.

The case of micellar solutions is very similar in terms of conditions and restrictions to the colloidal complexes as will be discussed below.

2.4.2.2 Crystal engineering: polymorphs, habits, particle sizes

For formulations using the solid drug as pure component solid-state properties may largely impact solubility and dissolution rates. Like formulations based on solutions, such an approach using, for example, "powder in the capsule" formulations are frequently used for early clinical trials. Again, these formulations are of limited complexity and can usually be developed quickly.

Polymorphs have different lattice energies. If this difference in lattice energy is large enough to be of practical relevance, they will show significantly different solubilities as well as dissolution rates. This makes the choice of the best polymorph decisive and metastable polymorphs might be useful for such formulations. A more in-depth discussion of solubilities of polymorphs and pseudo-polymorphs is provided in Chapter 9.

Crystallization from different solvents can lead to different habits of the same polymorphs that have altered relative surface areas of the different crystal faces. Different crystal faces have different surface properties, which promote differences in solvent–solids surface interactions. Wetting properties as well as dissolution rates above different crystal faces can differ widely. A closer discussion of this topic is given in Chapter 9.

The smaller the particles, the larger becomes the specific surface area of the material. This improves dissolution rate and thus bioavailability for poorly soluble drugs where exposure after oral intake limits absorption. Top-down processing by milling does not only lead to smaller particles, but it may also lead to a high proportion of less hydrophilic crystal faces because the crystals most likely breaking plane is that of weakest interactions. These typically are non-polar and less hydrophilic. Change in habit may occur by milling and result in a gain of the proportion of the hydrophobic crystal faces. An example of surface property alteration by milling a drug is described in [31]. Thus, milling can both increase dissolution rate by increased surface area but also decrease it by increased proportions of hydrophobic faces.

2.4.2.3 Pharmaceutical salts

For all APIs that are acids or bases, the use of pharmaceutical salts has classically been a means to increase solubility. Typically salts crystallize better than free acids and bases, have more reproducible particle and bulk properties, and are therefore also preferred for processing. There is a wide choice of pharmaceutically acceptable counter ions. The choice of pharmaceutical salts has many practical aspects beyond solubility and dissolution rate, for example, habit, stability, propensity for electrostatic charging, tabletability, and so on [32].

Pharmaceutical salts dissolve much faster than the corresponding free acids or bases. Even at the "wrong" pH, pharmaceutical salts act as "springs" because the dissolution process starts at the very surface of the crystals where the more hydrophilic salt form is present. The ions are released into the diffusion layer. The buffering effect of the released ions in the close proximity of the solid alters the local pH and allows more salt to dissolve. The pH jump into the bulk medium can lead to true supersaturation for a transitional period and enhance absorption.

2.4.2.4 Co-crystals

Co-crystals are basically very similar to both pharmaceutical salts and "pseudo-polymorphs", namely by being composed of the active drug molecules and at least one other compound type in the same crystal. While for solvates and hydrates the guest molecules of the drug crystals are substances that are liquid at room conditions, the co-formers for co-crystals are solids. The difference between a co-crystal and pharmaceutical salt is that for the latter, a proton is transferred between the two compounds and ions are generated. As a rule of thumb, this is the case if the pK_a-values of the two components differ by more than two. For co-crystals, on the other hand, the proton is not transferred between the molecules but shared due to a smaller pK_a-value difference. From that point of view, co-crystals appear closely related to "pseudo-polymorphs" such as hydrates and solvates. However, in contrast to hydrates, that typically are less soluble than their respective anhydrates, co-crystals are usually better soluble due to the typical choice of hydrophilic and well soluble co-formers. Their solubility and dissolution rates may be higher or lower than for salts [33].

The solubility advantage of co-crystals may be compromised if the parent drug precipitates. This is most critical if high degrees of supersaturation are created. In that case it would be a valuable strategy to limit the dissolution rate and thus also limit the degree of supersaturation that will be reached. An approach is to use excess of the co-former: By the "common co-former effect", the solubility of the co-crystal and the dissolution rate are reduced. This leads to a lower degree of supersaturation [34]. Furthermore, the pH partitioning in the presence of micelles will also play a role and affect solubility, precipitation, and dissolution rates [35].

2.4.2.5 Nanocrystals

For nanocrystal systems the diameters of the drug particles are below 1,000 nm, and typically not below 100 nm. Increased solubility is probably not the main reason for their frequent use in drug formulation. Increased dissolution rates – above the predicted values from larger surface area – are discussed to be due to surface energies that increase with decreasing particle size, especially if nanocrystals are prepared by top-down methods, for example, milling. Defects and (partial) amorphization of the particle surfaces may be induced by the processing. Therefore, in most cases additives are used to protect the surfaces of the nanocrystals by a stabilizing layer of surfactants or other type of surface modifier to prevent crystal growth based on Ostwald Ripening, aggregation, and sedimentation [8, 36].

Furthermore, presence of surface modifiers has been argued to possibly also act as wetting agents, precipitation inhibitors and even hydrotropic agents. It is not necessary to distinguish these effects from one another, because all of them help to reach the observed high dissolution rates and possible supersaturation, which is a common feature of nanocrystal systems.

2.4.2.6 Amorphous and partly amorphous solid dispersions (ASD)

In solid dispersions the drug is dispersed in the form of small particles in a hydrophilic matrix typically being polymers or polymer mixtures. The dispersed particles might be amorphous or partly amorphous. Their amorphous state is stabilized or their recrystallization is hindered by the polymer. The dissolution rate and the apparent solubility are increased due to good wetting and dissolution properties of the hydrophilic matrix.

Typically, surfactants are added to the formulations to reduce the glass transition temperature of the polymers and to improve the processing properties. These excipients, both the polymers and the surfactants, are the same classes of polymers that are found denoted as "crystallization inhibitors". They may therefore act as precipitation inhibitors as well (spring and parachute principle) and stabilize the amorphous state of the dispersions. Examples of kinetically stabilized ASDs have been reported to retain their amorphous state even after storage under ambient conditions in closed containers for as long as 25 years [37]. The principle of the ASDs is successfully utilized in a number of drug products on the market for which ASDs are typically processed by hot melt extrusion, spray drying, or bead layering.

Upon dissolution of the ASDs, possibly true supersaturated states can be reached. However, in the case of interactions between drug and polymers or surfactants (depending on their quality), apparent solubility enhancement by solubilization may be more pronounced. Even if precipitation occurs, it may lead to amorphous nano- or microparticles which re-dissolve fast [24].

Overviews on the ASD formulation principle can be found in reviews [38, 39].

2.4.2.7 Solid solutions and co-amorphous systems

The ideal case of solid dispersions is the concept of solid solutions where the drug is distributed in the carrier as single molecules. If the fraction of drug in the carrier is kept below its solubility level in the carrier, the preparation is thermodynamically stable. Upon dissolution, supersaturation will be reached. The extent and duration of supersaturation depends on the co-dissolving matrix material. However, these ideal conditions of miscibility need to be carefully analysed [40].

In contrast to solid solutions, that traditionally contain polymers as excipients, co-amorphous drug formulations are based on binary systems of small molecules, for example, comprising two different APIs or using excipients of the sugar type or amino acids. Co-amorphous systems are typically prepared in laboratory scale by melt quench, spray drying, or extended ball milling procedures at low temperatures. If the temperature during ball milling is high, the material will be activated risking re-crystallization in the form of a less stable polymorph [41]. The principles regarding amorphization, crystallization, stabilization, solid-state characterization, as well as dissolution behaviour is similar to the ASDs.

2.4.2.8 Mesoporous silica

Ordered mesoporous silica (SiO_2) particles are sub-micron particles with pores of intermediate size – width 2 –50 nm – in hexagonal arrangement [42]. Drug molecules are deposited from a solution onto the large inner surface area of the pores.

The dimensions of the pores shall be as small as to prevent nucleation and crystallization inside. On the other hand, the high mobility of the drug molecules in the system promotes assembly of crystalline nuclei and thus crystallization. Therefore, the critical amount of drug relative to silica has to be experimentally determined. The formulations increase the dissolution rate and reach supersaturation. Precipitation inhibitors may be useful additives.

2.4.2.9 Complexes: inclusion complexation with cyclodextrins

CDs are "basket"-shaped cyclic glucose oligomers that yield stoichiometric drug: CD complexes. The CDs shield lipophilic parts of API molecules in the interior of their "basket" while the outside of CD molecules is quite hydrophilic. Thus, CD formulations increase the apparent solubility of the drug. Depending on the drug, CD complexing constant, and local concentrations of competing molecules (e.g., bile salts), the complexes may rapidly release free drug molecules and possibly reach supersaturated states. By using amounts of CD above the stoichiometric fraction needed for full complexation the extent and persistence of supersaturation can be tuned: Slowing down the release rate may avoid too high degrees of supersaturation and thus minimize the risk of spontaneous precipitation (also see Chapter 8).

2.4.2.10 Polymeric micelles

The principle of polymeric micelles is very much the same as for complexes: There is hydrophobic–hydrophobic interaction between the micelle-former and the drug molecules. Colloidal associates of stoichiometric or non-stoichiometric compositions are formed. Polymeric surfactants are built from a diversity of building blocks to tailor-make the balance between hydrophilic and lipophilic domains. Polymeric surfactants have typically lower CMCs and better tolerability as compared to small-molecule surfactants. Such surfactants are typically not used for oral administration but for intravenous administration where the priority is on the prevention of precipitation rather than on supersaturation [43].

2.4.2.11 Lipid-based formulations (LBFs)

There are several approaches to take advantage of the better solubility of lipophilic "grease ball"-type drugs in LBFs. For all of them the digestive processes are key to the understanding of bioavailability upon oral intake [44].

Liquid formulations in which the drug is pre-dissolved are composed of mixtures of oils, surfactants, and hydrophilic co-solvents. An example is self-emulsifying drug delivery systems [45]. LBFs avoid the solid-to-liquid phase transition process of the drug substance. Upon dispersion into the digestive system the formulations meet other dietary lipids, bile salts, and phospholipids as well as digestive enzymes. The composition of the dispersion changes widely, as well as its pH, both locally with transit and during time. Differential solubility and a complex interplay between different types of micelles, mixed micelles, and oil droplets occur. The absorption rate for drug molecules may be increased by local supersaturation in terms of molecularly dissolved drug when lipid metabolites are absorbed. Local precipitation may also occur. However, re-dissolution of drug in the dynamic environment replenishes the media with drug molecules and promotes further absorption.

Formulations based on phospholipids include solid phospholipid nanoparticles, pro-liposomes, and liposome dispersions. Similar conditions regarding digestion and absorption apply.

Solid formulations are prepared in the form of solid lipid nanoparticles. Only few examples of drug-loaded solid lipid nanoparticles have been studied in vivo so far. Their colloidal size is hoped to qualify for transcytosis, for example, lymphatic uptake in M-cells of Peyer´s patches. The lipid digestion process takes advantage of increased surface areas by their small particle size [46].

2.4.2.12 Prodrugs

Molecular structure alterations such as prodrug approaches reveal new substances that are created with the aim to widely alter their physico-chemical properties. One of the most obvious aims is to increase solubility of the poorly soluble "mother" drug. In such a case, the solubility and supersaturation aspects become less prominent. On the other hand, for such prodrugs the in vivo transition reaction to the original API is the key, including questions about the kinetics of the respective chemical or enzymatic reactions.

2.5 Conclusion

Supersaturating drug delivery systems achieve high dissolution rates and high local drug concentrations. Thus, faster passive drug transport and better oral bioavailability can be expected. This makes supersaturation an attractive formulations approach in the development of drugs with limited oral bioavailability. Several promising drug delivery systems based on supersaturation as a principle have intensively been studied and quite a number of them have entered the drug market [47].

For a consistent and meaningful description of the mechanism behind such drug delivery systems, the correct use of nomenclature is necessary to differentiate true supersaturation from stable systems with increased solubility. It is proposed to clearly distinguish true solubility from apparent solubility, as well as true, freely dissolved drug molecules and their apparently dissolved and solubilized states. If so, the term "supersaturation" also needs to distinguish the unstable and stable states, implying their potential to increase bioavailability.

Drug formulation principles to increase bioavailability have briefly been summarized with respect to their general principles regarding true and apparent solubility, ditto supersaturation states, and their potential and limitations for bioavailability enhancement.

References

[1] IUPAC. Solubility Compendium of Chemical Terminology, 2nd ed. (the "Gold Book"). Compiled by A. D. McNaught and A. Wilkinson. Blackwell Scientific Publications, Oxford (1997). XML on-line corrected version: http://goldbook.iupac.org (2006-) created by M. Nic, J. Jirat, B. Kosata; updates compiled by A. Jenkins. ISBN 0-9678550-9-8. https://doi.org/10.1351/goldbook. Last update:2014-02-24; version: 2.3.3. DOI of this term: https://doi.org/10.1351/goldbook.S05740.

[2] FDA Guidance for Industry: ANDAs: Pharmaceutical Solid Polymorphism, U.S. Department of Health and Human Services Food and Drug Administration Center for Drug Evaluation and Research (CDER) July 2007 https://www.fda.gov/downloads/Drugs/GuidanceComplianceRegulatoryInformation/Guidances/UCM072866.pdf, accessed 04.10.2018.

[3] Perlovich, GL., Surov, AO., & Bauer-Brandl, A. Thermodynamic properties of flufenamic and niflumic acids – specific and non-specific interactions in solution and in crystal lattices, mechanism of solvation, partitioning and distribution. J. Pharm. Biomed. Analysis. 2007, 45, 679–687.

[4] Wassvik, C.M., Holmén, A.G., Draheim, R., Artursson, P., & Bergström, C.A.S. Molecular characteristics for solid-state limited solubility. J. Med. Chem. 2008, 51, 3035–3039.

[5] Bergström, C.A.S., Wassvik, C.M., Johansson, K., & Hubatsch, I. Poorly soluble marketed drugs display solvation limited solubility. J. Med. Chem. 2007, 50, 5858–5862.

[6] IUPAC Ostwald Ripening IUPAC. Compendium of Chemical Terminology, 2nd ed. (the "Gold Book"). Compiled by A. D. McNaught and A. Wilkinson. Blackwell Scientific Publications, Oxford (1997). XML on-line corrected version: http://goldbook.iupac.org (2006-) created by M. Nic, J. Jirat, B. Kosata; updates compiled by A. Jenkins. ISBN 0-9678550-9-8. https://doi.org/10.1351/goldbook. Last update:2014-02-24; version: 2.3.3. DOI of this term: https://doi.org/10.1351/goldbook.O04348.

[7] Grant, D.J.W., & Brittain, H.G. Solubility of Pharmaceutical Solids, Physical Characterization of Pharmaceutical Solids, Brittain H.G., ed., Marcel Dekker Inc, New York, 1995, 229–237.

[8] van Eerdenbrugh, B., Vermant, J., Martens, JA., Froyen, L., van Humbeeck, J., van Den Mooter, G., & Augustijns, P. Solubility increases associated with crystalline drug nanoparticles: Methodologies and significance. Mol. Pharm. 2010, 7, 1858–1870.

[9] Pharmacopoia Europaeia 9.8; 2019, monograph 2. 9.61: Characterization of crystalline solids by microcalorimetry and solution calorimetry, EDQM, Strasbourg.

[10] Mosharraf, M., & Nyström, C. Apparent solubility of drugs in partially crystalline systems. Drug. Dev. Ind. Pharm. 2003, 29, 603–622.

[11] Morrison, J.: A tale of two drug delivery strategies: simple solubilization or sophisticated supersaturation?, American Pharmaceutical Review, March 31, 2016; available at: https://www.americanpharmaceuticalreview.com/Featured-Articles/184363-A-Tale-of-Two-Drug-Delivery-Strategies-Simple-Solubilization-or-Sophisticated-Supersaturation/; approached 26.02.2019.
[12] Serajuddin, A. T. M., & Jarowski, C. I. Effect of diffusion layer pH and solubility on the dissolution rate of pharmaceutical acids and their sodium salts. II: salicylic acid, theophylline, and benzoic acid. J. Pharm. Sci. 1985, 74, 148–154.
[13] Dokoumetzidis, A., Papadopoulou, V., Valsami, G., & Macheras, P. Development of a reaction-limited model of dissolution: application to official dissolution tests experiments. Int. J. Pharm., 355, 2008, 114–125.
[14] Frank, KJ., Rosenblatt, KM., Westedt, Ulrich., Holig, P., Rosenberg, J., Magerlein, M., Fricker, G., & Brandl, M. Amorphous solid dispersion enhances permeation of poorly soluble ABT-102: True supersaturation vs. apparent solubility enhancement. Int. J. Pharm. 2012b, 437, 288–293.
[15] Bevernage, J., Forier, T., Brouwers, J., Tack, J., Annaert, P., & Augustijns, P. Excipient-mediated supersaturation stabilization in human intestinal fluids. Mol. Pharmaceut., 8, 2011, 564–570.
[16] Price, DJ., Ditzinger, F., Koehl, NJ., Jankovic, S., Tsakiridou, G., Nair, A., Holm, R., Kuentz, M., Dressman, JB., & Saal, C. Approaches to increase mechanistic understanding and aid in the selection of precipitation inhibitors for supersaturating solutions – A PEARRL review. J. Pharm. Pharm. 2018. doi 10.1111/jphph.12927.
[17] Xu, S., & Dai, W.-G. Drug precipitation inhibitors in supersaturable formulations, Int. J. Pharmaceutics, Vol. 453, 2013, 36–43.
[18] Gao, Y., & Olsen, K.W. Drug-polymer interactions at water-crystals interfaces and implications for crystallization inhibition. J. Pharm. Sci 2015, 104, 2132–2141.
[19] Ting, JM., Tale, S., Purchel, AA., Jones, SD., Widanapathirana, L., Tolstyka, ZP., Guo, L., Guillaudeu, SJ., Bates, FS., & Reineke, TM. High-throughput excipient discovery enables oral delivery of poorly soluble pharmaceuticals. ACS Cent. Sci. 2016, 2, 748–755.
[20] Koziolek, M., Grimm, M., Becker, D., Iordanov, V., Zou, H., Shimizu, J., Wanke, C., Garbacz, G., & Weitschies, W. Investigation of pH and temperature profiles in the gi tract of fasted human subjects using the intellicap system. J. Pharm. Sci. 2015, 104(9), 2855–2863.
[21] Rippie, E.G., Lamb, D.J., & Romig, P.W. Solubilization of weakly acidic and basic drugs by aqueous solutions of polysorbate 80. J. Pharm. Sci., 53(1964), 1346–1348.
[22] Levy, G., & Reuning, R.H. Effect of complex formation on drug absorption I. J. Pharm. Sci., 53, 1964, 1471–1475.
[23] Buckley, ST., Frank, KJ., Fricker, G., & Brandl, M. Biopharmaceutical classification of poorly soluble drugs with respect to "enabling formulations". Eur. J. Pharm. Sci. 2013, 50, 8–16.
[24] Frank, KJ., Westedt, U., Rosenblatt, KM., Rosenberg, J., Maegerlein, M., Fricker, G., & Brandl, M. The amorphous solid dispersion of the poorly soluble ABT-102 forms nano/microparticulate structures in aqueous medium: Impact on solubility. Int. J. Nanomedicine 2012a, 7, 5757–5768.
[25] Sironi, D., Rosenberg, J., Bauer-Brandl, A., & Brandl, M. Dynamic dissolution-/permeation-testing of nano- and microparticle formulations of fenofibrate. Euro. J. Pharm. Sci., 96, 2017, 20–27.
[26] Williams, H.D., Trevaskis, N.L., Charman, S.A., Shanker, R.M., Charman, W.N., Pouton, C.W., & Porter, C.J. Strategies to address low drug solubility in discovery and development. Pharmacol. Rev. 2013, 65(1), 315–499.
[27] Brouwers, J., Brewster, ME., & Augustijns, P. Supersaturating drug delivery systems: the answer to solubility-limited oral bioavailability?. J. Pharm. Sci. 2009, 98(8), 2549–2572.
[28] Friedgeirsdottir, GA., Harris, R., Fischer, PM., & Roberts, CJ. Support tools in formulation development for poorly soluble drugs. J. Pharm. Sci., 105, 2016, 2260–2269.

[29] Waterman, KC. Understanding and predicting pharmaceutical product shelf-life, Handbook of stability testing in pharmaceutical development, Huynh-Ba K, Ed., Springer, New York, 2009, 115–135.
[30] Strickley, R.G., & Reza, O. Solubilizing vehicles for oral formulation development, Chapter 9 in, Augustijns, P, Brewster ME, eds., Solvent Systems and Their Selection in Pharmaceutics and Biopharmaceutics, 2007, 257–308.
[31] Heng, J.Y. Y., Thielmann, F., & Williams, D.R. The effects of milling on the surface properties of form i paracetamol crystals pharmaceutical research, Vol. 23, No. 8, August 2006 1918–1927.
[32] Stahl H.P., Wermuth C.G. eds, Handbook of Pharmaceutical Salts, IUPAC Wiley –VCH, 2nd. ed., Zürich, 2011.
[33] Canavagh, K.L., Maheshwhari, C., & Rodríguez-Hornedo, N. Understanding the differences between cocrystal and salt aqueous solubilities. J. Pharm. Sci, 107, 2018, 113–120.
[34] Yamashita, H., & Sun, C.C. Harvesting potential dissolution advantages of soluble cocrystals by depressing precipitation using the common coformer effect. Cryst. Growth Des. 2016, 16, 6719–6721.
[35] Cao, F., Amidon, G.L., Rodriguez-Hornedo, N., & Amidon, G.E. Mechanistic analysis of cocrystal dissolution as a function of pH and micellar solubilization. Mol. Pharm. 2016, 13, 1030–1046.
[36] Ghosh, I., Schenck, D., Bose, S., & Ruegger, C. Optimization of formulation and process parameters for the production of nanosuspension by wet media milling technique: effect of vitamin e TPGS and nanocrystal particle size on oral absorption. Euro. J. Pharm. Sci. 2012, 47, 718–728.
[37] Theil, F., Anantharaman, S., Kyeremateng, SO, van Lishaut, H., Dreis-Kühne, SH., Rosenberg, J., Mägerlein, M., & Woehrle, GH. Frozen in Time: Kinetically Stabilized Amorphous Solid Dispersions of Nifedipine Stable after a Quarter Century of Storage. Mol. Pharmaceutics 14, 2017, 183–192.
[38] Baghel, S., Cathcart, H., & O´Reilly, N.J. Polymeric amorphous solid dispersions: A review of amorphization, crystallization, stabilization, solid-state characterization, and aqueous solubilization of biopharmaceutical classification system class II drugs. J. Pharm. Sci. 2016, 105, 2527–2544.
[39] van Den Mooter, G.,. The use of amorphous solid dispersions: A formulation strategy to overcome poor solubility and dissolution rate. Drug. Discov. today: Technol. 2012, 9, e80–e85.
[40] Marsac, P.J., Shamblin, S.L., & Taylor, L.S. Theoretical and practical approaches for prediction of drug-polymer miscibility and solubility. Pharm. Res. 2006, 23, 2417–2426.
[41] Xu, K., Xiong, X., Zhai, Y., Wang, L., Li, S., Yan, J., Wu, D., Ma, X., & Li, H. Effect of milling conditions on solid-state amorphization of glipizide, and characterization and stability of solid forms. J. Pharm. Bio. Analys. 2016, 129, 367–377.
[42] van Speybroeck, M., Mellaerts, R., Mols, R., Thi, T.D., Martens, J.A., Van Humbeeck, J., Annaert, P., Van Den Mooter, G., & Augustijns, P. Enhanced absorption of the poorly soluble drug fenofibrate by tuning its release rate from ordered mesoporous silica. Euro. J. Pharm. Sci., 41, 2010, 623–630.
[43] Strickley, R.G. Solubilizing excipients in oral and injectable formulations. Pharm. Res., 21, 2004, 201–230.
[44] Crum, MF., Trevaskis, NL., Pouton, CW., & Porter, CJH. Transient supersaturation supports drug absorption from lipid based formulations for short periods of time, but ongoing solubilisation is required for longer absorption periods. Mole. Pharm. 2017, 14, 394–405.

[45] Feeney, OM., Crum, MF., & Porter, CJH. 50 years of oral lipid –Based formulations: Provenance, progress and future perspectives. Adv. Drug. Deliv. Rev. 2016, 101, 167–194.
[46] Mehnert, W., & Mäder, K. Solid lipid nanoparticles: production, characterization and applications. Adv. Drug. Deliv. Rev. 64 (Suppl.) 2012, 83–101.
[47] Jermain, S.V., Brough, C., & Williams, R.O. Amorphous solid dispersions and nanocrystal technologies for poorly water-soluble drug delivery – an update. Int. J. Pharm., 535, 2018, 379–392.

James L McDonagh, John BO Mitchell, David S Palmer and Rachael E Skyner

3 In Silico methods to predict solubility

3.1 Solubility: What is all the fuss about?

Solubility is a vital parameter in many different scenarios like for the direct question "*will compound A dissolve in solvent B?*" to questions such as "*what is the toxicity of a substance?*" [1], "*what environmental impacts may a substance have?*" [2], and "*how sensitive are humans to a particular odor?*" [3]. As a result, many industries have an interest in determining solubility at an early stage in molecular discovery and development, as well as in later stages of chemical and product formulation. Some may even wish to optimize a molecular design to maximize or minimize solubility depending on the use case [4]. As a result, the computational prediction of solubility is an attractive idea as no chemicals are required to be used. Hence, many predictions can be run in parallel at minimal cost in terms of human research time, chemical usage and chemical disposal. Computational methods, therefore, offer the potential to dramatically shorten the time to solution, minimize the environmental impact of molecular discovery and reduce research costs. Nonetheless, the general aim in applying computational methods is not to completely replace laboratory experiments, but rather to guide investigators towards focusing their experimental resources on the most promising areas of research.

Solubility is a particularly important property in the pharmaceutical industry, playing a critical role in determining *pharmacokinetics* – the mechanism by which a substance is transported around the body and excreted; and *pharmacodynamics* – an active substance's pharmaceutical action in vivo.

Pharmacokinetics covers the absorption, distribution, metabolism, excretion, and toxicity (ADMET) properties, which are all influenced by a substance's *bioavailability*, that is, in what concentration and how fast an active compound will be available under physiological conditions at the point of action. For orally administered pharmaceuticals, active substances are required to pass through gastric fluids, cell membranes, and blood to reach their sites of action. This covers a wide

James L McDonagh, IBM Research UK, Hartree Centre, STFC Daresbury Laboratory, Sci-Tech Daresbury, Keckwick Lane, Daresbury
John BO Mitchell, EaStCHEM School of Chemistry and Biomedical Sciences Research Complex, University of St Andrews, St Andrews, United Kingdom
David S Palmer, Department of Pure and Applied Chemistry, University of Strathclyde, Glasgow, Scotland
Rachael E Skyner, Diamond Light Source Ltd., Diamond House, Harwell Science and Innovation Campus, Fermi Ave, Didcot.

https://doi.org/10.1515/9783110559835-003

variety of environments: hydrophobic, hydrophilic, acidic stomach, and slightly acidic to neutral intestinal tract. Many drug molecules are weak acids or bases, meaning that ionization can occur in response to such environmental changes, affecting the probability of absorption at different sites within the gastrointestinal tract. Because of these factors, if a substance has a low solubility, or a significant shift in its solubility due to these changes in environment, then the quantity of that substance available within the gastrointestinal tract and bloodstream may consequently be low. This leads to difficulties in formulation as there may be a high variance in the quantity of the active substance available in different patients [4].

Pharmacodynamics describes the pharmaceutical's therapeutic action. Solubility can affect how easily an active substance can reach a target, and hence contributes to a substance's pharmacological activity. Solubility has become a major source of attrition in the development of new chemical entities (NCEs), and the subject of regulatory conditions for low solubility active substances [5].

3.2 Definitions and concepts

A solution is a homogenous mixture of solute(s) and solvent(s) in any physical state, solid, liquid, or gas, where the solute is the substance dissolved in the solvent. A substance's solubility describes the extent to which the substance can be dissolved in a given solvent, resulting in a solution. Solubility is a thermodynamic property, related to the equilibrium between the solute and solvent. From these general ideas, we can begin to discuss the finer details of solubility and the process of solvation.

3.2.1 Solubility data and experimental determinations

Solubility data, particularly in the pharmaceutical industry, tend to be of two distinct types: *equilibrium solubility* measurements, also called thermodynamic solubility, and *kinetic solubility* measurements. In the following paragraphs we provide a short overview of how solubility data can be measured. A more in-depth discussion of this topic is provided in the chapter 7 "The role of solubility to optimize drug substances – a medicinal chemistry perspective" of this book.

3.2.1.1 Equilibrium solubility

Equilibrium solubility is the concentration of solute in equilibrium with its saturated solution. These measurements are often made at the later stages of pharmaceutical research or during development [6, 7].

The classical method for determining equilibrium solubility is the *shake flask method* [8], which involves mixing a sample of a solute with a solvent. The suspension is then shaken until equilibrium is reached and a final solubility determined by the dissolved concentrations after liquid-solid phase separation. This is often achieved using high-pressure liquid chromatography. This method relies on long shaking times, as it is difficult to determine when the solution has reached equilibrium.

A more modern approach is the *CheqSol* (*ch*asing *eq*uilibrium *sol*ubility) method [9–12]. The procedure involves acid-base titrations between concentrations slightly above and slightly below that where precipitation occurs. A major advantage of the CheqSol method over others is its speed. The CheqSol experiment takes approximately 20–80 min [10], whereas traditional shake-flask methods can take days, and other titration-based methods [13] can take up to 10 h.

Another commonly used method is the *synthetic* method [8], which is particularly useful for viscous solutions. This method uses a laser and detector to determine the equilibrium point by a significant drop in laser light reaching the detector, signifying that the solute is no longer entering solution.

3.2.1.2 Kinetic solubility

Kinetic solubility is the solubility at which an induced precipitate is first detected. The kinetic solubility value is attributable to a metastable state, which results from a supersaturated solution, a solution in which the concentration of the solute is greater than its equilibrium value. Hence, typically these values suggest a compound's solubility to be higher than the true equilibrium solubility. As mentioned in 3.2.1.1, equilibrium measurements often require a long time, and thus kinetic solubility measurements have emerged as an alternative due to their speed, allowing fast screening in the early stages of research. Kinetic solubility values can be helpful in guiding the experimental design of a NCE towards an optimal solubility [6, 7]. On the other hand, they have to be considered with care as results frequently refer to the amorphous solid-state form and accordingly they can also be misleading.

Kinetic solubility is often determined using *turbidimetric assays*. Turbidimetry is a process that measures the loss of transmitted light intensity due to the scattering effect of suspended particles. This usually involves mixing the solute with a solvent and using UV spectroscopy for detection when precipitation occurs. The solubility is then determined based on the concentration that has been added to the solvent. Experimental solubility determinations have been discussed in much more detail by Lipinski et al. [14], and by Alsenz & Kansy [15], and in chapter 7 of the current volume.

3.2.2 Intrinsic solubility

An important definition in the solubility literature is *intrinsic solubility*. The intrinsic solubility of an ionizable molecule is defined as *the equilibrium solubility of the unionized form at a given set of thermodynamic conditions* [16, 17].

This is a significant quantity in many industries where it is used to indicate the bioavailability of a substance. This is important for pharmaceuticals – where bioavailability determines how effective an API can be and the dose required – as well as in industries like agrochemicals where environmental concerns are critical for pesticide and insecticide development. Several well-established models also link the intrinsic solubility to pH dependent solubility and the dissolution process through the *Noyes-Whitney equation* [18] (eq. (3.1)) and *Henderson-Hasselbalch equation* [19] (eqs. (3.4a) & (3.4b)), as discussed in more detail below. Intrinsic solubility is generally referred to using the notation S_0 or, for its base 10 logarithm, log S_0. The solubility is most often referred to units of moles per litre (M), though when considered as an equilibrium constant it is technically unitless.

3.2.3 The solvation process and factors which affect solvation

3.2.3.1 Solvation: equilibrium solubility and dissolution

The process of solvation involves two clear and distinct concepts, the first is equilibrium solubility (described in 3.2.1.1) and the second is *dissolution*. Dissolution is a kinetic property and describes the rate at which molecules become available for dispersal from the solid solute into solution. Dispersal of solute molecules through a solvent continues until a constant equilibrium concentration is achieved. Solubility and dissolution are important concepts, particularly in the context of pharmaceuticals, as drug delivery is impacted by the dissolution rate and solubility whilst drug activity is impacted by a solute's equilibrium solubility [8]. The dissolution rate can be described by the Noyes-Whitney equation [18]:

$$\frac{dW}{dt} = \frac{kA(C_s - C)}{L} \tag{3.1}$$

Equation (3.1). *Noyes-Whitney equation: dW/dt is the rate of dissolution, A stands for the solute surface area that is in contact with the solvent, C represents the instantaneous solute concentration in the bulk solvent, C_s is the diffusion layer solute concentration (given from the solubility of the molecule with the assumption that the diffusion layer is saturated), k is the diffusion coefficient, and L is the diffusion layer thickness.*

3.2.3.2 Thermodynamic effects on solubility

There are several physical and chemical factors that affect a substance's solubility. As discussed above, solubility is a thermodynamic property, thus thermodynamic variables such as temperature and pressure influence a substance's solubility.

Temperature affects a substance's solubility in accordance with the second law of thermodynamics: *for an isolated system a spontaneous change will occur in the direction of increasing entropy*. The Gibbs free energy of solvation is composed of enthalpic and entropic terms:

$$\Delta G^{*}_{sol} = \Delta H^{*}_{sol} - T\Delta S^{*}_{sol} \tag{3.2}$$

where the asterisks refer to the 1 M standard state (see below). The entropic term plays an increasingly important role as the temperature increases, since molecular motion increases at higher temperatures, leading to a more disordered system. For example, a gas has a much greater available volume, is more dispersed, and therefore has a higher entropy than a liquid or solution. Hence, dissolving a gas in a liquid is entropically unfavourable, and a gas will generally be less soluble at a higher temperature. Therefore, gas solvation must be enthalpically driven. The opposite is true of a solid, since breaking up an ordered crystal lattice is an entropically favourable process. Thus, a solid generally becomes more soluble at higher temperature, an observation so familiar that it can be considered common sense, with the solution representing a higher entropy state. Solid solvation can therefore be entropically driven.

The partial pressure of a gas *(the pressure that a single gas component of a mixture would have if it alone occupied the same volume at the same temperature as the mixture)* is another example of a thermodynamic variable that affects a substance's solubility. The solution is in equilibrium with the surroundings. So, if the composition, temperature, or pressure of the surrounding gas changes, the partial pressure also changes, and the equilibrium responds giving a change in solubility. This is the thermodynamic explanation of the process which, for example, occurs when one opens a carbonated drink and carbon dioxide escapes from the solution.

When discussing the solubility of a solute material, molecular interactions that exist between the solute molecules need to be accounted for. Where the solute is a solid, due consideration of the nature of the solid-state form is required, whether that be amorphous or crystalline, as discussed in chapter 9 of this book.

The polymorphic form of a crystalline material must also be considered. Closer packing of molecules within a crystal lattice, and stronger interactions, lead to energetically favourable lattice energies. More enthalpy is required to dissociate the energetically favourable crystal, which makes it less soluble. Therefore, the lowest energy polymorph of a compound is the least soluble. A more weakly bound structure, such as a less stable polymorph or especially an amorphous state, is enthalpically easier to break up, and thus the solubility is higher as these structures can dissociate more easily. Polymorphism is highly relevant in the pharmaceutical industry, as polymorphic transitions can substantially change the solubility, pharmacokinetics, and other physicochemical properties of a substance. This is perhaps most famously presented pharmaceutically in the case of the HIV protease inhibitor Ritonavir [20–22].

Ionization also plays an important role in determining a substance's solubility. Many biologically active ingredients fall into a category of being a weak acid or weak base, meaning the active molecules ionization changes, depending on environmental pH. This ionization occurs due to a reaction with the solvent, which forms an equilibrium. An example of the equilibrium states, where water is the solvent, is given in Eqs. (3.3a) & (3.3b) (for an acid – 3a, and for a base – 3b):
Acid:

$$\mathrm{HA} + \mathrm{H_2O} \rightleftharpoons \mathrm{A^-} + \mathrm{H_3O^+} \quad \therefore \quad pKa = -log_{10}\left[\frac{[A^-][H_3O^+]}{[HA]}\right] \tag{3.3a}$$

Base:

$$\mathrm{B} + \mathrm{H_2O} \rightleftharpoons \mathrm{BH^+} + \mathrm{OH^-} \quad \therefore \quad pKb = -log_{10}\left[\frac{[BH^+][OH^-]}{[B]}\right] \tag{3.3b}$$

Equations (3.3a) and (3.3b). *(a) Acid and (b) base equilibria and definitions of* pK_a *and* pK_b*. HA represents an acidic molecule and B represents a basic molecule, the ionized species are then represented by the charged forms.*

The strength of an acid in solution is measured by $pK_{a,}$ which is the negative base 10 logarithm of the acid dissociation constant K_a. A more positive pK_a value represents a smaller extent of dissociation at a given pH, as shown by the Henderson-Hasselbalch equation – eqs. (3.4a) and (3.4b). Generally, as ionization increases, the solubility of electrolytes (substances that dissociate upon solvation into ions and enable the solution to conduct electricity) also increases and the solubility of non-electrolytes decreases. As a result, the pH at which experimental measurements are made is an important factor when assessing a substance's solubility. The total solubility of an ionizable substance is calculable by the Henderson-Hasselbalch equation (for an acid – eq. (3.4a), for a base – eq. (3.4b)) by consideration of the intrinsic solubility, pH of the environment, and pK_a of a substance. Note that the exponent of the rightmost term differs in the case of an acidic or basic solute.

$$log_{10}S^{Acid} = log_{10}S_O + log_{10}\left(1 + 10^{pH - pKa}\right) \tag{3.4a}$$

$$log_{10}S^{Base} = log_{10}S_O + log_{10}\left(1 + 10^{pKa - pH}\right) \tag{3.4b}$$

Equations (3.4a) and (3.4b). *The Henderson-Hasselbalch equation.* S_O *is the intrinsic solubility,* pK_a *is defined in eq. (3.3a), and pH is the acidity or basicity of the solution.*

Solubility is influenced by intermolecular interactions between molecular species. This means the interactions between the solute molecules, interactions between solvent molecules, and the cross interactions between solute and solvent molecules all need to be accounted for when performing solubility predictions. In addition, thermodynamic variables and ionization need to be considered. This leads to many degrees of freedom, hence demonstrating the difficulty faced in making accurate predictions *in silico* [8, 23].

3.3 Computational prediction of solubility

Methods typically applied to solubility prediction broadly fall into two categories: *first principles* calculations and *chemoinformatics.*

First principles calculations generally apply physical modelling methods such as coarse-grained simulations, molecular dynamics (MD) [24], and quantum chemistry. Whilst considerably closer to physics than are chemoinformatics approaches, such methods are rarely *first principles* in the sense of doing fully *ab initio* quantum chemistry on both solvent and solute. Nonetheless, these methods look to solve real physical equations to elucidate the physicochemical processes that are occurring.

Chemoinformatics, in contrast, seeks methods to correlate so called *features* with a property of interest, in our current case solubility. These features range in complexity from simple counts (for example the number of carbon atoms in a molecule), to more complex descriptors such as those representing the topology of a molecule, for example, shape indices [25, 26].

Both first principles and chemoinformatics approaches have their advantages and disadvantages. Chemoinformatics is a data-driven discipline whose models have been shown to provide accurate results quickly once they have been suitably trained. However, these methods usually provide little in the way of phenomenological or mechanistic information and can perform poorly outside of the domain they were trained for. First principles calculations have also been shown to provide good results but are usually a little less accurate and require much more computational time and hence higher cost. Such methods do, however, provide chemical and physical insights into the phenomena and mechanisms that physically transpire.

In first principles and some chemoinformatics models, it is common to apply a thermodynamic cycle to predict the solubility of a molecule. There are two frequently employed cycles: the *fusion cycle* and the *sublimation cycle.* These are shown in Figure 3.1.

The fusion cycle describes the transition from the solid-state to solution-state through an intermediate supercooled liquid-state. From a physical perspective, this provides two free energy changes: the free energy of fusion and the free energy of transfer. The sublimation cycle couples the solid-state to the solution-state through a gaseous state, which again consists of two free energy changes: free energy of sublimation and free energy of solvation or hydration. These cycles, or variations

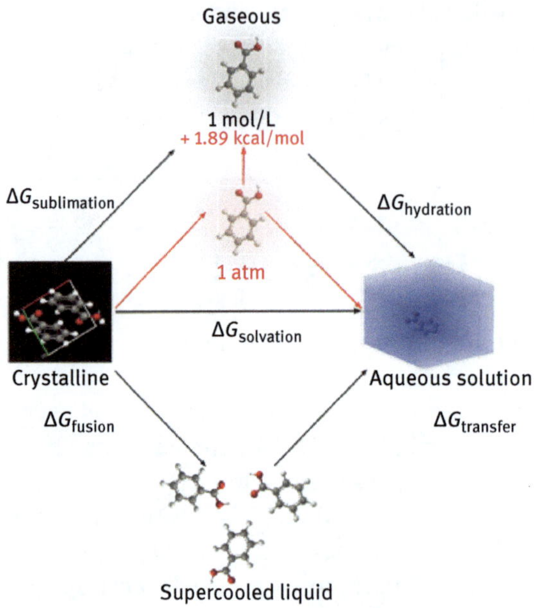

Figure 3.1: Sublimation and fusion cycles used to predict solubility. Representation of the 1 atm and 1M standard states with their difference in energies calculated at 298 K.

on them, have been used for first principles models [17, 27–31], chemoinformatics models [32], and model equations [33].

3.3.1 Standard state conventions

Sublimation energies are typically quoted or calculated in the *1 atm standard state*. In contrast to this, solvation free energies are often given using the *1 M or Ben-Naim* standard-state [34, 35]. This must be accounted for in modelling solubility by a thermodynamic cycle. In this chapter, ΔG^{o} will be used to indicate the 1 atm standard state. ΔG^{*} will be used to indicate the 1 M standard state [34].

The energetic difference across these two standard states is calculated from the work for isothermal expansion or compression of a gas between its initial volume V_i and final volume V_f as:

$$\Delta G = RT\ ln\left(\frac{V_f}{V_i}\right) \tag{3.5}$$

Taking the initial condition to be 1 M and the final as 1 atm, this becomes RT ln (24.46) at 298 K, since the molar volume of an ideal gas at 1 atm is 24.46 L. The corresponding energy difference at 298 K is therefore 1.89 kcal/mol or 7.91 kJ/mol.

When calculating the sublimation energy, the energetic correction is positive in the conversion from 1 atm to 1 M and negative in the conversion 1 M to 1 atm, this is shown diagrammatically in Figure 3.1.

3.3.2 Solubility from first principles

It is very striking that a wide diversity of methods exists for computing solubility on the basis of chemical and physical theory. We will refer to such methods as *first principles* approaches, notwithstanding the fact that to varying degrees they all depend in one way or another on empirical parameters.

3.3.2.1 Computational models of the solvent for first principles calculations

Across computational chemistry, it has been shown that the chemical environment often needs to be modelled to accurately reproduce the chemical and physical characteristics of a system [36]. In some cases, the environment takes an active role in chemical processes, meaning that the environment needs to be explicitly accounted for when modelling the system [36–39].

Models for solvents can be split into two distinct categories: *explicit* and *implicit*, although some hybrid models [40, 41] have also been generated. Implicit models are computationally more efficient but lack the explicit insights into the role of the solvent. Implicit models are particularly common in quantum chemistry [42–44], and some coarse grained simulation is common in chemical engineering. Depending on the property one is considering, the additional detail afforded by explicit models may not be strictly required. Implicit models tend to provide bulk response properties such as polarization, but do not explicitly represent discrete solvent molecules. Implicit models model a solvent as a continuous field, which interacts with a bounding surface around the solute. The reaction field, induced by the charge distribution of the solute polarizing the solvent field, is then evaluated on this surface. This is shown in Figure 3.2. Notable versions of implicit solvation models include the Polarizable Continuum Model (PCM) [45, 46], the Solvation Model based on Density (SMD) [47]. A notable variant of the dielectric continuum models is the Conductor-like Screening Model (COSMO) [48].

Explicit models directly model solvent molecules, thereby accounting for spatial and orientational degrees of freedom of the solvent. Explicit models pay additional overheads for accounting directly for all the degrees of freedom but provide greater insight into the role solvents play in a chemical system. Explicit models are most common in molecular dynamics [24] – a simulation method for studying the physical motion of atoms and molecules, where the trajectories of atoms and molecules are determined from Newton's equations. Explicit models for various solvents have been generated over decades, which has allowed the tuning of efficient parameterized

Figure 3.2: Implicit solvation by a continuum model. A probe shown in yellow is used to trace a surface known as the solvent accessible surface defining a surface area which is accessible for solvent interactions with the solute. A reaction field is then induced as a result of the solute molecule's electrostatic potential polarizing the continuum field.

models to represent the physical processes that are occurring. Genuinely, first principles quantum chemical methods have also advanced to a state at which a limited number of explicit solvent molecules can be included in a calculation [49, 50], in these examples only one or two water molecules are included. In addition, coupled multi-scale methods such as quantum mechanics/molecular mechanics (QM/MM) [51] have allowed quantum chemical calculations of reactions in the presence of a larger solvent environment. Quantum mechanics is used to treat the area in which chemical processes take place at an appropriate level of quantum theory, and molecular mechanics is used to model the rest of the system with a force field. Such advances have enabled computational studies of enzyme reactions [52–54].

There are some emerging hybrid models which offer averaged spatial distribution information without the extra computational cost of explicitly representing the solvent molecules, an important example being the 3-Dimensional Reference Interaction Site Model (3D-RISM) [40, 55–58]. 3D-RISM stands at a level intermediate between the respective kinds of solvent representation used in explicit and implicit solvent models. 3D-RISM uses three-dimensional solvent density distributions, but not spatial coordinates of individual solvent molecules. Although 3D-RISM is based on rigorous statistical mechanics, carefully chosen 3D-RISM functionals are required to obtain numerically accurate results.

3.3.2.2 Modelling water

Aqueous solution is one of the most critical chemical environments, and the subject of much computational as well as experimental research. Therefore, many models have been generated to represent water in various computational methods. In this section we outline some of the common water models. An exhaustive list of water models in computational chemistry is outside the scope of this text, but we provide references to extended discussions on that topic [40, 41, 43, 44, 59, http://www1.lsbu.ac.uk/water/water_models.html accessed 12 August 2019].

The water molecule is composed of two hydrogen atoms covalently bound to a single oxygen atom in a distorted tetrahedral geometry. The oxygen lone pairs above and below the plane of the molecule cause a deviation from the ideal 109.5 ° tetrahedral bond angle, the experimentally determined value being 104.5 °. Water is a polar molecule, having point group C_{2v} and a permanent dipole moment along its principal axis. Water can also act as both a hydrogen bond donor and hydrogen bond acceptor. It can form up to four hydrogen bonds with neighboring molecules. This leads to highly ordered structures being formed; particularly in its crystalline solid-state form, ice, which has many different polymorphic forms that differ in the regular 3D arrangements of the water molecules and can be formed under different conditions of temperature and pressure. One of the notable anomalous properties of water is that water's maximum density is not in the solid phase, as is the case with most substances. In fact, water's maximum density occurs as a liquid around 4 °C.

Generally, explicit water models can be categorized through three characteristics: 1) the number of interaction sites each molecule has, 2) whether the molecule is treated as a rigid or a flexible body, and 3) whether polarizability is included or not.

The most minimal models in common use are the three site models. These models generally have a rigid structure and place interaction sites at the atomic locations within the water molecule. The hydrogen sites are typically represented by a point charge. The oxygen site is represented by a point charge and a Lennard-Jones repulsion-dispersion potential. Regularly utilized models of this type are: *Simple-Point-Charge* (*SPC*) model [60], *SPC-Extended* (*SPC/E*) [61], and *Transferable-Intermolecular-Potential with 3 Points* (*TIP3P*) [37]. The SPC model was the first of these models to be published, followed by TIP3P. Both models are rigid with the SPC model having an ideal tetrahedral HOH angle of 109.5 °, whilst the TIP3P HOH angle is the experimental 104.5 ° [37, 60]. These models were generally fitted to reproduce experimental density and heat of vaporization and typically have a dipole moment marginally larger than gaseous water. The SPC/E model was derived using a heat of vaporization that accounted for the energy needed to polarize the model beyond the gas phase dipole moment. Consequently, this model has a larger dipole moment than SPC. A further addition to the SPC model is the flexible-SPC model, which has been shown to be a very accurate three-site model [62–64]. The flexible-SPC model enables O-H bond distance and angle variations during a simulation. Three site models are very computationally efficient owing to their relative simplicity.

A two-site model also exists, and is based on the SPC model, maintaining the dipole moment, size, and charge separation of the three-site SPC model. This model has been shown to capture the solvation properties of water well for apolar solutes and the bulk solvent [65].

Beyond these models are the four-site models. These models add a fictitious negatively charged interaction site along the principal axis of the water molecule. The additional site better represents the electrostatic charge distribution over the water molecule. Of the four-site models, TIP4P [37] is the most commonly applied, and is used in a variety of forms that have been optimized for different scenarios. One of the most notable optimizations of the TIP4P is the TIP4P/Ew model [66] that has been optimized for use in simulations that employ Ewald summation techniques for treating the long-range electrostatic interactions. The TIP4P/Ew model has been shown to have significantly improved water property reproduction in comparison to the TIP4P model [66]. TIP4P/Ice [67] is a model optimized for simulating and recovering properties of ice structures, with charge and Lennard-Jones parameters used similarly to the TIP4P/Ew model, and fitting equations of state for different forms of ice. TIP4P/2005 [68] is a general-purpose parameterization for the condensed phases of water which shows promising reproductions of experimental properties.

Further models have been constructed adding additional sites to the water molecule with five-site models typically including interaction sites for the oxygen lone pairs and removing the fictitious site from the four-site models. TIP5P is an example of this type of model [38, 69], and has been shown to provide some improvements in reproducing the structure of water clusters.

Most of the models mentioned in the preceding paragraphs are rigid, non-polarizable models, although many have extensions that aim to add parameters to model electronic polarization, for example, TIP3P/Fw and SPC/Fw [64]. However, when introduced as a solvent in biomolecular simulations, or for dilute solutions, their performance decreases. Polarizable water models aim to account for the instantaneous electrostatic environment of each water molecule in solution – that is to say, individual water molecules in solution are inequivalent, as opposed to classical water models, which try to produce an "average" representation of a water molecule.

Generally, fixed point-charge models tend to over-stabilize the water dimer, in comparison to polarizable models [70]. One method to include polarizability is the Drude oscillator model, where a classical charged Drude particle is attached to the water oxygen by a harmonic spring. One such model is SWM4 [71], which represents the permanent charge distribution of water by three point-charges; two on the hydrogen atoms, and an additional point at the HOH bisector. There are five charged sites in total. Other Charge-On-Spring (COS) [72] models include the COS/B1-B2 [73], COS/G2-G3 [74] – based on the TIP4P geometry, and COS/D [75] models. COS/G2 and COS/G3 models are very similar in performance, and arguably better than the COS/D model.

BK3 [76] is a polarizable model with Gaussian spatial distributions of atomic charges as opposed to point charges. Multipolar representations of water molecules can be used in conjunction with MD. These methods go beyond point charges by including dipoles, quadruples and sometimes higher multipoles to improve the representation of the anisotropy of the molecular charge distribution [77, 78], a notable example being Amoeba [79]. Electronic coarse graining now enables very accurate water models to be applied to research problems [80].

Coarse-grained techniques have also been produced. These methods enable larger simulations to be carried out using simplified models. Methods such as coarse-grained MD and dissipative particle dynamics (DPD) are now becoming common in industrial and academic settings. Some efforts are now underway to construct optimal coarse-grained models and investigate the transferability of such models [81, 82].

3.3.2.3 Computing sublimation energies

Those methodologies using the sublimation cycle (Figure 3.1) require the computation of the sublimation free energy. There are three main approaches to this which are used either in solubility calculations, or in crystal structure prediction (CSP). These are firstly the ψ_{mol} approach, secondly the ψ_{crys} method, and thirdly the Einstein Crystal technique.

In chemistry and materials science, for systems where a compound's crystal structure is unknown, prediction methods are widely used. These typically involve generating possible crystal packings and identifying those with the most favourable lattice energies. These methods fall under the category of CSP. Conveniently, lattice energy corresponds almost exactly to the sublimation energy described in Figure 3.1, albeit with a sign reversal. Thus, techniques originally developed for CSP can be incorporated into first principles solubility computation.

One approach [83], which has been popular in CSP, is to obtain the lattice energy with a model potential for the repulsion and dispersion terms, plus distributed multipole analysis (DMA) [78] for the electrostatics. This approach to CSP or lattice energy computation is sometimes known as the ψ_{mol} approach. The name indicates that the wavefunction, or charge density for density functional theory (DFT) methods, is calculated explicitly only for the isolated molecule. From this charge density, one obtains the DMA, giving an atom-atom anisotropic description of the electrostatic interactions. The lattice energy is then minimized by relaxing the lattice parameters and the positions and orientations of the molecules within the unit cell in the lattice minimization programme DMACRYS [83]. This approach is used in simulation-free solubility calculation [17, 30, 84], with the entropic components then being approximated using statistical thermodynamics [85].

A second possibility is to compute the lattice energy by quantum chemistry, which approaches genuine first principles accuracy. The state of the art is periodic DFT, which however requires some correction for the missing dispersion energy. Hoja & Tkatchenko [86], used the PBE0 functional [87] along with a many body dispersion correction and mode-by-mode analysis of the vibrational contributions for their most accurate CSP results. The approach is known as ψ_{crys} since it obtains the wavefunction or charge density for the periodic crystalline system. Curtis et al. [88], similarly used a dispersion-corrected periodic DFT ψ_{crys} approach as part of their genetic algorithm approach to CSP.

Buchholz et al. [84], obtained results for both the ψ_{mol} and ψ_{crys} approaches, and both the melt and sublimation cycles, in their simulation-free study of the relative solubilities of racemic and enantiopure organic crystals. They also discussed the benefits of going beyond the statistical thermodynamic approximation of the vibrational contributions by explicitly computing the contribution of each phonon or molecular vibrational mode. Iuzzolino et al. [89], used ψ_{crys} and ψ_{mol} approaches successively in their CSP protocol.

The performance of modern CSP methods is covered in the literature describing blind tests held periodically by the Cambridge Crystallographic Data Centre (CCDC). There have been six such tests to date, covering a multitude of different models from various researchers [90–95].

A third approach to the sublimation energy, more popular in solubility prediction than in CSP, is to obtain the free energy difference between an Einstein crystal, which is a simple hypothetical model of a solid, and the real crystal by simulation. This approach has been used by Li et al. [31], and by Sanz & Vega [96]. Simulation-based approaches to this problem are discussed at greater length in Moustafa et al. [97].

3.3.2.4 Computing hydration energies

The computation of hydration energies has been reviewed at length by Skyner et al. [41], and will only be summarized here. There exists a wide diversity of approaches. Alchemical free energy simulations with explicit solvent models have been used to calculate hydration free energies by Mobley et al. [98], and Westergren et al. [99], amongst others. QM/MM approaches [52] are also possible. Moving down the scale of accuracy and cost, we can consider solvent density without the need for simulations of explicit water molecules by utilizing an integral equation theory model with a suitable free energy functional, for example, 3D-RISM with the UC, PC, or PC + functional [57, 100–102], or classical molecular density functional theory [103], which describes the density of molecules in a fluid rather than the more familiar use of DFT to describe the density of electrons in a molecule. This can provide an excellent compromise between cost and precision. Hydration energies can also be calculated at relatively low computational cost using implicit solvent continuum models [104, 105]. Finally, the COSMO-RS model can also generate good results [42, 106].

3.3.2.5 First principles routes to solubility

Figure 3.3 shows the variety of different thermodynamic cycles used to obtain the chemical potential or free energy difference between crystalline solute and aqueous solution. As discussed earlier, a simple cycle where the total solid-to-solution free energy change is obtained from considering a route via the gas phase is known as a sublimation cycle. The archetypical sublimation cycle is the simple crystal-gas-solution route shown in yellow in Figure 3.3, and as the top half of Figure 3.1. A cycle proceeding via a liquid, often a notional supercooled liquid at room temperature, is known as a melt cycle and corresponds to the bottom half of Figure 3.1.

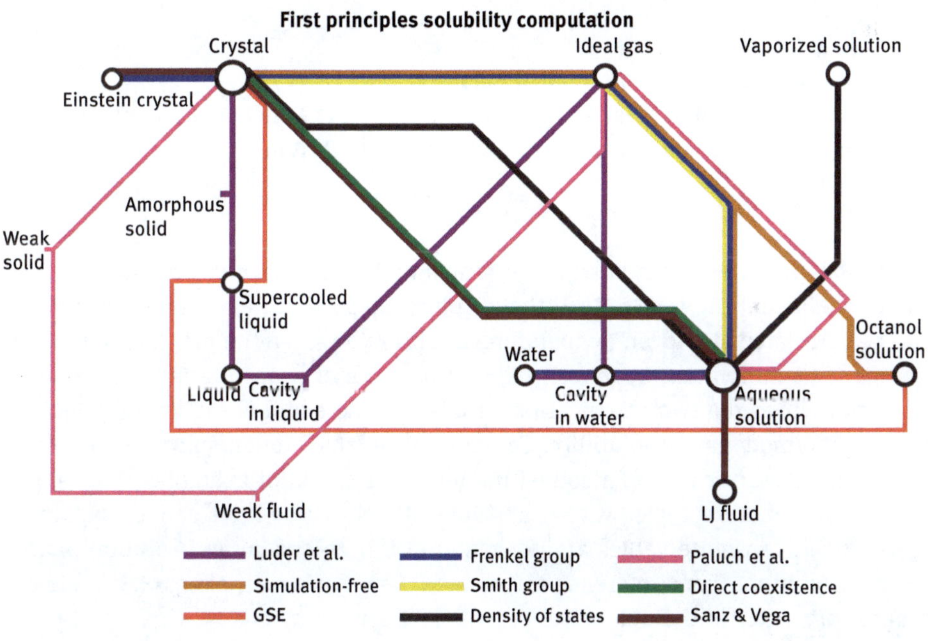

Figure 3.3: States considered in a sample of nine first principles approaches to solubility calculation. All such methods require a suitable route, or thermodynamic cycle, linking the crystal and aqueous solution. This route must permit the change in free energy between crystal and aqueous solution, and therefore the equilibrium constant describing solubility, to be deduced from contributions which can all be computed with available techniques and models. States are included here if they are considered explicitly, visited in simulations, or used as a reference for the calculation of chemical potential or free energy. Some states correspond to real systems, others are hypothetical.

The general solubility equation (GSE) method of Yalkowsky's group which is also discussed in chapters 1 and 7 of this book, [33] uses a cycle of this general kind, though also considering solution in octanol, and corresponds to the red line in Figure 3.3.

3.3.2.5.1 Direct coexistence

One approach is to run an MD or similar simulation of a solute and solvent until equilibrium is reached. This equilibrium will then represent the solubility limit of this system, and the concentration can then be obtained simply by counting the number of solvent and solute particles in the solution phase in the simulation. Kolafa [107], used both the SPC/E model of water [61] and the BK3 model [76] to simulate a slab of crystalline NaCl in direct contact with brine, and obtained solubilities around 3.7 mol/kg, rather less than the experimental 6.1 mol/kg. Such direct coexistence methods tend to be expensive and on occasion are described rather harshly as "brute force" approaches, despite the substantial technical and scientific expertise involved.

3.3.2.5.2 Chemical potentials from simulation

Another popular approach is based on the equality of the chemical potentials of the crystalline solid and aqueous solution phases at equilibrium:

$$\mu_{aq}^{solute} = \mu_{solid}^{solute} \tag{3.6}$$

where the chemical potentials (μ) are functions of temperature, pressure, and, in solution, concentration [23, 27]. If the absolute chemical potentials of both phases can be calculated at a given temperature and pressure, then the problem reduces to finding the solution chemical potential as a function of concentration. The concentration at which the two chemical potentials become equal is the limiting equilibrium or thermodynamic solubility. Computing the absolute chemical potentials of these phases requires comparison with a reference state, for example, an ideal gas as in a sublimation cycle, whose absolute chemical potential can be obtained. Alternatively, absolute values are not required if both crystal and solution phases are referred to the same reference state, since the requirement of eq. (3.6) is simply that the *difference* in their chemical potentials is zero.

Recent work in the Frenkel group [31, 108] calculated the chemical potential of the crystalline phase by using MD to simulate thermodynamically reversible paths between the Einstein crystal and the full crystalline solute. The solution phase chemical potential was obtained by MD simulation of the processes of growing a cavity in SPC water [60], inserting a solute molecule into the cavity, and then finally shrinking the cavity away to leave the molecule in aqueous solution [23]. Although Li *et al.*, in fact, simplified their analysis by assuming high dilution, such that only the insertion of a single isolated solute molecule into pure water need be considered, their method is easily extensible to higher solubilities. They applied their approach to naphthalene and obtained $\log S_0 = -5.32$, almost identical to the experimental value of −5.36 [31]. Whilst they suggest that this essentially perfect agreement may be somewhat fortuitous, it provides a tantalizing hint that their methodology may be highly effective. More recently [108], they calculated the solubility of phenanthrene as $\log S_0 = -6.51$ compared with the experimental −6.96, an error of only 0.45 $\log S_0$ units. However, their calculations underestimated the solubility of caffeine by approximately two orders of magnitude. To investigate the accuracy of their method properly, a much larger set of predictions on several tens of druglike compounds would be required.

The Smith group's work, Moučka et al. [28], similarly seeks conditions where eq. (3.6) is satisfied. They used an Osmotic Ensemble Monte Carlo Approach to calculate the chemical potential of NaCl solutions as a function of concentration, carrying out a series of Monte Carlo (MC) simulations with different numbers of ion pairs solvated in water. They also used MC simulations to compute the chemical potential of solid NaCl. Since both solid and solution chemical potentials were calculated relative to the ideal gas reference state, this work is also an example of the so-called sublimation cycle; see Figure 3.1. Like any approach intended for highly soluble compounds, their methodology is required to operate without any assumptions of high dilution. Like most NaCl solubility calculations, their results both depended quite strongly on the force field used and were underestimated relative to experiment. Their best result was a solubility of 3.6 mol/kg using a potential function from Joung & Cheatham [109], a factor of nearly two smaller than the experimental value of 6.1 mol/kg and similar to the results of Kolafa [107]. Other choices of force field led to values in the range 0.8–1.0 mol/kg. The inadequacies of potential energy functions are likely to be a major cause of underestimation of NaCl solubility in simulation studies.

Paluch et al. [27] also computed the solubility of NaCl. Their work contains a particularly useful discussion of the role of reference states in calculating both relative and absolute chemical potentials. They also obtain μ_{aq}^{solute} as a function of concentration and seek the conditions where it is equal to μ_{solid}^{solute}. The solid chemical potential is found by performing MD simulations along a pathway in which the crystal is transformed via two hypothetical states, a weakly interacting ordered solid and a weakly interacting liquid, into an ideal gas. Path integration is performed along this transformation pathway, so that the solid chemical potential can be found relative to the ideal gas reference state. For the solution, they carry out an analogous process using what they term an expanded ensemble method, linking together the ideal gas and the solution in SPC/E water [61] via solutions of varying concentrations, all simulated with an MC method. Like Moučka et al. [28], they underestimate the NaCl solubility significantly, obtaining 0.8 mol/kg.

Sanz & Vega [96], had earlier used MC simulations and thermodynamic integration to link the solid chemical potential to that of an Einstein crystal, and the solution chemical potential to a hypothetical Lennard-Jones fluid. Both were also related to an ideal gas reference state. They found the solid and solution chemical potentials of NaCl to be equal at a concentration of 5.4 mol/kg in SPC/E water [61], which translates to 4.8 M. Their result is significantly closer to the experimental value of 6.1 mol/kg, or 5.4 M, than many other simulation-derived solubilities for NaCl. Possible reasons for this are discussed at some length by Paluch et al. [27].

3.3.2.5.3 Free energy change via amorphous phases

In a series of four publications [29, 99, 110, 111], a Swedish team computed the solubility of druglike molecules using simulations via an elaborately planned route that visited crystalline and amorphous solids, supercooled liquid, liquid melt, and ideal gas on its way to the aqueous solution. Rather than explicitly modelling the crystalline phase, they linked crystalline to amorphous solubility by the empirical relationship

$$S_0^{amorph} \approx S_0^{crys} \exp\left(\frac{\Delta S_m}{R} \ln\left(\frac{T_m}{T}\right)\right) \tag{3.7}$$

where T_m is the melting point and ΔS_m is the entropy of melting. Their work is notable for its use of a simple linear response approximation, whereby the free energy required to transfer a single molecule from the vapour into an amorphous phase is:

$$\Delta G_{va} = \Delta G_{cav} + E_{LJ} + \frac{E_{QQ}}{2} \tag{3.8}$$

This is the sum of the free energy required to form a cavity in TIP4P water (ΔG_{cav}) [37], the Lennard-Jones energy of interaction of the molecule with the amorphous phase (E_{LJ}), and only half the Coulombic interaction energy, since the other half is assumed to be cancelled out by a corresponding entropy change. Their most accurate free energies were obtained by an expensive free energy perturbation method, but they demonstrated that good results could also be achieved by simpler and cheaper models which greatly reduced the cost of the simulations.

3.3.2.5.4 Simulation-free approaches

An alternative to simulation is to calculate the free energy changes in a static manner, without use of dynamics. Two related publications [17, 30] illustrate this approach, both computing solubility as an equilibrium constant derived from free energy changes calculated under standard conditions. This contrasts with the approach based on eq. (3.6) and used by many simulation methods, which seeks the non-standard conditions under which the change in chemical potential or free energy on solvation is zero. In Palmer *et al.*'s work a sublimation cycle is used, the crystal-gas leg being computed by the kind of ψ_{mol}-based lattice energy minimization common in crystal structure prediction [83]. In the 2008 publication, the gas-solution leg was computed via a simple quantum chemical model of the solution state. The authors looked both at a direct gas-aqueous route, and at one via octanol solution. The ease of computing logP, and hence the relevant partition equilibrium constant P, facilitated accurate modelling of transfer between the two solution environments. The authors found that significant improvements in accuracy could be obtained by allowing the contributions to the free energy change to be scaled by parameters fitted from training data. A regression model including three descriptors – first principles lattice energy, estimated logP, and the number of rotatable bonds – achieved an excellent root mean squared error (RMSE) of 0.71 $\log S_0$ units over an unseen test set of 26 druglike molecules [17]. Adding the computed hydration energy to the model was found not to improve the regression statistics.

Modest accuracy in the hydration energy was to be expected, since continuum solvation models contain many approximations. Solvent structure features from the solvation shell structure are missing in continuum models, and non-electrostatic energy terms are not represented in a first principles manner. In Palmer et al. [30], the more sophisticated 3D-RISM/UC model of Palmer et al. [57], parameterized specifically to yield numerically accurate hydration free energies, was used for the hydration leg.

Combining the two legs of the cycle, the solubility S_0 can be obtained from the free energy of dissolving the solid into aqueous solution, which is given by the sum

$$\Delta G^*_{\text{solu}} = \Delta G^*_{\text{sub}} + \Delta G^*_{\text{hyd}} = -RT\ln(S_0 V_m) \tag{3.9}$$

where S_0 is the intrinsic solubility in moles per litre and V_m is the crystal's molar volume. The molar volume V_m (in litres) appears due to the use of the Ben-Naim approach with the molecular centres of mass fixed and a standard state (denoted by *) of 1 M concentration. S_0V_m is the ratio of the compound's concentrations in aqueous solution and in the crystal, which Ben-Naim and Marcus [35], treat like a partition coefficient. A useful formula [30, 41] allows one to calculate the sublimation free energy at its usual 1 atm standard state and the solvation term similarly at the common 1 M, whilst eliminating the crystalline molar volume:

$$S_0 = -\frac{p_0}{RT}\exp\left(\frac{\Delta G_{sub}^{1\,atm} + \Delta G_{solv}^{1mol/L}}{RT}\right) \tag{3.10}$$

Palmer et al. [30], achieved a RMSE of 1.45 $\log S_0$ units from first principles across a set of 25 druglike molecules, good enough to be a useful prediction though with a larger error than is typical of informatics approaches to similar problems.

Also borrowing from CSP, Buchholz et al. [84], considered both ψ_{mol} and ψ_{crys} approaches to the sublimation energy in their investigation of the relative solubilities of racemic and enantiopure crystals of druglike organic molecules. For the gas-to-solution leg, they utilized the COSMO-RS model [42]. Their objective was to evaluate the feasibility of using differential solubility for enantiomer separation.

3.3.2.5.5 Solubility from density of states

A rather different method has been described by the Anwar group [112]. This involves computation of the density of states of the solution, which is carried out by means of MC simulations on NaCl solutions of different concentrations in SPC/E water [61], using the force field from Joung and Cheatham [109]. These simulations visit both solution and gaseous states, to facilitate particle insertion. The solid phase can be simulated similarly to other methods discussed above, or one can simply import an externally simulated value of μ_{solid}^{NaCl}. Restricting themselves to conditions at which the solid and solution chemical potentials were equal, the authors then looked at the computed density of states, in the form of the probability distribution function. If the density of states is known, scanning the probability distribution function in temperature at a given pressure determines the phase coexistence condition, and vice versa for pressure. The single-component probability distribution function contains two peaks, one corresponding to pure solute and the other corresponding to the saturated solution; the mole fraction of this peak is the limiting solubility. Like other simulation-based studies, they found a NaCl solubility around half the experimental value and noted a counterfactual decrease of computed solubility with increasing temperature.

3.3.3 Solubility from informatics

Informatics methods, in contrast to first principles ones, are designed with the simple objective of accurate numerical prediction. One seeks any method that will link the inputs – in this case, representations of molecular structures, to the required outputs – accurate estimates of experimental solubility, without any requirement to incorporate real-world physics or chemistry. Any interpretability or mechanistic insight from the model would be a secondary consideration at best. For a property where we believe that the underlying processes are unknown, difficult, or expensive to compute accurately, this approach of letting the data speak for themselves has much to commend it. Properties like bioactivity, logP and melting point are therefore generally computed with informatics methods. On the other hand, some other properties, such as dipole moments and infrared absorption frequencies, are relatively easy to compute accurately from first principles, and not generally predicted by chemoinformatics. Solubility is an intermediate property, where first principles computation is tractable, but currently informatics provides both much faster and more numerically accurate predictions.

3.3.3.1 Test sets, comparison, and experimental design

The traditional experimental design in informatics-based property prediction is to train the model on a training set. Optimization of a model on this training set would ultimately lead to overfitting, and thus any model must be validated on an independent external test set which has not in any way been used in the model's construction. Alternatively, multiple models are built using different subsets of the data for training and for testing. Techniques of this kind include cross-validation, bootstrapping, and jack-knifing, and again depend on the test data for a particular model being independent of its training. Where an algorithm has variable parameters, these may be optimized using an internal validation set, once more this needs to be entirely independent of the final external testing. For small datasets, tuning model parameters against a single validation set may induce model bias, which can sometimes be avoided by using cross-validation rather than simple training and validation to optimize these parameters, see Figure 3.4a. For instance, in a tenfold cross-validation exercise with an internal validation fold, the roles of the eight training, one internal, and one external validation set are permuted cyclically, and ten separate models generated, as shown in Figure 3.4b.

Validation methods, such as in the cyclical process described above, are often used for model selection. In these cases, the validation method needs to be sensitive enough to estimate differences between models, as discussed by Gütlein et al. [113]. For example, if the validation method has a bias in calculating predictivity, the method can still be applied, providing the systematic error applies across all models. Validation is also sensitive to dataset size, and the distribution of the data. The data in the training set will only produce a model that is good for predicting values from the same distribution. Validation estimates therefore only hold true for unseen data that fall within the same distribution as the training data. Sample selection bias using data biased towards a particular distribution can be reduced or avoided by using larger datasets for testing, by repeating the splitting of data, or by using stratified splitting where the folds of data are designed to have a similar distribution of property values, or sometimes coverage of chemical space, to the overall dataset. For example, if the dataset contains a few large hydrophobic and probably insoluble compounds, the splitting is designed as far as possible to put them in different folds so that each fold is representative of the distribution of these selected groups in the complete dataset.

Unfortunately, almost every published study uses somewhat different data. This makes it almost impossible to compare the quality of prediction between separate studies. Although the relative performance and ranking of methods within a given study may be meaningful, the examples below will show that these relative rankings are in fact often not maintained when studies are compared. Although several standard datasets exist, such as the Huuskonen set [114] and the DLS-100 set [115], typically studies adopt a pick-and-mix approach to dataset construction. It

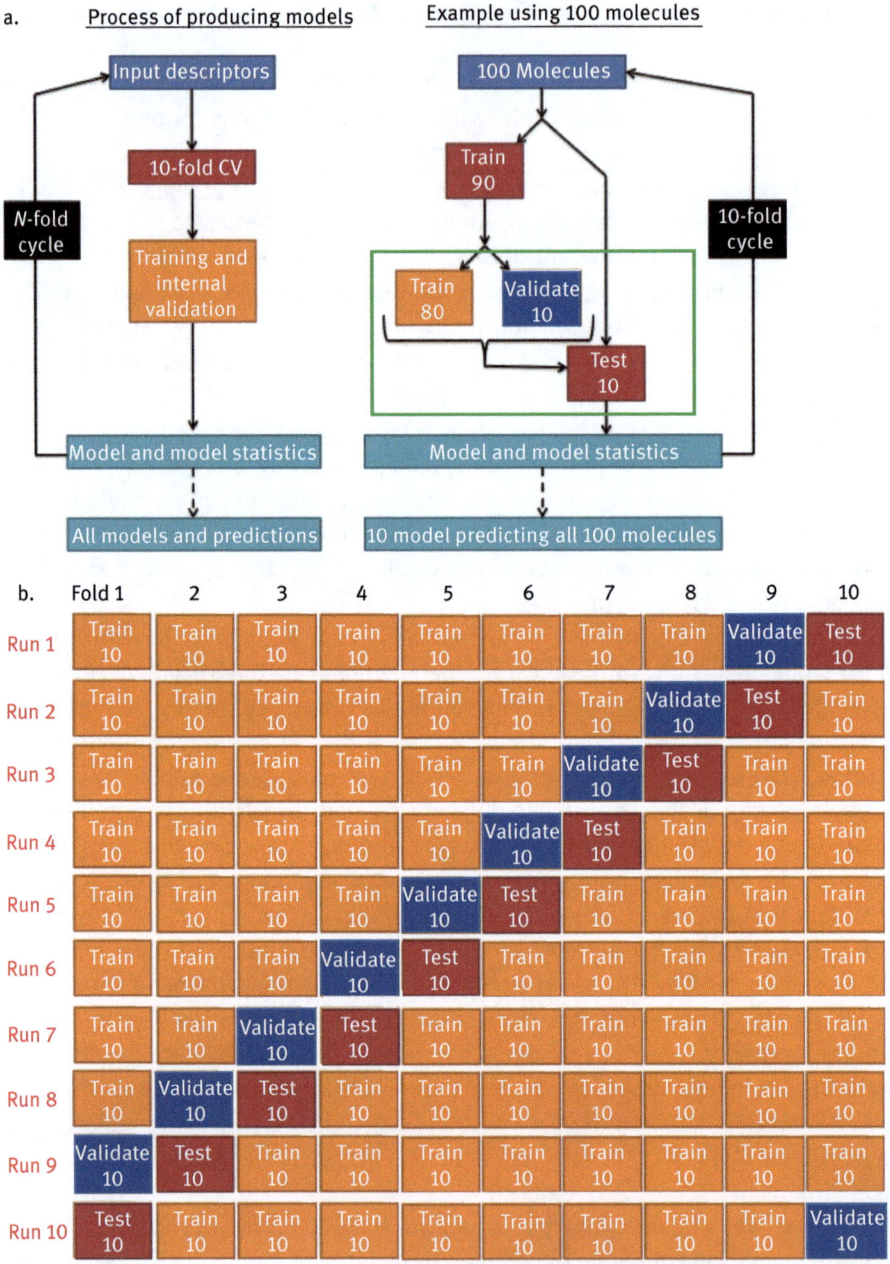

Figure 3.4: (a) Structure of a tenfold cross-validation experiment. Each model makes unseen predictions of solubility for ten molecules, such that each molecule is predicted once by the ensemble of ten models. (b) A detailed view of the tenfold cross-validation experiment, showing the roles of each fold in the ten different runs. This corresponds to the region outlined in green in Figure 3.4a.

is well known that some molecules are harder to predict solubility for than others [116, 117], and by extension predictivity cannot easily be compared between datasets drawn from different regions of chemical space or different degrees of molecular diversity. This means that prediction statistics from different studies can provide only a rough guideline for comparison.

A rare attempt to standardise solubility prediction was the solubility challenge [12, 117] where the idea was that 100 accurate solubilities would be provided as a training set and a further 32 would be held back as a test set. Research groups around the world were challenged to predict intrinsic aqueous solubilities for these 32 compounds. Of these, 94 training and 28 test compounds had usable numerical solubility data. The other compounds were either too soluble for CheqSol to detect any precipitate, in which case no numerical solubility is available, or found to decompose or react during the experiment; as in the cases of aspirin and, it was later discovered [118], indomethacin. There were 99 entries received, each of these being a set of predicted solubilities submitted by one participating research group. The best were high-quality models, whilst some of the others were not predictively useful. The organizers reported only R^2 on the test set, and number of answers correct to within the exacting margin of ± 0.5 $\log S_0$ units. On these criteria, only 18 entries obtained an R^2 above 0.5 on the 28 test compounds, and only 19 predicted at least half of the solubilities correctly. Fuller reporting of results, and especially a description of the methods employed by each entrant, would have made the solubility challenge even more valuable to the academic and industrial communities interested in solubility prediction. Nonetheless, these 122 compounds and their solubility data provide a standard training and test set than can be used to evaluate new and existing methods. A second Solubility Challenge was released by Llinas & Avdeef in 2019, comprising of two separate test sets [119].

3.3.3.2 General solubility equation

Before discussing purely informatics approaches to solubility prediction, we should mention one method that falls somewhere between informatics and first principles. Ran & Yalkowsky [33], give a theoretical justification of their general solubility equation (GSE) in terms of a version of the melt cycle, where the melting point determines the ratio of the solubilities of the solid and liquid phases, whilst the octanol-water partition coefficient logP describes the difference between solubility in an ideal solvent and in water. This leads to an equation

$$logS_0 = 0.5 - 0.01(MP - 25°C) - logP \tag{3.11}$$

where MP is the melting point in degrees Celsius, with the melting point term set to zero for compounds which are liquid at room temperature. Notwithstanding its theory-related derivation and its parameters being conveniently round numbers, this equation shares much in common with purely empirical relationships. Where experimental melting points and partition coefficients are available, the GSE can be quite accurate. Ran and Yalkowsky obtained a best root mean squared error (RMSE) of 0.52 $\log S_0$ units for a dataset including both druglike and smaller organic molecules. Yalkowsky's group later reported RMSEs between 0.53 and 0.86 on a variety of test sets [120]. McDonagh et al. [121], found an RMSE of 0.77 for 30 druglike molecules using a version of the GSE that incorporated experimental melting points and *in silico* predicted logP values, and 0.86 when both melting points and logP were predicted computationally. Although melting point is hard to predict accurately [122–124], each 1 K error in the melting point affects the GSE's predicted solubility by only 0.01 $\log S_0$ units, so even modest quality predicted melting points are acceptable for this purpose. Ali et al. [125], fitted a version of the GSE using training data; the coefficients, as expected, changed upon fitting and the RMSE therefore is slightly better than that of the Ran & Yalkowsky GSE on the same dataset. Ali *et al.* also looked at models involving topographical polar surface area (TPSA) as an alternative to melting point that is more easily predictable for unsynthesised virtual library compounds; this change has very little effect on the RMS error. Alternatively, TPSA can be used as well as melting point, in which case the error falls from 0.71 to 0.61 $\log S_0$ units with the inclusion of the one extra parameter.

3.3.3.3 Quantitative structure-property relationships *(QSPR)*

The idea that there exist property-to-property and structure-to-property correlations amongst molecules goes back to at least the 1860s [126, 127]. As early as 1863, Cros related solubility to toxicity in a PhD thesis [128]. Such relationships are predicated on molecular structure being both physically meaningful and capable of being represented, notions only gaining widespread acceptance in the mid-nineteenth century. Edinburgh chemist Alexander Crum Brown was an early advocate of the representation of molecular structure as bonds between atoms, using deliberately two-dimensional topological diagrams showing interatomic connections and bond orders. His 1868 paper [129] linked "the mutual relations of the atoms in the substance" to its physiological effect. QSPRs relating structure specifically to solubility included the work of Fühner [130], and Erickson [131], both of whom observed that adding extra CH_2 units reduces solubility by an approximately constant factor. QSPR pioneer Hansch demonstrated that solubility could be predicted by assuming a linear relationship between $\log S_0$ and logP [132].

3.3.4 Specific techniques in QSPR and machine learning

A substantial number of different mathematical and computational techniques have been used in the construction of QSPR models of solubility and other physicochemical properties [133]. The more sophisticated and typically non-linear algorithms that have more recently become mainstream in the field are generally categorized as supervised machine learning methods. That is, the computer learns the relationship between chemical structure and solubility by being trained on available data and generates a predictive model. Below we describe some of the interesting and important techniques, though this list is not exhaustive.

3.3.4.1 Linear techniques

Many historically important and mathematically compact approaches to the QSPR problem are linear. At its simplest, such a model uses a linear combination of input properties describing chemical structure and multiplies them by fitted coefficients to predict the value of the output property, here solubility. This can be simply visualized as an extension of the idea of interpolation from a line of best fit, although the number of variables used in the model is usually rather greater than one.

3.3.4.1.1 Group contribution and multi-linear regression methods

The constant effect of a CH_2 moiety described in Section 3.3.3 is a rudimentary example of a group contribution. This concept can be used to build models where each molecule is broken down into fragments, often similar to conventional functional groups. By using a suitable training set each such group is assigned a numerical parameter defining a transferable contribution to solubility, which it is presumed to make whenever it occurs in a molecule. For each test molecule, $\log S_0$ is calculated as the sum of the contributions from each group in the compound [134–136] – this is known as an additive group contribution method. Klopman et al. [134], carried out several predictive tests on different models. Their most generally applicable model gave an RMSE of 1.25 on 21 organic molecules. Hou et al. [137], defined contributions per atom, rather than per molecule, and obtained an RMSE of 0.79 $\log S_0$ units over 120 test compounds. Wang et al. [138], modified the group contribution idea by calculating an accessible surface area associated with each fragment, rather than just counting the occurrences, and added descriptors for other key properties into their model. They validated their model thoroughly and obtained an RMSE of 0.705 on their 120-molecule test set, compared with between 1.23 and 2.06 on external databases of druglike compounds. The UNIFAC (UNIversal quasi-chemical Functional-group Activity Coefficients) approach is effectively a group contribution method but proceeds via estimation of activity coefficients. Its use to predict organic solubilities was described by Gracin et al. [139], for nine different organic solutes in a variety of polar and non-polar solvents. Abraham's group [140] similarly used linear regression, but the quantities in their equation were designed to have specific physicochemical meanings: refractivity, polarizability, hydrogen bond acidity and basicity, and a characteristic volume. Their work gave an impressive RMSE of 0.5 over 65 test compounds. The same concepts can be extended to parameterize solvents, both allowing solubility to be predicted in different solvents and permitting similarities between solvents to be identified [141].

Multi-linear regression (MLR) is also applicable based on molecular, rather than groupwise, descriptors. For instance, Hewitt et al. [142], obtained an RMSE of 0.95 for the 28 usable test compounds of the solubility challenge [12, 117] using an MLR model based on only three descriptors. Catana et al. [143], implemented an MLR model, amongst other linear and non-linear methods, and obtained an impressive RMSE of 0.57 over 177 test compounds.

3.3.4.1.2 Partial least squares

Partial least squares (PLS), or projection to latent structures, is a linear regression method that uses latent variables to project both the input and output variables into a new space. PLS is essentially modelling covariance, seeking the linear combination of input features that explains the maximum proportion of the variance in the output variable, here solubility. Since it is effectively seeking a single, maximally explanatory, direction in the input space, PLS is robust against redundancy and mutual correlation amongst the input variables. It can be used similarly to multilinear regression, but without the need for prior aggressive feature selection. Catana et al. [143], implemented PLS and achieved RMSE values of around 0.5 over 177 test compounds for their PLS models. Hughes et al. [122], found that PLS was almost as good as Support Vector Machine (SVM) and a little better than random forest (RF) in terms of RMSE over 87 test compounds, with RMSE values around 0.95 depending on the exact descriptor set used. However, Palmer et al. [144], had previously observed that PLS was slightly less effective than either SVM or RF with an RMSE of 0.773 for 330 test molecules. Boobier et al. [116], had PLS as seventh best of ten machine learning algorithms, with an RMSE of 1.265 for 25 molecules. Cao et al. [145], got an RMSE of 0.769 on 45 test set molecules, marginally better than artificial neural networks (ANN) but behind SVM.

3.3.4.2 Non-linear machine learning methods

Whilst powerful enough often to obtain good empirical fits to experimental solubility data, the methods discussed above are essentially limited to linear relationships between descriptors and solubility, although Abraham and Le [140] did consider taking products of two descriptors as a pragmatic if inelegant means of capturing inter-descriptor interactions. However, machine learning methods provide a more natural way of accounting for more complicated, non-linear, QSPRs. In fact, although chemistry has often been rather slow to adopt techniques from computer science, several different machine learning approaches have now found application in solubility prediction and related QSPR problems.

3.3.4.2.1 Artificial neural networks

Neural networks have been studied since the 1950s, with the invention of backpropagation by Werbos [146], being a breakthrough. An artificial neural network (ANN) is a mathematical approach to pattern recognition and machine learning problems. Whilst the ANN is inspired by the structure of the human brain, it is not realistically attempting to simulate or reproduce the way the brain works. Indeed, the ANN is typically smaller even than the 302-neuron brain of the nematode worm *C. elegans* [147]. The ANN consists of nodes known by biological analogy as neurons, which are joined together by weight-carrying connections and arranged in an input layer, a hidden layer or layers, and an output layer. The mathematical weights are varied during training and adjusted through backpropagation. Although there is a significant risk of overfitting if ANN training is not stopped at an appropriate stage, the approach can produce good results.

ANNs have been used to predict solubility by several groups. Hewitt et al. [142], implemented a number of different models, including ANN, as part of their participation in the Solubility Challenge [12, 117]. They limited their models to no more than five input descriptors, albeit chosen by a genetic algorithm from an available pool of 426. Their best ANN model was a multilayer perceptron with only two input descriptors. The first was logP and the second a hard-to-interpret size-and-connectivity feature known as R2e+, or more fully as the R maximal autocorrelation of lag 2 weighted by atomic Sanderson electronegativities. Rather disappointingly, their ANN performed worse than multi-linear regression, with an RMSE of 1.51 $\log S_0$ units on the 28-compound Solubility Challenge test set. Catana et al. [143] found an RMSE of 0.608 over 130 test molecules using a multi-layer perceptron (MLP), a variety of ANN, with a single hidden layer. Louis et al. [148] used a backpropagation network to obtain an RMSE of 0.738 on a small 14-compound test set. Cao et al. [145] used the same methodology to obtain an RMSE of 0.789 on 45 molecules. Boobier et al. [116] also found an MLP to do very well, being the best of 10 assorted machine learning methods with an RMSE of 0.985 over a challenging test set of 25 druglike molecules. Erić et al. [149] used a counter-propagation neural network designed for interpretability, obtaining an RMSE of 0.679 on a 94-compound test set and interpreting their model in terms of the computed importances of the seven input descriptors of which logP was the most significant. Palmer et al. [144] reported that ANN did a little less well than RF in their study, with an RMSE of 0.751 over 330 test set molecules. Bhat et al. [123] designed an ensemble technique with 50 neural networks and implemented it to predict the melting points of organic molecules. This approach could be combined with logP prediction to obtain solubility *via* the GSE.

3.3.4.2.2 Random forest

Random Forest (RF) creates an ensemble of many diverse decision or regression trees, based on distinct samples from the same pool of data [150, 151]. For numerical prediction of solubility, regression rather than decision trees are used. Each tree is grown by recursive partitioning of the training set compounds, based on their descriptor representations. The resulting forest of regression trees is described as random for two reasons. Firstly, each tree is built from a new bootstrap sample of the training data, a sample of N out of N compounds chosen with replacement. Secondly, at each node a tree must make its partition by considering only a random subset of the descriptors, the size of which subset is a parameter known as m_{try}. A fresh set of m_{try} descriptors is selected for decision-making at each node as the tree is grown. A Gini-optimal [152] split of the training data is made at each mode, so that the compounds are grouped into increasingly homogeneous sets down the tree. Thus, the collection of molecules assigned to each terminal leaf node will share similar values of solubility, or of whatever other property is being predicted. Once built, the Random Forest consisting of a total of n_{tree} regression trees can be used to predict solubilities of previously unseen test compounds. The consensus of the different trees that forms the overall prediction of the forest is based simply on the mean of the individual trees' predictions. The probability of a given molecule not being selected for the bootstrap sample of a particular tree is $(1–1/N)^N$, which tends to the limiting value of 1/e as N becomes large. This means that for each tree approximately 37 % of the training data are unused, and these can be adopted as a so-called out-of-bag validation set. RF tends to cope well with the presence of correlated descriptors and is generally robust against overfitting [150]. Indeed, many different descriptors will play at least some role in an RF model, given that only a modestly sized sample is available to be selected at any one node of any one tree.

RF has been applied to various aqueous solubility datasets by different authors. Schroeter et al. [153] obtained an RMSE of 0.855 on an external test set of 536 compounds. Palmer et al. [144], reported the RMSE of 0.690 for 330 test molecules, and subsequently [154] compared RF models built firstly on literature solubilities and secondly on new CheqSol [9] experiments for 80 compounds. Perhaps surprisingly, they found that the new experiments, despite having a consistent methodology and reporting scheme for all 80 compounds, generated a model with an almost identical cross-validated RMSE to that obtained from solubilities harvested from a variety of literature sources and methodologies, around 0.88 $\log S_0$ units. RF methods have been applied by Hughes et al. [122], Kovdienko et al. [155], McDonagh et al. [32], and by Boobier et al. [116], who found that it was the joint second best amongst 10 machine learning predictors tested and of similar quality to the second best of a panel of 22 human predictors. Kew et al. [156] observed RF to generate an RMSE of 1.02 in a tenfold cross-validation using 262 molecules from Hughes et al. [122], and thus to be essentially joint best alongside Support Vector Machine of 15 methods for solubility prediction.

3.3.4.2.3 Support Vector Machine

Support Vector Machine (SVM) [157, 158] is a popular machine learning method which transforms the input data into a high dimensional space, by means of a typically non-linear kernel function. For binary classification, SVM seeks a hyperplane which optimally separates the data between the two classes, ideally such that they lie almost entirely on opposite sides of it. This is achieved by maximizing the margin between the closest points, known as support vectors, and the hyperplane. For solubility, however, it is usual to generate quantitative predictions, rather than a binary soluble-insoluble classification, and thus SVM is adapted for regression with the hyperplane now playing the role of a regression line.

Numerous studies have addressed solubility in this way. Lind et al. [159] reported cross-validated RMSEs between 0.57 and 0.77 on different datasets. Palmer et al. [144] found that SVM obtained an RMSE of 0.72 on a 330 compound test set, slightly worse than RF. In contrast Hughes et al. [122] in the same research group found SVM to somewhat outperform RF over 87 molecules. Comparison of these two studies emphasizes that the performance of these machine learning methods is similar and that there is unlikely to be a universally best-performing algorithm. Louis et al. [148] found SVM to do a little better than ANN on a small test set of 14 molecules with an RMSE of 0.832. Cao et al. [145] found a better RMSE, 0.731 over 45 compounds, with SVM than with two other machine learning methods. Kew et al. [156] observed SVM to get an RMSE of 1.01 in a tenfold cross-validation over 262 compounds taken from Hughes et al. [122], and thus to be essentially joint best with RF of 15 methods for solubility prediction. Boobier et al. [116] however, found SVM to be only the eighth best out of 10 methods for a 75-25 training-test split of the DLS-100 dataset [115] with an RMSE of 1.280 for 25 test compounds.

3.3.4.2.4 k-nearest neighbours

In k-Nearest Neighbours (kNN), the solubility of a query molecule is predicted from the solubilities of its k nearest neighbours in chemical space amongst the available dataset. In general, k is a small integer so that only the local variation of the property in chemical space affects the kNN prediction. This distinguishes it from other machine learning approaches, which attempt to generate global models. Although the predictivity is essentially local, the coverage of diverse chemical structures is as broad as the composition of the dataset. A quantitative prediction is made by averaging the solubilities of the k neighbours. In some applications the average can be weighted by distance [124]. The methodology requires a robust measure of distance in chemical space, so descriptors should be appropriately scaled. Hughes et al. [122] found that kNN was less effective than SVM, PLS or RF over 87 test compounds, with RMSE values around 1.10 depending on the descriptor set used. Kew et al. [156] found kNN to be only 12th best of 15 methods for solubility prediction. Boobier et al. [116] in contrast, found kNN to be the fourth best of ten machine learning algorithms, slightly ahead of PLS and SVM amongst others, with an RMSE of 1.204 for 25 molecules. Kühne et al. [160] employed kNN in a more cryptic way, using it to select the most appropriate model for each compound rather than directly to predict solubility.

3.3.4.2.5 Gaussian processes

Gaussian processes are Bayesian methods commonly applied in various fields of machine learning. A Gaussian process is a stochastic process, which extends a multivariate normal distribution to an infinite number of random variables. As a result, a Gaussian process effectively represents a probability distribution over functions. The Gaussian process is completely specified by a mean and covariance function [161]. The covariance function can be selected to represent previous knowledge about the data such as periodic patterns or how sharply a function may change between points. Common choices of covariance function include the squared exponential and Martérn covariance functions [161]. Gaussian processes have over the past decade begun to be applied to a range of chemical problems with success [162–165].

Gaussian processes have been used effectively to predict ADMET properties and solubility. Obrezanova and Seagall [166] applied Gaussian process classification to build QSPR models of a variety of ADMET and bioactivity properties, including blood-brain barrier penetration and hERG inhibition. They compared the results to a range of other machine learning classification methods, including RF and SVM, and found that, whilst no method was notably more successful than the others, Gaussian processes were often the best performing. Schwaighofer et al. [167] developed QSPR solubility models for a range of datasets of electrolyte molecules, achieving promising results, and outperforming many of the commercially available solubility prediction packages.

3.3.4.2.6 Deep learning

The notion of deep learning [168] in the machine learning field comes from the use of large multi-layer neural networks, which can perform abstractions from data and hence intrinsically define features, as opposed to traditional ANNs which correlate fixed feature representations with a given property. The hidden layers are where the decisions are made in a network. Deep learning for pharmaceutical property prediction has seen widening use over recent years. Deep learning specifically for solubility prediction has, however, received less attention, though one notable example is the work of Lusci et al [169]. In this publication the authors produced a novel two-step approach, using a first ANN to determine the optimal molecular representation encoding chemical structure *in silico*, and a second ANN to find the best mapping function between this representation and solubility. They achieved good predictive statistics over a range of datasets, including cross-validated RMSEs of 0.60–0.92 $\log S_0$ units on the 1026-molecule Huuskonen dataset [114] and RMSEs between 1.00 and 1.41 on the solubility callenge set [12, 117] for variants of their method.

3.3.4.2.7 Consensus methods

There is potential to improve the performance of machine learning by taking a consensus of different predictors. RF is itself by design a consensus of different trees, whilst Bhat et al. [123] created an analogous ensemble of ANNs. Going beyond this, it is equally possible to combine different machine learning approaches to generate a single prediction. Kew et al. [156] showed that a Greedy Ensemble incorporating several other machine learning predictors performed similarly to the best individual algorithm with an RMSE of 0.83 $\log S_0$ units on the Solubility Challenge dataset and 1.13 on the 262 molecule Hughes dataset. Boobier et al. [116] similarly observed that a median-based consensus was essentially as good as the best single method, whose identity would not be apparent in advance. Thus, there is scope to leverage the power of diverse algorithms and take advantage of the wisdom of crowds [170, 171] with consensus approaches.

3.4 Conclusions

This chapter has covered a wide range of methods from physical simulations to chemoinformatics. Currently, informatics methods provide the best quantitative predictions of solubility in terms of the lowest RMSE and highest R^2. Nonetheless, they are very limited in the physical insight they can offer. Some methods, such as RF, have inbuilt measures of descriptor importance. However, since descriptors are often correlated with one another, the list of the most numerically significant descriptors for predictivity may not be a good guide to physical or chemical importance. Further, these kinds of models demonstrate correlation not causation, and thus do not directly inform us about the physicochemical reasons or mechanisms for some molecules being more soluble than others.

There is no clear or repeatable pattern as to which machine learning methods are most effective at solubility prediction. Indeed, looking at the various references cited herein, RF, SVM and ANN are all competitive with one another, but it is not possible to identify in advance which will be most accurate for any given dataset. However, all the informatics methods presented herein are data driven and thus have some dependence on the accuracy and reliability of the experimental measurements used to acquire the data. To this end, reliable and carefully curated datasets for solubility and related quantities are of great value to the computational and modelling community. More laboratory data enabling better estimation of the appropriate error bars for experimental solubility values would help us to understand whether informatics methods are now close their limiting accuracy, or whether there remains substantial scope for improvement from new machine learning algorithms, from novel descriptors encoding different information, or from better feature selection.

In terms of physical models, we have presented methods which make good solubility predictions and provide varying levels of physical insight. Much work by those predicting solubility has focused on modelling of the solvent and solution. This has provided a rich choice of models which can be selected for a particular problem. The best choice of model depends upon the physics one is interested in representing. Implicit models are efficient but provide little information on solvent structure. Hybrid methods such as RISM are also efficient and provide limited statistical information on the solvent structure. Explicit solvent representation can be expensive but will provide maximal information on the solvent's structure and response to a solute.

Solid-state modelling has not received the same level of attention by those predicting solubility as solution phase modelling has, but there is considerable expertise and experience available from CSP. There are many options here also, which range from lattice simulation methods to periodic quantum mechanical calculations. Again, the choice depends on balancing compute time against the level of detail needed to describe the physics of interest. There is much scope in this area to investigate the effects of different solid-state modelling techniques on solubility predictions.

Both the solid-state and solution-states require suitable representation for accurate predictions via thermodynamic cycles. Predictions from both the fusion cycle and the sublimation cycles have shown similar levels of accuracy, as described by the references presented herein. Over recent years, physics-based methods have shown improvements and begun to offer much improved solubility predictions. Amongst the most pressing priorities is for these methods to be tested on datasets of sufficient size to assess their accuracy and compare their quantitative performance with that of chemoinformatics.

Overall, we foresee a bright future for solubility prediction using both physics-based models and informatics methods. Informatics provides excellent opportunities for fast virtual screening, with deep learning offering new opportunities for automated feature extraction. Physical models will continue to provide deeper insights into the chemical and physical phenomena which define a substance's solubility, enabling us to discover rules for designing molecules intelligently to enhance or reduce solubility. The combination of these methods could yield a particularly powerful tool in the future.

Acknowledgement: Dr William Swope of IBM for useful discussions on water models.

References

[1] Hutchinson, T. C., Hellebust, J. A., Mackay, D., Tarn, D., & Kauss, P. Relationship of hydrocarbon solubility to toxicity in algae and cellular membrane effects. Int. Oil Spill Conf. Proc. 1979, 1, 541–547.

[2] Doerr-MacEwen, N. A., & Haight, M. E. Expert stakeholders' views on the management of human pharmaceuticals in the environment. Environ. Manage. 2006, 38(5), 853–866.
[3] Sell, C. S. The Mechanism of Olfaction, *Chemistry and the Sense of Smell*, John Wiley & Sons, Inc., 2014, 32–187.
[4] Leach, AG., Jones, HD., Cosgrove, DA., Kenny, PW., Ruston, L., MacFaul, P. et al. Matched molecular pairs as a guide in the optimization of pharmaceutical properties; a study of aqueous solubility, plasma protein binding and oral exposure. J. Med. Chem. 2006, 49(23), 6672–6682. http://dx.doi.org/10.1021/jm0605233.
[5] FDA (US Food and Drug Administration), The Biopharmaceutics Classification System (BCS) guidance, 2014.
[6] Narasimham, L. Y. S., & Barhate, V. D. Kinetic and intrinsic solubility determination of some beta-blockers and antidiabetics by potentiometry. J. Pharm. Res. 2011, 4(2), 532–536.
[7] Saal, C., & Petereit, A. C. Optimizing solubility: kinetic versus thermodynamic solubility temptations and risks. Eur. J. Pharm. Sci. 2012, 47(3), 589–595.
[8] Jouyban, A., & Fakhree, M. Experimental and Computational Methods Pertaining to Drug Solubility, *Toxicity and Drug Testing*, Acree W, Ed, Intech Open, London, 2012.
[9] Box, K., Comer, JE., Gravestock, T., & Stuart, M. New ideas about the solubility of drugs. Chem. Biodivers. 2009, 6(11), 1767–1788. http://dx.doi.org/10.1002/cbdv.200900164.
[10] Stuart, M., & Box, K. Chasing equilibrium: measuring the intrinsic solubility of weak acids and bases. Anal. Chem. 2005, 77, 983–990.
[11] Etherson, K., Halbert, G., & Elliott, M. Determination of excipient based solubility increases using the cheqsol method. Int. J. Pharm. 2014, 465(1–2), 202–209.
[12] Llinàs, A., Glen, RC., & Goodman, JM. Solubility challenge: can you predict solubilities of 32 molecules using a database of 100 reliable measurements?. J. Chem. Inf. Model. 2008, 48, 1289–1303. http://dx.doi.org/10.1021/ci800058v.
[13] Avdeef, A. pH-metric solubility. 1. solubility-ph profiles from Bjerrum plots, Gibbs buffer and pKa in the solid state. Pharm. Pharmacol. Commun. 1998, *4(3)*, 165–178.
[14] Lipinski, C.A., Lombardo, F., Dominy, B. W., & Feeney, P. J. Experimental and computational approaches to estimate solubility and permeability in drug discovery and developmental settings. Adv. Drug Deliv. Rev. 1997, 23, 3–25.
[15] Alsenz, J., & Kansy, M. High throughput solubility measurement in drug discovery and development. Adv. Drug Deliv. Rev. 2007, 59(7), 546–567.
[16] Hörter, D., & Dressman, J. Influence of physicochemical properties on dissolution of drugs in the gastrointestinal tract. Adv. Drug Deliv. Rev. 2001, 46(1–3), 75–87.
[17] Palmer, DS., Llinàs, A., Morao, I., Day, GM., Goodman, JM., Glen, RC., & Mitchell, JBO. Predicting intrinsic aqueous solubility by a thermodynamic cycle. Mol. Pharm. 2008, 5(2), 266–279. http://dx.doi.org/10.1021/mp7000878.
[18] Noyes, A. A., & Whitney, W. R. The rate of solution of solid substances in their own solutions. J. Am. Chem. Soc. 1897, 19(12), 930–934.
[19] Po, HN., & Senozan, NM. The Henderson-Hasselbalch equation: its history and limitations. J. Chem. Educ. 2001, 78(11), 1499.
[20] Chemburkar, S. R., Bauer, J., Deming, K., Spiwek, H., Patel, K., Morris, J., Henry, R., Spanton, S., Dziki, W., Porter, W., Quick, J., Bauer, P., Donaubauer, J., Narayanan, B. A., Soldani, M., Riley, D., & McFarland, K. Dealing with the impact of ritonavir polymorphs on the late stages of bulk drug process development. Org. Process Res. Dev. 2000, 4(5), 413–417.
[21] Neumann, M., & van de Streek, J. How many ritonavir cases are there still out there? Faraday Discuss. 2018, 211, 441–458. https://dx.doi.org/10.1039/C8FD00069G.
[22] Bauer, J., Spanton, S., Henry, R., Quick, J., Dziki, W., Porter, W., & Morris, J. Ritonavir: an extraordinary example of conformational polymorphism. Pharm. Res. 2001, 18(6), 859–866.

[23] Bergström, CAS., & Larsson, P. Computational prediction of drug solubility in water-based systems: qualitative and quantitative approaches used in the current drug discovery and development setting. Int. J. Pharm. 2018, 540, 185–193. https://dx.doi.org/10.1016/j.ijpharm.2018.01.044.
[24] Jorgensen, WL., & Tirado-Rives, J. Monte Carlo vs molecular dynamics for conformational sampling. J. Phys. Chem. 1996, 100(34), 14508–14513. http://dx.doi.org/10.1021/jp960880x.
[25] Hall, LH., & Kier, LB. The molecular connectivity chi indexes and kappa shape indexes in structure-property modeling. Rev. Comput. Chem. 2007, 2, 367–422.
[26] Sharma, V., Goswami, R., & Madan, A. K. Eccentric connectivity index: a novel highly discriminating topological descriptor for structure–property and structure–activity studies. J. Chem. Inf. Comput. Sci. 1997, 37(2), 273–282.
[27] Paluch, AS., Jayaraman, S., Shah, JK., & Maginn, EJ. A method for computing the solubility limit of solids: application to sodium chloride in water and alcohols. J. Chem. Phys. 2010, 133(12), 124504. http://dx.doi.org/10.1063/1.3478539.
[28] Moučka, F., Nezbeda, I., & Smith, WR. Chemical potentials, activity coefficients, and solubility in aqueous nacl solutions: prediction by polarizable force fields. J. Chem. Theor. Comput. 2015, 11(4), 1756–1764. http://dx.doi.org/10.1021/acs.jctc.5b00018.
[29] Lüder, K., Lindfors, L., Westergren, J., Nordholm, S., & Kjellander, R. In silico prediction of drug solubility: 2. Free energy of solvation in pure melts. J. Phys. Chem. B 2007a, 111(7), 1883–1892. http://dx.doi.org/10.1021/jp0642239.
[30] Palmer, DS., McDonagh, JL., Mitchell, JBO., van Mourik, T., & Fedorov, MV. First-principles calculation of the intrinsic aqueous solubility of crystalline druglike molecules. J. Chem. Theory. Comput. 2012, 8(9), 3322–3337. http://dx.doi.org/10.1021/ct300345m.
[31] Li, L., Totton, T., & Frenkel, D. Computational methodology for solubility prediction: application to the sparingly soluble solutes. J. Chem. Phys. 2017, 146, 214110. http://dx.doi.org/10.1063/1.4983754.
[32] McDonagh, JL., Nath, N., De Ferrari, L., van Mourik, T., & Mitchell, JBO. Uniting chemoinformatics and chemical theory to predict the intrinsic aqueous solubility of crystalline druglike molecules. J. Chem. Inf. Model. 2014, 54, 844–856. http://dx.doi.org/10.1021/ci4005805.
[33] Ran, Y., & Yalkowsky, SH. Prediction of drug solubility by the general solubility equation (GSE). J Chem Inf Comput Sci 2001, 41, 354–357. http://dx.doi.org/10.1021/ci000338c.
[34] Ben-Naim, A. Standard thermodynamics of transfer. Uses and misuses. J. Phys. Chem. 1978, 82(7), 792–803.
[35] Ben-Naim, A., & Marcus, Y. Solvation thermodynamics of nonionic solutes. J. Chem. Phys. 1984, 81(4), 2016–2027. http://dx.doi.org/10.1063/1.447824.
[36] Luccarelli, J., Michel, J., Tirado-Rives, J., & Jorgensen, W. L. Effects of water placement on predictions of binding affinities for p38α map kinase inhibitors. J. Chem. Theory Comput. 2010, 6(12), 3850–3856.
[37] Jorgensen, W. L., Chandrasekhar, J., Madura, J. D., Impey, R. W., & Klein, M. L. Comparison of simple potential functions for simulating liquid water. J. Chem. Phys. 1983, 79(2), 926–935.
[38] Jorgensen, W. L., & Tirado-Rives, J. Molecular modeling of organic and biomolecular systems using BOSS and MCPRO. J. Comput. Chem. 2005, 26(16), 1689–1700. https://dx.doi.org/10.1002/jcc.20297.
[39] Bhat, T. N., Bentley, G. A., Boulot, G., Greene, M. I., Tello, D., Dall'Acqua, W., Souchon, H., Schwarz, F. P., Mariuzza, R. A., & Poljak, R. J. Bound water molecules and conformational stabilization help mediate an antigen-antibody association. Proc. Natl. Acad. Sci. U. S. A. 1994, 91(3), 1089–1093.

[40] Ratkova, E. L., Palmer, D. S., & Fedorov, M. V. Solvation thermodynamics of organic molecules by the molecular integral equation theory: approaching chemical accuracy. Chem. Rev. 2015, 115(13), 6312–6356.
[41] Skyner, R. E., McDonagh, J. L., Groom, C. R., van Mourik, T., & Mitchell, J. B. O. A review of methods for the calculation of solution free energies and the modelling of systems in solution. Phys. Chem. Chem. Phys. 2015, 17(9), 6174–6191. https://dx.doi.org/10.1039/c5cp00288e.
[42] Klamt, A. The COSMO and COSMO-RS solvation models. WIREs Comput Mol Sci 2011, 1(5), 699–709. http://dx.doi.org/10.1002/wcms.56.
[43] Tomasi, J., Mennucci, B., & Cammi, R. Quantum mechanical continuum solvation models. Chem. Rev. 2005, 105(8), 2999–3093.
[44] Cramer, C. J., & Truhlar, D. G. Implicit solvation models: equilibria, structure, spectra, and dynamics. Chem. Rev. 1999, 99(8), 2161–2200.
[45] Miertuš, S., Scrocco, E., & Tomasi, J. Electrostatic interaction of a solute with a continuum. A direct utilization of ab initio molecular potentials for the prevision of solvent effects. Chem. Phys. 1981, 55(1), 117–129.
[46] Cammi, R., & Tomasi, J. Remarks on the use of the apparent surface charges (ASC) methods in solvation problems: iterative versus matrix-inversion procedures and the renormalization of the apparent charges. J. Comput. Chem. 1995, 16(12), 1449–1458. http://dx.doi.org/10.1002/jcc.540161202.
[47] Marenich, AV., Cramer, CJ., & Truhlar, DG. Universal solvation model based on solute electron density and on a continuum model of the solvent defined by the bulk dielectric constant and atomic surface tensions. J. Phys. Chem. B. 2009b, 113(18), 6378–6396.
[48] Klamt, A., & Schüürmann, G. COSMO: A new approach to dielectric screening in solvents with explicit expressions for the screening energy and its gradient. J. Chem. Soc. Perkin Trans. 2 1993, 0(5), 799–805.
[49] Hogan, SW., & van Mourik, T. Competition between hydrogen and halogen bonding in halogenated 1-methyluracil: water systems. J. Comput. Chem. 2016, Mar 30, 37(8), 763–70. https://doi.org/10.1002/jcc.24264.
[50] Huan, G., Xu, T., Momen, R., Wang, L., Ping, Y., Kirk, SR., Jenkins, S., & van Mourik, T. A QTAIM exploration of the competition between hydrogen and halogen bonding in halogenated 1-methyluracil: water systems. Chem. Phys. Lett. 2016, 662, 67–72. https://doi.org/10.1016/j.cplett.2016.09.031.
[51] Warshel, A., & Levitt, M. Theoretical studies of enzymic reactions: dielectric, electrostatic and steric stabilization of the carbonium ion in the reaction of lysozyme. J. Mol. Biol. 1976, 2(15), 227–249.
[52] Mulholland, A. J. Modelling enzyme reaction mechanisms, specificity and catalysis. Drug Discov. Today 2005, 10(20), 1393–1402.
[53] Alderson, RG., De Ferrari, L., Mavridis, L., McDonagh, JL., Mitchell, JBO., & Nath, N. Enzyme informatics. Curr. Top. Med. Chem 2012, 12(17), 1911–1923.
[54] Senn, HM., O'Hagan, D., & Thiel, W. Insight into enzymatic C-F bond formation from QM and QM/MM calculations. J. Am. Chem. Soc. 2005, 127, 13643–13655. https://dx.doi.org/10.1021/ja053875s.
[55] Kovalenko, A., & Hirata, F. Self-consistent description of a metal–water interface by the Kohn–Sham density functional theory and the three-dimensional reference interaction site model. J. Chem. Phys. 1999, 110(20), 10095–10112. http://dx.doi.org/10.1063/1.478883.
[56] Genheden, S., Luchko, T., Gusarov, S., Kovalenko, A., & Ryde, U. An MM/3D-RISM approach for ligand binding affinities. J Phys Chem B 2010, 114(25), 8505–8516. http://dx.doi.org/10.1021/jp101461s.

[57] Palmer, DS., Frolov, AI., Ratkova, EL., & Fedorov, MV. Towards a universal method for calculating hydration free energies: A 3D reference interaction site model with partial molar volume correction. J. Phys. : Condens. Matt. 2010, 22(49), 492101. http://dx.doi.org/10.1021/jp111271c.
[58] Chandler, D., McCoy, JD., & Singer, SJ. Density functional theory of nonuniform polyatomic systems. II. rational closures for integral equations. J. Chem. Phys. 1986, 85, 5977–5982. https://doi.org/10.1063/1.451511.
[59] Mark, P., & Nilsson, L. Structure and dynamics of the TIP3P, SPC, and SPC/E water models at 298 K. J. Phys. Chem. A 2001, 105(43), 9954–9960. http://dx.doi.org/10.1021/jp003020w.
[60] Berendsen, H. J. C., Postma, J. P. M., van Gunsteren, W. F., & Hermans, J. Interaction Models for Water in Relation to Protein Hydration, Springer, Dordrecht, 1981. 331–342.
[61] Berendsen, H. J. C., Grigera, J. R., & Straatsma, T. P. The missing term in effective pair potentials. J. Phys. Chem. 1987, 91(24), 6269–6271.
[62] Toukan, K., & Rahman, A. Molecular-dynamics study of atomic motions in water. Phys. Rev. B 1985, 31(5), 2643. https://dx.doi.org/10.1103/PhysRevB.31.2643.
[63] Praprotnik, M., Janezic, D., & Mavri, J. Temperature dependence of water vibrational spectrum: a molecular dynamics simulation study. J Phys Chem A 2004, 108(50), 11056–11062. http://dx.doi.org/10.1021/jp046158d.
[64] Wu, Y., Tepper, H.L., & Voth, G.G. Flexible simple point-charge water model with improved liquid state properties. J. Chem. Phys. 2006, 124, 024503.
[65] Dyer, KD., Perkyns, JS., Stell, G., & Pettit, MB. Site-renormalised molecular fluid theory: on the utility of a two-site model of water. Mol. Phys. 2009, 107, 423–431.
[66] Horn, H. W., Swope, W. C., Pitera, J. W., Madura, J. D., Dick, T. J., Hura, G. L., & Head-Gordon, T. Development of an improved four-site water model for biomolecular simulations: TIP4P-Ew . J. Chem. Phys. 2004, 120(20), 9665–9678.
[67] Abascal, J. L. F., Sanz, E., García Fernández, R., & Vega, C. A potential model for the study of ices and amorphous water: TIP4P/Ice. J. Chem. Phys. 2005, 122(23), 234511.
[68] Abascal, J. L., & Vega, C. A general purpose model for the condensed phases of water: TIP4P/2005. J. Chem. Phys. 2005, 123(23), 234505.
[69] Mahoney, M. W., & Jorgensen, W. L. A five-site model for liquid water and the reproduction of the density anomaly by rigid, nonpolarizable potential functions. J. Chem. Phys. 2000, 112(20), 8910.
[70] Baranyai, A., & Bartók, A. Classical interaction model for the water molecule. J Chem Phys 2007, 126(18), 184508. https://dx.doi.org/10.1063/1.2730510.
[71] Lamoureux, G., MacKerrel Jr., A.D., & Roux, B. A simple polarizable model of water based on classical Drude oscillators. J. Chem. Phys. 2003, 119(10), 5185–5197.
[72] Straatsma, TP., & McCammon, JA. Molecular dynamics simulations with interaction potentials including polarization development of a noniterative method and application to water. Mol. Simulat. 1990, 5, 181–192.
[73] Yu, H., Hansson, T.,& van Gunsteren, W.F. Development of a simple self-consistent polarizable model for water. J. Chem. Phys 2003, 118, 221–234.
[74] Yu, H., Hansson, T., & van Gunsteren, W.F. Charge-on-spring polarizable water models revisited: from water clusters to liquid water to ice. J. Chem. Phys. 2004, 121, 9549–9564.
[75] Bachmann, S.J., & van Gunsteren, W.J. An improved simple polarisable water model for use in biomolecular simulation. J. Chem. Phys. 2014, 141, 22D515.
[76] Kiss, PT., & Baranyai, A. A systematic development of a polarizable potential of water. J Chem Phys 2013, 138(20), 204507. https://dx.doi.org/10.1063/1.4807600.

[77] Handley, C. M., Hawe, G. I., Kell, D. B., & Popelier, P. L. Optimal construction of a fast and accurate polarisable water potential based on multipole moments trained by machine learning. Phys. Chem. Chem. Phys. 2009, 11(30), 6365–6376.
[78] Stone, AJ. Distributed multipole analysis, or how to describe a molecular charge-distribution. Chem. Phys. Lett. 1981, 83(2), 233–239. https://dx.doi.org/10.1016/0009-2614(81)85452-8.
[79] Ren, P., & Ponder, JW. Polarizable atomic multipole water model for molecular mechanics simulation. J. Phys. Chem. B 2003, 107(24), 5933–5947.
[80] Jones, A., Cipcigan, F., Sokhan, V. P., Crain, J., & Martyna, G. J. Electronically coarse-grained model for water. Phys. Rev. Lett. 2013, 110(22), 227801.
[81] Bejagam, KK., Singh, S., An, Y., Berry, C., & Deshmukh, SA. PSO-assisted development of new transferable coarse-grained water models. J. Phys. Chem. B 2018, 122(6), 1958–1971.
[82] Yesylevskyy, S. O., Schäfer, L. V., Sengupta, D., & Marrink, S. J. Polarizable water model for the coarse-grained MARTINI force field. PLoS Comput. Biol. 2010, 6(6), e1000810.
[83] Price, SL., Leslie, M., Welch, GWA., Habgood, M., Price, LS., Karamertzanis, PG., & Day, GM. Modelling organic crystal structures using distributed multipole and polarizability-based model intermolecular potentials. Phys Chem Chem Phys 2010, 12(30), 8478–8490. http://dx.doi.org/10.1039/c004164e.
[84] Buchholz, H. K., Hylton, R. K., Brandenburg, J. G., Seidel-Morgenstern, A., Lorenz, H., Stein, M., & Price, S. L. Thermochemistry of racemic and enantiopure organic crystals for predicting enantiomer separation. Cryst. Growth. Des. 2017, 17(9), 4676–4686. http://dx.doi.org/10.1021/acs.cgd.7b00582.
[85] McQuarrie, DA., & Simon, JD. Physical Chemistry: A Molecular Approach, University Science Books, Sausalito, California, 1997.
[86] Hoja, J., & Tkatchenko, A. First-principles stability ranking of molecular crystal polymorphs with the DFT + MBD approach. Faraday. Discuss. 2018, 211, 253–274. http://dx.doi.org/10.1039/C8FD00066B.
[87] Adamo, C. Toward reliable density functional methods without adjustable parameters: the PBE0 model. J. Chem. Phys. 1999, 110:6158. https://doi.org/10.1063/1.478522.
[88] Curtis, F., Rose, T., & Marom, N. Evolutionary niching in the gator genetic algorithm for molecular crystal structure prediction. Faraday Discuss. 2018, 211, 61–77. http://dx.doi.org/10.1039/C8FD00067K.
[89] Iuzzolino, L., McCabe, P., Price, SL., & Brandenburg, JG. Crystal structure prediction of flexible pharmaceutical-like molecules: density functional tight-binding as an intermediate optimisation method and for free energy estimation. Faraday Discuss. 2018. https://dx.doi.org/10.1039/c8fd00010g.
[90] Lommerse, JP. *et al.* A test of crystal structure prediction of small organic molecules. Acta. Cryst. B. 2000, B56, 687–714.
[91] Bardwell, DA. *et al.* Towards crystal structure prediction of complex organic compounds – A report on the fifth blind test. Acta. Cryst. B. 2011, B67, 535–551.
[92] Day, G. *et al.* A third blind test of crystal structure prediction. Acta. Cryst. B. 2005, B61, 511–527.
[93] Day, G. *et al.* Significant progress in predicting the crystal structures of small organic molecules – A report on the fourth blind test. Acta. Cryst. B. 2009, B65, 107–125.
[94] Motherwell, WD. *et al.* Crystal structure prediction of small organic molecules: A second blind test. Acta. Cryst. B. 2002, B58, 647–661.
[95] Reilly, A.M. et al. Report on the sixth blind test of crystal structure prediction methods. Acta. Cryst. B. 2016, B72, 439–459.
[96] Sanz, E., & Vega, C. Solubility of KF and NaCl in water by molecular simulation. J. Chem. Phys. 2007, 126(1), 014507. http://dx.doi.org/10.1063/1.2397683.

[97] Moustafa, SG., Schultz, AJ., & Kofke, DA. A comparative study of methods to compute the free energy of an ordered assembly by molecular simulation. J. Chem. Phys. 2013, 139, 084105. https://doi.org/10.1063/1.4818990.
[98] Mobley, DL., Bayly, CI., Cooper, MD., & Dill, KA. Predictions of hydration free energies from all-atom molecular dynamics simulations. J. Phys. Chem. B 2009, 113(14), 4533–4537. http://dx.doi.org/10.1021/jp806838b.
[99] Westergren, J., Lindfors, L., Höglund, T., Lüder, K., Nordholm, S., & Kjellander, R. In silico prediction of drug solubility: 1. Free energy of hydration. J. Phys. Chem. B 2007, 111(7), 1872–1882. http://dx.doi.org/10.1021/jp064220w.
[100] Misin, M., Fedorov, M. V., & Palmer, D. S. Accurate hydration free energies at a wide range of temperatures from 3D RISM. J. Chem. Phys. 2015, 142, 091105. http://dx.doi.org/10.1063/1.4914315.
[101] Sergiievskyi, V., Jeanmairet, G., Levesque, M., & Borgis, D. Solvation free-energy pressure corrections in the three dimensional reference interaction site model. J. Chem. Phys. 2015, 143, 184116. https://doi.org/10.1063/1.4935065.
[102] Misin, M., Fedorov, M. V., & Palmer, D. S. Hydration free energies of molecular ions from theory and simulation. J. Phys. Chem. B 2016, 120, 975–983. http:/dx.doi.org/10.1021/acs.jpcb.5b10809.
[103] Wu, J. Classical Density Functional Theory for Molecular Systems, Wu J., eds, Variational Methods in Molecular Modeling. Molecular Modeling and Simulation (Applications and Perspectives), Springer, Singapore, 2017.
[104] Sulea, T., Wanapun, D., Dennis, S., & Purisima, EO. Prediction of SAMPL-1 hydration free energies using a continuum electrostatics-dispersion model. J. Phys. Chem. B 2009, 113(14), 4511–4520. http://dx.doi.org/10.1021/jp8061477.
[105] Marenich, AV., Cramer, CJ., & Truhlar, DG. Performance of SM6, SM8, and SMD on the SAMPL1 test set for the prediction of small-molecule solvation free energies. J. Phys. Chem. B 2009a, 113(14), 4538–4543. http://dx.doi.org/10.1021/jp809094y.
[106] Klamt, A., Eckert, F., & Diedenhofen, M. Prediction of the free energy of hydration of a challenging set of pesticide-like compounds. J Phys Chem B 2009, 113(14), 4508–4510. http://dx.doi.org/10.1021/jp805853y.
[107] Kolafa, J. Solubility of NaCl in water and its melting point by molecular dynamics in the slab geometry and a new BK3-compatible force field. J. Chem. Phys. 2016, 145, 204509. http://dx.doi.org/10.1063/1.4968045.
[108] Li, L., Totton, T., & Frenkel, D. Computational methodology for solubility prediction: application to sparingly soluble organic/inorganic materials. J. Chem. Phys. 2018, 149, 054102. https://dx.doi.org/10.1063/1.5040366.
[109] Joung, IS., & Cheatham, TE. Determination of alkali and halide monovalent ion parameters for use in explicitly solvated biomolecular simulations. J. Phys. Chem. B 2008, 112, 9020–9041. https://doi.org/10.1021/jp8001614.
[110] Lüder, K., Lindfors, L., Westergren, J., Nordholm, S., & Kjellander, R. In silico prediction of drug solubility. 3. Free energy of solvation in pure amorphous matter. J. Phys. Chem. B. 2007b, 111(25), 7303–7311. http://dx.doi.org/10.1021/jp071687d.
[111] Lüder, K., Lindfors, L., Westergren, J., Nordholm, S., Persson, R., & Pedersen, M. In silico prediction of drug solubility: 4. Will simple potentials suffice?. J. Comput Chem. 2009, 30 (12), 859–1871. http://dx.doi.org/10.1002/jcc.21173.
[112] Boothroyd, S., Kerridge, A., Broo, A., Buttar, D., & Anwar, J. Solubility prediction from first principles: A density of states approach. Phys. Chem. Chem. Phys. 2018, 20, 20981–20987. http://dx.doi.org/10.1039/C8CP01786G.

[113] Gütlein, M., Helma, C., Karwath, A., & Kramer, S. A large-scale empirical evaluation of cross-validation and external test set validation in (Q)SAR. Mole. Info. 2013, 32(5-6), 516–528. http://dx.doi.org/10.1002/minf.201200134.
[114] Huuskonen, J. Estimation of aqueous solubility for a diverse set of organic compounds based on molecular topology. J. Chem. Info. Comp. Sci. 2000, 40, 773–777. http://dx.doi.org/10.1021/ci9901338.
[115] Mitchell, JBO., McDonagh, JD., & Boobier, S. DLS-100 solubility dataset, University of St Andrews 2017. http://dx.doi.org/10.17630/3a3a5abc-8458-4924-8e6c-b804347605e8
[116] Boobier, S., Osbourn, A., & Mitchell, JBO. Can human experts predict solubility better than computers? J. Cheminformat. 2017, 9, 63. http://dx.doi.org/10.1186/s13321-017-0250-y.
[117] Hopfinger, AJ., Esposito, EX., Llinàs, A., Glen, RC., & Goodman, JM. Findings of the challenge to predict aqueous solubility. J. Chem. Info. Model. 2009, 49, 1–5. https://dx.doi.org/10.1021/ci800436c.
[118] Comer, J., Judge, S., Matthews, D., Towers, L., Falcone, B., Goodman, J., & Dearden, J. The intrinsic aqueous solubility of indomethacin. ADMET & DMPK 2014. http://dx.doi.org/10.5599/admet.2.1.33.
[119] Llinas, A. & Avdeef, A. Solubility Challenge Revisited after Ten Years, with Multilab Shake-Flask Data, Using Tight (SD ~ 0.17 log) and Loose (SD ~ 0.62 log) Test Sets. J. Chem. Inf. Model. 2019, 59(6), 3036–3040. doi: 10.1021/acs.jcim.9b00345.
[120] Sanghvi, T., Jain, N., Yang, G., & Yalkowsky, S. Estimation of aqueous solubility by the general solubility equation (GSE) the easy way. QSAR Comb. Sci. 2003, 22, 258–262. http://dx.doi.org/10.1002/qsar.200390020.
[121] McDonagh, JL., van Mourik, T., & Mitchell, JBO. Predicting melting points of organic molecules: applications to aqueous solubility prediction using the general solubility equation. Mol Inf. 2015, 34, 715–724. http://dx.doi.org/10.1002/minf.201500052.
[122] Hughes, LD., Palmer, DS., Nigsch, F., & Mitchell, JBO. Why are some properties more difficult to predict than others? a study of QSPR models of solubility, melting point, and log P. J. Chem. Info. Model. 2008, 48, 220–232. http://dx.doi.org/10.1021/ci700307p.
[123] Bhat, AU., Merchant, SS., & Bhagwat, SS. Prediction of melting points of organic compounds using extreme learning machines. Ind. Eng. Chem. Res. 2008, 47, 920–925. http://dx.doi.org/10.1021/ie0704647.
[124] Nigsch, F., Bender, A., Buuren, B., Tissen, J., Nigsch, E., & Mitchell, JBO. Melting point prediction employing k-nearest neighbor algorithms and genetic parameter optimization. J. Chem. Inf. Model. 2006, 46, 2412–2422. https://dx.doi.org/10.1021/ci060149f.
[125] Ali, J., Camilleri, P., Brown, MB., Hutt, AJ., & Kirton, SB. Revisiting the general solubility equation: in silico prediction of aqueous solubility incorporating the effect of topographical polar surface area. J. Chem. Inf. Model. 2012, 52, 420–428. http://dx.doi.org/10.1021/ci200387c.
[126] Dearden, JC. In silico prediction of aqueous solubility. Expert. Opin. Drug. Disc. 2006, 1, 31–52.
[127] Yousefinejad, S., & Hemmateenejad, B. Chemometrics tools in QSAR/QSPR studies: A historical perspective. Chem. Intel. Lab. Sys. 2015, 149, 177–204. http://dx.doi.org/10.1016/j.chemolab.2015.06.016.
[128] Cros, AFA. *Action de l'alcool amylique sur l'organisme.* PhD Thesis, University of Strasbourg, 1863.
[129] Crum Brown, A., & Fraser, TR. On the connection between chemical constitution and physiological action. part. I.–On the physiological action of the salts of the ammonium bases, derived from strychnia, brucia, thebaia, codeia, morphia, and nicotia. Trans. R. Soc. Edinb. 1868, 25, 151–203. http://dx.doi.org/10.1017/S0080456800028155.
[130] Führer, H. Water-solubility in homologous series. Ber. Dtsch. Chem. Ges. 1924, 57B, 510–515.

[131] Erickson, L. The solubility of homologous series of organic compounds. Naturwiss 1952, 39, 41–42.
[132] Hansch, C., Quinlan, JE., & Lawrence, GL. Linear free-energy relationship between partition coefficients and the aqueous solubility of organic liquids. J. Org. Chem. 1968, 33, 347–350. http://dx.doi.org/10.1021/jo01265a071.
[133] Mitchell, JBO. Machine learning methods in chemoinformatics. WIREs Comput. Mol. Sci. 2014, 4(5), 468–481. http://dx.doi.org/10.1002/wcms.1183.
[134] Klopman, G., Wang, S., & Balthasar, DM. Estimation of aqueous solubility of organic molecules by the group contribution approach. application to the study of biodegradation. J. Chem. Info. Comp. Sci. 1992, 32, 474–482. http://dx.doi.org/10.1021/ci00009a013.
[135] Marrero, J., & Gani, R. Group-contribution-based estimation of octanol/water partition coefficient and aqueous solubility. Ind.Eng. Chem. Res. 2002, 41, 6623–6633. http://dx.doi.org/10.1021/ie0205290.
[136] Klopman, G., & Zhu, H. Estimation of the aqueous solubility of organic molecules by the group contribution approach. J Chem Inf Comput Sci 2001, 41, 439–445. http://dx.doi.org/10.1021/ci000152d.
[137] Hou, TJ., Xia, K., Zhang, W., & Xu, XJ. ADME evaluation in drug discovery. 4. Prediction of aqueous solubility based on atom contribution approach. J. Chem. Info. Comp. Sci. 2004, 44, 266–275. http://dx.doi.org/10.1021/ci034184n.
[138] Wang, J., Krudy, G., Hou, T., Zhang, W., Holland, G., & Xu, X. Development of reliable aqueous solubility models and their application in druglike analysis. J. Chem. Info. Model. 2007, 47, 1395–1404. http://dx.doi.org/10.1021/ci700096r.
[139] Gracin, S., Brinck, T., & Rasmuson, ÃC. Prediction of solubility of solid organic compounds in solvents by UNIFAC. Indus. Eng. Chem. Res. 2002, 41, 5114–5124. http://dx.doi.org/10.1021/ie011014w.
[140] Abraham, MH., & Le, J. The correlation and prediction of the solubility of compounds in water using an amended solvation energy relationship. J. Phar. Sci. 1999, 88, 868–880. http://dx.doi.org/10.1021/js9901007.
[141] Bradley, JC., Abraham, MH., Acree, WE., & Lang, A. Predicting Abraham model solvent coefficients. Chem. Cent. J. 2015, 9, 12. http://dx.doi.org/10.1186/s13065-015-0085-4.
[142] Hewitt, M., Cronin, MTD., Enoch, SJ., Madden, JC., Roberts, DW., & Dearden, JC. In silico prediction of aqueous solubility: the solubility challenge. J. Chem. Inf. Model. 2009, 49, 2572–2587. https://dx.doi.org/10.1021/ci900286s.
[143] Catana, C., Gao, H., Orrenius, C., & Stouten, PFW. Linear and nonlinear methods in modeling the aqueous solubility of organic compounds. J. Chem. Inf. Model. 2005, 45, 170–176. http://dx.doi.org/10.1021/ci049797u.
[144] Palmer, DS., O'Boyle, NM., Glen, RC., & Mitchell, JBO. Random forest models to predict aqueous solubility. J. Chem. Info. Model. 2007, 47, 150–158. http://dx.doi.org/10.1021/ci060164k.
[145] Cao, DS., Xu, QS., Liang, YZ., Chen, X., & Li, HD. Prediction of aqueous solubility of druglike organic compounds using partial least squares, back-propagation network and Support Vector Machine J. Chem. 2010, 24, 584–595. http://dx.doi.org/10.1002/cem.1321.
[146] Werbos, PJ. Beyond regression: new tools for prediction and analysis in the behavioral sciences, PhD thesis, Harvard University, 1975.
[147] Connors, BW., & Long, MA. Electrical synapses in the mammalian brain. Annu. Rev. Neurosci. 2004, 27, 393–418.
[148] Louis, B., Agrawal, VK., & Khadikar, PV. Prediction of intrinsic solubility of generic drugs using MLR, ANN and SVM analyses. Eur. J. Med. Chem. 2010, 45, 4018–4025. http://dx.doi.org/10.1016/j.ejmech.2010.05.059.

[149] Erić, S., Kalinić, M., Popović, A., Zloh, M., & Kuzmanovski, I. Prediction of aqueous solubility of drug-like molecules using a novel algorithm for automatic adjustment of relative importance of descriptors implemented in counter-propagation artificial neural networks. Int. J. Pharm. 2012, 437, 232–241. http://dx.doi.org/10.1016/j.ijpharm.2012.08.022.
[150] Svetnik, V., Liaw, A., Tong, C., Culberson, JC., Sheridan, RP., & Feuston, BP. Random forest: A classification and regression tool for compound classification and QSAR modeling. J. Chem. Inf. Comput. Sci. 2003, 43, 1947–1958. http://dx.doi.org/10.1021/ci034160g.
[151] Breiman, L. Random forests. Mach. Learn. 2001, 45, 5–32. http://dx.doi.org/10.1023/a:1010933404324.
[152] Raileanu, L., & Stoffel, K. Theoretical comparison between the Gini index and information gain criteria. Ann. Math. Artif. Intell. 2004, 41, 77–93. https://doi.org/10.1023/B:AMAI.0000018580.96245.c6.
[153] Schroeter, TS., Schwaighofer, A., Mika, S., Ter Laak, A., Suelzle, D., Ganzer, U. *et al.* Estimating the domain of applicability for machine learning QSAR models: A study on aqueous solubility of drug discovery molecules. J. Comp.-Aid. Mol. Des. 2007, 21, 485–498. http://dx.doi.org/10.1007/s10822-007-9125-z.
[154] Palmer, DS., & Mitchell, JBO. Is experimental data quality the limiting factor in predicting the aqueous solubility of druglike molecules? Mole. Pharm. 2014, 11, 2962–2972. http://dx.doi.org/10.1021/mp500103r.
[155] Kovdienko, NA., Polishchuk, PG., Muratov, EN., Artemenko, AG., Kuz'min, VE., Gorb, L. *et al.* Application of random forest and multiple linear regression techniques to QSPR prediction of an aqueous solubility for military compounds. Mole. Info. 2010, 29, 394–406. http://dx.doi.org/10.1002/minf.201000001.
[156] Kew, W., & Mitchell, JBO. Greedy and linear ensembles of machine learning methods outperform single approaches for QSPR regression problems. Mol. Inf. 2015, 34, 634–647. 2015 http://dx.doi.org/10.1002/minf.201400122.
[157] Noble, WS. What is a Support Vector Machine? Nat. Biotech. 2006, 24, 1565–1567. http://dx.doi.org/10.1038/nbt1206-1565.
[158] Vapnik, VN. Statistical Learning Theory, Wiley, New York, 1998.
[159] Lind, P., & Maltseva, T. Support Vector Machine for the estimation of aqueous solubility. J. Chem. Inf. Comput. Sci. 2003, 43, 1855–1859. http://dx.doi.org/10.1021/ci034107s.
[160] Kühne, R., Ebert, RU., & Schuurmann, G. Model selection based on structural similarity-method description and application to water solubility prediction. J. Chem. Inf. Model. 2006, 46, 636–641. http://dx.doi.org/10.1021/ci0503762.
[161] Rasmussen, CE., & Williams, CKI. Gaussian Processes for Machine Learning, MIT Press, 2006, ISBN, 026218253X. http://www.gaussianprocess.org/gpml/chapters/RW.pdf.
[162] Popelier, PLA. Molecular simulation by knowledgeable quantum atoms. Physica. Scripta. 2016, 91:033007. http://dx.doi.org/10.1088/0031-8949/91/3/033007.
[163] Obrezanova, O., Csányi, G., Gola, JMR., & Segall, MD. Gaussian processes: a method for automatic QSAR modeling of ADME properties. J. Chem. Info. Model. 2007, 47(5), 1847–1857. http://dx.doi.org/10.1021/ci7000633.
[164] Pyzer-Knapp, EO., Simm, GN., & Aspuru Guzik, A. A Bayesian approach to calibrating high-throughput virtual screening results and application to organic photovoltaic materials. Mater. Horiz. 2016, 3, 226–233. http://dx.doi.org/10.1039/C5MH00282F.
[165] McDonagh, JL., Silva, AF., Vincent, MA., & Popelier, PLA. Machine learning of dynamic electron correlation energies from topological atoms. J. Chem. Theo. Comput. 2018, 14(1), 216–224. http://dx.doi.org/10.1021/acs.jctc.7b01157.

[166] Obrezanova, O., & Segall, MD. Gaussian processes for classification: QSAR modeling of ADMET and target activity. J. Chem. Info. Model. 2010, 50(6), 1053–1061. http://dx.doi.org/10.1021/ci900406x.
[167] Schwaighofer, A., Schroeter, T., Mika, S., Laub, J., Ter Laak, A., Sülzle, D. *et al.* Accurate solubility prediction with error bars for electrolytes: A machine learning approach. J. Chem. Info. Model. 2007, 47(2), 407–424. http://dx.doi.org/10.1021/ci600205g.
[168] LeCun, Y., Bengio, Y., & Hinton, G. Deep learning. Nature 2015, 521(7553), 436–444. http://dx.doi.org/10.1038/nature14539.
[169] Lusci, A., Pollastri, G., & Baldi, P. Deep architectures and deep learning in chemoinformatics: the prediction of aqueous solubility for drug-like molecules. J. Chem. Inf. Model 2013, 53(7), 1563–1575. http://dx.doi.org/10.1021/ci400187y.
[170] Surowiecki, J. The wisdom of crowds: why the many are smarter than the few and how collective wisdom shapes business, economies, societies, and nations, Doubleday, New York, 2004.
[171] Galton, F. Vox populi. Nature 1907, 75, 450–451. https://dx.doi.org/10.1038/075450a0.

Bernard Faller, Sandrine Desrayaud, Joerg Berghausen,
Marc Laisney and Stephanie Dodd

4 How solubility influences bioavailability

Bioavailability is the result of a complex interplay between the intrinsic properties of the molecule, the physiology of the species considered, the formulation of the drug, and the dose. This chapter focuses mainly on oral bioavailability with a specific paragraph on non-oral routes at the end. The link between solubility and bioavailability goes back to Fick's laws of diffusion. For oral delivery, solubility and permeability are the main contributors defining the fraction of dose absorbed (F_a). The flux (FL) through the gastro-epithelial tract is governed by the relationship as follows:

$$\mathrm{FL} = P_e\ (\Delta C) \tag{4.1}$$

where P_e is the effective permeability and ΔC the concentration gradient across the epithelium. ΔC is defined by the dose if the compound is sufficiently soluble and by the solubility if only a fraction of the dose is in solution. In this case, absorption will be limited by solubility.

In the context of bioavailability, the notion of high and low solubility has to be considered together with the intended dose. As a first approach, the concept of dose number [1] helps to define when solubility is likely a limiting absorption

$$D_O = \frac{M_O/V_O}{C_s} \tag{4.2}$$

where M_O is the dose, V_O the administered volume, and C_s the solubility of the API. In our experience, when $D_O > 20$, the compound is considered poorly soluble, and solubility often limits oral absorption.

Another approach that was proposed to give guidance in the discovery and early development phase is the concept of maximum absorbable dose (MAD), as defined by Curatolo [2]:

$$\mathrm{MAD} = S \times K_a \times \mathrm{SIWV} \times \mathrm{SITT} \tag{4.3}$$

where S is the solubility in mg/mL at pH 6.5, K_a the absorption rate constant, SIWV the small intestine water volume (mL), and SITT the small intestine transit time (min). The absorption rate constant K_a is obtained directly from intestinal perfusion measurement or can be estimated from Caco-2 or MDCK permeability experiments [3]. More recently, it has been suggested that MAD tends to under-predict oral absorption [4].

Oral absorption is rarely impacted by solubility in the low dose (<5 mg/kg p.o.) rodent studies performed in early discovery. The interpretation of solubility

Bernard Faller, Sandrine Desrayaud, Joerg Berghausen, Marc Laisney, Stephanie Dodd, Novartis, Institutes for BioMedical Research

https://doi.org/10.1515/9783110559835-004

limitation from a single PK study is difficult and most often identified from the observation of an under-proportional dose exposure relationship or a comparison of the AUC obtained with a suspension versus a solution formulation. The relationship between dose and exposure can be defined as an exposure proportionality (EP), a dimensionless value that is described by the following equation:

$$\text{Exposure proportionality(EP)} = \frac{\text{AUC}_{(0-t)}\text{highest dose}/\text{AUC}_{(0-t)},\text{lowest dose}}{\text{Highest dose(mg)}/\text{lowest dose(mg)}} \quad (4.4)$$

The key question at the candidate selection stage is more how to predict which molecules will fail to achieve exposure multiples in toxicology studies or simply fail to reach the efficacious plasma concentration. Another challenge is to extrapolate from pre-clinical species to human, especially for difficult molecules that were dosed using a specific formulation in pre-clinical species. Wuehling et al. [5] analysed a large dataset of more than 900 experiments performed at Merck in an attempt to predict when EP will become an issue.

In this work, the authors used the pre-clinical dose number (PD_0), parameter that is simply the dose in mg/kg divided by the fasted state simulated intestinal fluid (FaSSIF) solubility in mg/mL. PD_0 values >10,000 in the rat and >5,000 in dog and monkey were indicative of a high risk of under-proportional dose exposure. Again, similar to our own observations, under-proportional dose exposure was rarely observed at doses lower than 10 mg/kg.

In early discovery, solubility is typically obtained from DMSO stock solutions as the handling of solutions is a lot easier than powders. The impact of DMSO on solubility in FaSSIF is modest at 2.5% DMSO and increasing with 5% solvent with ca. 20% of the molecules showing a significant higher solubility in the presence of DMSO. Typically, solubility values from solubility screening assays tend to overestimate the true thermodynamic solubility and should therefore be taken as maximum solubility values.

In early development, one can refine the impact of solubility on absorption considering the three key parameters defined by Amidon: the dose number (D_o), the absorption number (A_n), and the dissolution number (D_n). When the dose is in suspension at the beginning of the intestine, the fraction absorbed is the result of a complex interplay between effective permeability, solubility, and dissolution rate. The fraction of the dose in solution at the beginning of the intestine is not always easy to define as supersaturation may happen due to solubilization in the stomach, particularly for basic compounds (See Chapter 2). If the dose is entirely dissolved at the beginning of the intestine the prediction of the fraction absorbed (F_a) becomes easier and the following equation can be used:

$$F_a = 1 - e^{-2An} \quad (4.5)$$

with $$A_n = P_{eff}\,\pi R L / Q$$

where P_{eff} is the effective permeability, R the radius of the intestine, L the length of the intestine, and Q the flow rate. Equation (4.5) largely applies to low dose (<5 mg/kg) experiments typically performed in early discovery even when the compound is given as a suspension generally, including carboxymethylcellulose to get a homogeneous and stable suspension and a wetting agent like Tween as most compounds will solubilize in the stomach in low-dose experiments. In this context, the notion of supersaturation stability appears useful and several teams have proposed to measure the solubility after 15 min starting from concentrated solutions in either organic solvent (solvent-shift method) or at a different pH value (pH-shift method) [6, 7]. The supersaturation solubility at 15 min gives an indication of what formulation will be able to achieve in the best-case scenario. Skolnik [7] reported that 80% of low supersaturation stability compounds had an EP < 0.8 while all molecules with a supersaturation solubility target >200 µM at the 15 min time point demonstrated an EP > 0.8. Brouwers [8] analysed the impact of delivery strategies using supersaturation principles on oral absorption. These authors suggested that for high permeable compounds, the production of a highly supersaturated solution was important, while for medium-to-low permeable compounds stabilization of the supersaturated solution for a prolonged time period was necessary to provide sufficient time for absorption.

If permeability nor dissolution is rate limiting, the fraction absorbed can be predicted from the following equation:

$$F_a = 2A_n / D_o \tag{4.6}$$

The dissolution rate can be modulated by the particle size of the drug as it changes the surface area in contact with the medium. Johnson and Swindell [9] looked at the impact of particle size on absorption for various absorption rate constants and dose. The largest effect is achieved with low soluble compounds given at a low dose (<10 mg in humans). For highly permeable compounds, the effect of particle size reduction is expected to be significant for a low soluble compound also at higher dose (up to 100 mg) compared to low permeable compounds. The effect is, however, maximal at lower doses.

4.1 Species differences (physiology)

There is currently no animal model that directly relates to the human physiology in the whole extension of the gastrointestinal (GI) tract. It is thus important to understand how the GI luminal environment changes along the GI tract of the different animal models. The GI tract size and anatomy can contribute in a profound fashion to absorption differences across species (Table 4.1).

Table 4.I: Comparison of the anatomical and physiological features of the GI tract of various animals and humans.

	Mouse	Rat	Rabbit	Monkey	Dog	Pig	Human
BW kg	**0.02**	**0.25**	**2.5**	**5**	**10**	**15–25**	**70**
Stomach capacity (L)	0.0003	0.005	0.10	0.10	1.00	7.00	1.30
Fluid content in (g)							
Stomach	0.09	2.29	54.6	100	500	278	118
Small intestine	0.31	3.89	19.9	794	300	476	212
Cecum	0.28	3.38	62.9	14	250	204	
Colon	0.12	1.34	15.3			587	187
Small intestine diameter (cm)		0.3–0.5		1.2–2	2–2.5	2.5–3.5	5
Small intestine length (m)							
Duodenum		0.10		0.05	0.25		0.25
Jejunum		0.9–1.35		1.00	3.74		3.00
Ileum		0.025–0.035		1.50	0.15		3.00
Total	0.34	1.02–1.5	3.56	2.55	4.14	18.29	6.25
Cecum length (m)		0.05–0.07	0.61	0.05–0.06	0.08	0.23	0.1–0.3
Colon length (m)	0.09–0.14	0.26	1.65	0.4–0.5	0.60	4.99	1.50
Rectum length (m)		0.08					0.15–0.19
Bile flow (mL/day/kg)	50	48–92	130	19–32	19–36		2.2–22.2

Adapted from Refs [10–14].

The gross morphology of the mammalian GI tract differs considerably among species although it exhibits some basic structural similarities. Indeed, the structural aspects such as the volume of stomach and small intestine, diameter, absorptive surface area, villi, and microvilli structures are greatly related to the diet and nutrient processing, as illustrated in the comprehensive literature review published by Hatton [10].

For example, carnivores (e.g. dog) possess a relatively simple colon but a well-developed small intestine (long villi). This is consistent with a diet that is low in fibre but high in fat and protein. Omnivores like human possess a well-developed small intestine but have a more complex lower intestine to compensate for their more diverse diet. The lower intestine of the pigs is differentiated enough to allow for fermentation.

The small intestine in turn constitutes the majority of gut length in mammals, and its large surface area is highly specialized and optimized for the absorption of nutrients. The small intestine of the dog is about half of that in humans, though the difference in absorptive surface area is to some extent compensated by the shape of dog intestinal villi. This is a predominantly dietary adaptive function. These structural parameters will be of great importance to extrapolate formulations tested in pre-clinical species to human.

The GI content is a complex mixture of different sources and its composition and characteristics are key for the solubilization of xenobiotics.

The gastric fluid is mainly composed of saliva, gastric secretion, dietary food, and liquid and refluxed liquid from the duodenum. The upper small intestine contains gastric mixture secretions from the liver, the pancreas, and the wall of the small intestine. The fluid composition is affected by fluid compartmentalization, mixing patterns, absorption of fluid into the intestinal wall, and transit down the intestinal tract. Secretions from the pancreas include bicarbonate as well as proteases (the major ones are trypsin and chymotrypsin), amylases, and lipases. The liver secretes bile, which contains bile salts, phospholipids, bicarbonate, cholesterol, bile pigments, and organic waste. The wall of the small intestine secretes mineral ions such as bicarbonate, sodium, and chloride, as well as water. Bicarbonate is secreted to neutralize gastric secretion in the GI lumen and by the duodenal epithelial cells to protect the duodenal epithelium from acid-related damage. The buffer capacity and pH in the GI media can significantly affect the dissolution rates of ionizable drugs.

The two most important properties affecting solubility include hydrogen ion concentration affecting the pH, and the bile salts that can be combined with lipid form mixed micelles, enhancing the solubility of some drugs and may also decrease surface tension and overcome dissolution limitations caused by poor wettability of the powder.

Solubilization can be enhanced by either native or co-ingested surfactants, binding to peptides or proteins and solubilization in lipid components of the meal.

4.2 Stomach fluids, intestinal fluids – fed versus fasted state

Modulation of solubility for drug absorption of oral dosage forms in vivo is largely determined by solubility in the GI tract. Besides the variation of total fluid volume available for compound dissolution, factors like pH gradient and fluid composition vary along the drug's journey through the GI tract. As emphasized in the previous paragraph, one major challenge for drug discovery and development arises from the differences between pre-clinical species GI tract morphology and fluid composition and the situation in humans.

Many additional factors are contributing to variability of fluid volumes and composition, like the permanent fluid recirculation and changes of composition induced by uptake the of food and beverages. Additionally, pH changes have been observed based on the circadian rhythm, without impact of food [15].

Besides the impact on pH, food intake and the resulting effect for drug absorption in humans have been widely studied, as discussed in the following sections.

4.2.1 Stomach: fasted–fed

The stomach is not the predominant site for drug absorption but plays a major role in drug dissolution. The role of stomach fluid for dissolution and thus bioavailability is much dependent on the compound's ionization profile. Absorption of drugs in a fasted stomach of humans [16] and rat [17] has been already studied in the 1950s. The authors revealed the role of ionization and typically found rapid absorption of acidic drugs. In contrast, basic drugs were found to get nearly not absorbed. It is important to note that acidic drugs in a basic solution were not absorbed by the stomach, thus emphasizing the importance of ionization and solubility for drug absorption. Similarly, absorption of basic drugs is facilitated in a basic solution. To study this aspect of absorption, rat models are still an area of active research [18].

The composition of human-fasted gastric fluid has been published by Vertzoni [19] and Kalantzi [20]. High variability of stomach pH in the fasted state is reported, which has been confirmed by more recent studies [21]. Bile salts are typically not present in significant quantities in the fasted state.

In the postprandial state, conditions vary over the course of gastric residence of the meal. For instance, after meal ingestion the pH in stomach was reported to decrease continuously from pH 6.4 to 2.7 in humans [20]. Even in fed state, bile salt concentrations typically remain very low in the stomach but may be refluxed after food intake. Further differences of fluid composition do contribute to drug absorption and are discussed in Chapter 5.

4.2.2 Intestine: fasted–fed

As the majority of low-molecular-weight drugs are absorbed in the small intestine, solubility in intestinal fluid is a critical factor for their bioavailability. Due to their ionization status, acidic molecules generally are better soluble in intestinal fluids than bases, zwitterions, or neutral compounds.

With increasing intestinal pH, the ionization of a weak base, and thus solubility, is typically decreasing. In case the compound has been dissolved in the stomach, the compound may remain soluble for a certain period of time, as supersaturated solution, and the stability of supersaturated solutions in the GI tract has been widely investigated in recent years [22, 23].

Besides ionization, one major component that influences solubility in the small intestine is bile salts. In humans, average bile salt content has been reported to be around 3 mM [24] in the fasted state. This value can increase significantly after food intake, reaching values of approximately 12 mmol in the fed state. These bile salt concentrations were found to decrease again and reach the fasted state value 180 min after food consumption [25]. In addition, significant differences in buffer capacity, surface tension, osmolality, and food components can be observed pre-/postprandially [26]. A study focusing on buffer capacity reported the influence of buffer capacity on solubility of the zwitterionic drug Mesalamine. It was found that solubility of this drug is strongly dependent on increasing buffer capacity down the GI tract, whereas prednisolone did not show this dependency [27].

The role of bile salt content for drug solubilization has been studied in the 1990s by Mithani [28]. The study revealed a direct relationship with log *P* of the dissolved compound. Using a more diverse set of proprietary compounds, we could show that the concentration-dependent solubilization power of bile salts may often deviate from this relationship [29] but confirmed the strong impact the bile salt concentration can have on solubilization of drugs for absorption. On the other hand, other studies have shown that attempts to correlate the solubility of lipophilic compounds with bile acid phospholipid content in human intestinal fluid have been less successful [30]. In our experience, about 25 % of discovery molecules have a substantially higher solubility (3 times or more) in FaSSIF compared to aqueous solubility.

Besides the fluid composition, the absolute amount of fluid available for drug dissolution needs to be considered, too. Magnetic resonance imaging studies revealed that fluid is not homogeneously distributed along the small intestine. In fact, fluid is distributed in separated pockets, which leads to highly variable drug concentration locally [31].

4.3 Absorption across species

4.3.1 Low-dose PK screening (<5 mg/kg)

An animal PK study typically includes two arms, such as intravenous and oral dosing, with 6–12 serial blood samplings for both. These PK screening experiments are generally performed at low dose, typically 0.1–1 mg/kg for i.v. and 0.3–3 mg/kg for oral administration, to provide robust estimates of PK parameters. This approach to use low doses for PK screening is dictated by different reasons:

- Variability is known to increase with dose as the interplay between animal physiology and physicochemical properties become more likely to impact absorption and disposition [32]. In particular, a higher PK variability has been reported in larger species, especially in dogs. This observation may reflect higher potential for absorption limitation in dogs and/or a higher inter-animal variability in physiological parameters (e.g., gastric pH, susceptibility to emesis) for that species.
- The synthesis of the compound at early stage is generally restricted to the minimum required to conduct key in vivo and in vitro assays (generally ≤50 mg).
- The cassette dosing approach, that is, simultaneous administration of several compounds per animal to increase the throughput of the in vivo rodent PK studies, calls for low doses to reduce potential for drug–drug interactions.

Given that the early PK screening aims at ranking and identifying the most promising candidate with a rapid turnaround time, a compound specific formulation development is often not conceivable at this stage. Moreover, compounds that do not show adequate oral exposure at low dose using a simple aqueous formulation principle will require more formulation development work to identify acceptable formulations for use in preclinical safety studies, clinical trials, and as the commercial drug product. Consequently, the general approach is to use a simple oral suspension with a good suspending agent such as methylcellulose (0.5% w/v) combined with a surfactant, typically Tween 80 at 0.1% (w/v), to ensure proper wetting and uniform drug dispersion. At low dose, the extent of oral absorption is rarely impacted by solubility and/or dissolution, so the oral exposure (AUC and C_{max}) obtained using suspension formulation does not often differ from the exposure obtained with a solution dosage form.

Finally, to ensure a reliable decision-making at the compound selection stage and homogeneity in the data set, a robust control of PK study parameters (age, sex, body weight, feeding regimen, surgical intervention, dosing and blood sampling procedures, etc.) is important.

4.3.2 High-dose PK (>10 mg/kg)

Understanding the impact of solubility on exposure at increasing doses is of pivotal importance to enable toxicology studies, where exposure multiples are required to establish a therapeutic index and progress a compound into early development. In general, doses greater than 10 mg/kg, and often exceeding 100 mg/kg are required for this purpose. Commonly, doses encompass a range greater than 10-fold between the lowest and highest dose required for a study. In order to select a formulation capable of delivering dose-EP in the range required (up to 2 g/kg), solubility in intestinal fluids has been found to be a useful surrogate prior to in vivo studies. Utilization of FaSSIF [33] to mimic human physiological conditions in the lumen has provided a reasonable estimation of in vivo solubility and can provide a relative guidance on expected EP in non-rodent species (i.e. dog), which are common choices for pre-clinical toxicology investigations.

Practically, the high-dose PK studies are performed in the selected toxicology species, using the same strain, gender, and body weight range as foreseen in the toxicology studies. The design of these high dose PK studies varies markedly according to the project, the properties of the molecule, and the formulation used. In general, one or two more sophisticated formulations (Tox-enabling formulation) are tested against the early simple formulation (e.g. suspension) in a cross-over fashion (i.e. same animals used in a consecutive manner) to demonstrate the benefit of the new formulation principle on the oral exposure (AUC and/or C_{max}). For molecules with low solubility, a large volume of application may be required for dosing but is limited by animal use and care guidance regulations and practicability (see Table 4.2). The food intake needs also to be carefully controlled as this factor can affect the outcome of the study. Large dose volume and food intake will overload the capacity of the stomach and as a result the drug could pass immediately into the small intestine and/or cause reflux into the oesophagus.

Table 4.2: Recommended volumes of oral application (by gavage) of compounds for common laboratory animals.

	Ideal volume (mL/kg) (single dose per day)	**Maximum volume* (mL/kg) (single dose per day)**
Mouse	10	20
Rat	10	20
Rabbit	10	20 (empty stomach)
Monkey	5	10
Dog	5	15
Minipig	10	15

Adapted from [34, 35].
*Larger volume may be envisaged when the dose is divided over time (e.g. 10 mL/kg administered 4 times a day to reach a total of 40 mL/kg).

Besides the volume, the tolerability of the formulation must be considered, as vehicles should be biologically inert and have no toxic effect on the animals. As such, the excipient biocompatibility (species-specificities exist, see [34]), the pH, the viscosity, the osmolality, buffering capacity, and stability are factors to consider.

The maximum dose (mg/kg) that can be applied with a specific formulation can be estimated by the ratio between the maximum drug load in the formulation in mg/mL and the dose volume applied (mL/kg).

In a recent evaluation, Skolnik et al. [7] were looking at the relationship between solubility and EP in 59 non-rodent (dog and monkey) pre-clinical pharmacokinetic studies. It was shown that in order to achieve EP (defined as EP > 0.8) using a conventional formulation (defined as a formulation that did not impact solubility, i.e. only dissolution controlled) an equilibrium solubility of minimally 0.1 mg/mL in FaSSIF is required. When FaSSIF solubility was >0.1 mg/mL, and a conventional formulation used, five of eight studies (63%) showed EP values >0.8. When equilibrium solubility was <0.1 mg/mL 22 of 29 studies (76%) had an EP < 0.8.

When solubility was low (<0.1 mg/mL), supersaturation was investigated as a surrogate for performance of a solubility-enhancing formulation (i.e. an amorphous solid dispersion or a stable solution). From the same set of 59 non-rodent studies, it was found that if the supersaturation solubility in FaSSIF is <0.1 mg/mL and an enabled formulation (i.e. solubility-enhancing formulation) was employed, despite this extra effort in formulation, seven of nine studies (78%) showed EP < 0.8. Conversely, if measured supersaturation solubility in FaSSIF was >0.1 mg/mL, and an enabled formulation was employed, 12 of 13 studies (92%) showed EP values > 0.8.

While FaSSIF can be a reasonable surrogate for in vivo solubility in dog and monkey, due to similar bile salt concentrations in the duodenum in the fasted state compared to humans (2.8 mM in humans and monkey, 5.0 mM in the dog), it could be a poor predictor for rats, the most commonly used rodent toxicology species employed in R&D. In general, rats exhibit similar bile salt concentrations in the duodenum in the fed and fasted states, about 20 mM, due to lack of a gall bladder, which regulates the release of bile in higher species. Therefore, rat in vivo solubility may be more appropriately reflected by fed state simulated intestinal fluid (FeSSIF), which has a bile salt concentration of approximately 10 mM. If a compound shows high sensitivity to bile salts for solubility, when FaSSIF solubility is <0.1 mg/mL, but FeSSIF solubility is >0.1 mg/mL, a conventional formulation approach (dissolution controlled) may be sufficient enough to provide a linear EP in rodents.

In a study conducted in rats with 11 compounds with solubility <0.1 mg/mL in FaSSIF, whereas FeSSIF solubility was >0.1 mg/mL (all dosed with a conventional suspension), five compounds showed dose proportionality >0.8 (45%). This is approximately double the percentage we typically see in dog. Therefore, in rats, due to the higher bile flow and hence higher solubility in the duodenum, a conventional suspension may be possible to support high-dose toxicology studies, even when solubility is limited in FaSSIF.

Situations in which this may not be successful could possibly include molecules with low permeability.

4.4 What can formulations do/not do?

Orally administered formulations may impact two critical aspects related to absorption: drug dissolution and drug solubility. A key aspect of formulation development is the determination of whether improving the dissolution rate will be sufficient to increase oral absorption, or whether a formulation must also improve the solubility of the drug in the GI tract. Dissolution rate improvements may be achieved through various approaches, including particle size reduction, modification of solid-state through salt formation or polymorphic control. Solubility improvements may be achieved through removal of the solid-state completely, such as a solution formulation (co-solvents, polymers, and cyclodextrins), or lessening of the impact of the solid-state through amorphorization (solid dispersions), emulsification (microemulsions), co-crystallization with inactive hydrophilic entities, nanosuspensions, or precipitation control (addition of polymers).

In previous paragraphs, it is highlighted that rats do not have a gall bladder, and hence high bile flow may make it feasible to formulate a suspension formulation (dissolution controlled) for a compound that exhibits poor solubility (<0.1 mg/mL) in FaSSIF media, but high solubility (>0.1 mg/mL) in FeSSIF media. For the same compound, a solubility-controlled formulation (i.e. a solution) may be required in dogs, due to low bile flow in the fasted state, resulting in poor absorption from a suspension caused by the poor solubility in the GI tract.

An example of this type of scenario is one of our recent kinase inhibitors, named compound A. Compound A is a phosphate salt and has a solubility in FaSSIF media of 0.6 µg/mL and a solubility in FeSSIF media of 1.25 mg/mL. When dosed as a suspension to rats the phosphate salt shows an EP of 0.97 (25–100 mg/kg dose), while the same phosphate salt dosed in a hard gelatine capsule to fasted dogs resulted in an EP of 0.42 only (20–60 mg/kg dose). The under-proportionality in dog can be attributed to the presence of much less bile present in the GI tract, compared to the rat. This was confirmed when dogs were administered the phosphate salt as a suspension at 20 mg/kg under fed and fasted conditions, as this led to an increase in exposure of approximately 2.4-fold (AUC, 0–*t*) in the fed dogs.

Further exploration of solubility versus dissolution rate limitation on absorption in the rat was conducted between a solution formulation, a suspension of the phosphate salt, and the suspension of the free base. All three formulations were dosed to fasted Sprague–Dawley rats, at a dose normalized to 50 mg/kg. The phosphate salt suspension improved the absorption compared to the free base suspension of approximately 2.6-fold (C_{max}), indicating the dissolution-limited absorption of the molecule. However, when the solution formulation was dosed, it did not provide any further improvement on absorption compared to the suspension of the phosphate salt (both formulations provided similar C_{max} +/−30%). Overall, this would indicate that compound A was essentially dissolution rate limited in rats, despite its low solubility in FaSSIF.

As an example of a compound which is limited by both solubility and dissolution, one can look at compound B, which blocks one component of the hedgehog pathway. The measured solubility in FaSSIF media is 1 µg/mL, and in FeSSIF media is 148 µg/mL. Another phosphate salt was developed for this compound, and when the pharmaceutical salt was dosed at 3 mg/kg in suspension compared to the free base in suspension at the same dose, the salt improved the AUC (0–*t*) by approximately 16-fold! Additionally, a solution formulation was dosed at 10 mg/kg, and the dose-normalized exposure improved the absorption of compound B by approximately 1.8-fold. Therefore, it could be concluded that compound B is heavily limited in oral absorption by dissolution rate, and to a much lesser extent by solubility. For compound B, the dose would play a critical role in absorption. A phosphate suspension was dosed in rats at 10 and 100 mg/kg, and the EP was found to be limited to 0.32. A solution formulation was dosed in rats at the same dose range, and the EP was found to be 0.59. Thus, while solubility-limited absorption at low dose was not too high, it became much more significant as the dose increased, and a solution formulation was able to improve the EP by almost twofold. In dog, the phosphate salt suspension provided an EP of 0.73 in a dose range 30–100 mg/kg. The slightly higher EP in dog than rat may be attributed to narrower dose range or potentially a higher solubility in the stomach of the dog as a fasted dog will have a stomach pH slightly lower than that of a fasted rat. Compound B being a weak base, there could be more compound dissolved in the dog compared to the rat stomach.

Other factors can also influence oral absorption, most notably permeability, pre-systemic degradation, and metabolism, and these parameters need to be understood to assess the formulation strategy. A less desirable space is the situation where a formulation attempts to address a perceived solubility or dissolution limitation, when in fact poor bioavailability may be due to a poor permeability.

In recent years, physiologically based pharmacokinetic (PBPK) modelling has been successfully used in the pre-clinical space in order to identify the rate-limiting step(s) for oral absorption across pre-clinical species. This information can help selecting the best formulation approach, especially in the context of high-dose PK studies needed to enable toxicological investigations. This is especially important as physiological differences in rodent and non-rodent species can have a large impact on oral absorption. The PBPK models that are commercially available from SimulationsPlus (GastroPlus) and Certara (SimCYP) allow for the incorporation of experimental solubility values measured in media with varying levels of bile salt. Species and prandial state (i.e. fed monkey) can be selected and the appropriate physiology is used to predict cross species differences. These simulation tools also incorporate pH-dependant solubility, which can be used in conjunction with the bile salt corrections across species. The models also allow for inclusion of parameters, which may influence permeability such as lipophilicity and ionization, and pre-systemic metabolism. The models, through parameter sensitivity analysis (PSA), may provide a reasonable estimation at which doses a dissolution-limited absorption or solubility-limited absorption may occur in a particular species and prandial state, or if neither solubility or dissolution will be limiting absorption.

To illustrate how modelling can be used to inform formulation decisions, we can revisit the example of compound A. Solubility data across the pH range of 1 to 7 for the phosphate salt were measured, and fed into a GastroPlus model, along with measured solubility in simulated physiological media. The physiology was chosen to be fasted monkey, and $\%F_a$ was determined for a 20 mg/kg dose and 60 mg/kg dose. The ratio of $\%F_a$ of high/low dose calculated is approximately 0.86. This is very close to the measured value of 0.72 observed in vivo. This model was then taken, and the physiology was changed from fasted monkey to fed rat. The simulation for $\%F_a$ was determined at 25 mg/kg dose and 100 mg/kg dose. The ratio of $\%F_a$ of high/low dose was calculated as approximately 0.90. This is also very close to the measured value of 0.97 observed in vivo. Finally, the physiology was changed from fasted monkey to fasted dog. The simulation for $\%F_a$ was performed for 20 and 60 mg/kg dose. The ratio of $\%F_a$ of high/low dose was calculated as approximately 0.48, which is also very close to the measured value of 0.42 observed in vivo. This example nicely demonstrates that available PBPK models can help translate in vitro measured parameters into in vivo solubility numbers, which allow for rapid understanding of when absorption may be high (in this case, in fed rats), moderate (in this case, fasted monkey), and low (in this case, fasted dog). However, connecting in vitro parameters with the animal physiology to predict in vivo exposure across dose and species does not always work that well.

4.5 When are we able to predict solubility-limited absorption with confidence?

As described in previous paragraphs, absorption is a very complex phenomenon for which solubility is only one component. During last decades, commercial software packages (GastroPlus, SimCYP, and PKSim) enabled to connect the intrinsic properties of the drug with the physiology of the GI tract (gut sections with different pH values, volume of fluid, permeability) to simulate the absorption of drug candidates in pre-clinical species and humans (e.g. the ADAM PBPK absorption model implemented GastroPlus [36–38]). An application of absorption modelling is to detect early on if a drug can be sufficiently absorbed to lead to an efficacious exposure in patients.

Typically, the assessment of drug solubility/absorption/bioavailability is done at several time points along the drug-development phases. In early discovery, an in silico prediction of solubility and pK_a(s) can be made based on molecular structure. This can be used to rank order virtual compounds and help chemists prioritize synthesis. Once a compound is available, the measured in vitro solubility can be used to predict exposure and then compare the predicted and observed systemic drug exposure in pre-clinical species. Our experience is that in vitro solubility measured in aqueous buffer tends to under-predict exposure at high dose in animals. In vitro solubility measured in biorelevant media (Simulated Gastric Fluid and FaSSIF) are usually higher than in aqueous buffer (due to media composition: e.g. bile-salt), which helps the prediction of low-dose exposure in animals where predictions are typically within two- to fourfold the observed $C_{\max}$ and AUC. In early development, simulations are performed to anticipate possible saturation of drug absorption in toxicology studies. At this stage, we usually only check if under-proportional exposure is anticipated in a qualitative assessment. When an under-proportional exposure is anticipated with a classical suspension, additional modelling is done to evaluate an alternative enabling formulation. The objective is then to boost as much as possible the release and dissolution of the active ingredient from the formulation (see previous paragraph). Our modelling approach is to make use of all measured solubility versus pH values and solubility values in biorelevant media to check which dataset/model best describes the exposure at high dose in animals and then adjust slightly the solubility versus pH curve to match as close as possible all observed in vivo profile(s). This deviation from measured values is possibly due to the fact that in vitro simulated fluids do not entirely reflect the in vivo situation. The optimized solubility versus pH that matches best the observed in vivo time–concentration profile is the starting point to predict exposure in humans. Under-proportional exposure in animals helps modelling the exposure versus dose anticipated in humans. Typically, for a compound with poor solubility (e.g. FaSSIF solubility <0.1 mg/mL) and under-proportional exposure in animals, we can relatively accurately assess the absorption starting from a low dose up to a predicted efficacious dose in humans. Again, anticipating a solubility-limited absorption in the dose-range of the single ascending dose (SAD) helps selecting the best formulation for the first in human study (FIH).

Our experience is that solubility-limited absorption in SAD of an FIH study is usually well predicted and then used to select an optimized formulation before further development to avoid additional bridging studies. One should also keep in mind that solubility is the starting point of a long sequence of events from drug dissolution in the gut until getting adequate exposure in the target organ. Modelling can effectively connect all parameters leading to exposure at the target and allow for a PSA. For example, one can factor in drug solubility (and permeability) across pH and gut sections and take into account that the available fluid volume decreases from small to large intestine.

Usually, the absorption of BCS class I and III drugs are well predicted. For BCS class II and IV, we do not often see under-proportional exposure in SAD study using conventional capsules likely because the in vivo solubility is higher than the solubility measured in vitro (fluid composition, dynamic forces in the gut, super-saturation).

A review paper on PK modelling to predict in human exposure has been published by the ORBITO consortium [39], showing rather moderate success rate of PBPK modelling to do a correct "a priori" prediction of the concentration versus time profile in humans. This is not our experience and the reason for this difference is unclear. Our perception of the ORBITO results is that it corresponds to a fully automated upload of the model parameters and using decision trees to decide on final model conditions to be used. This is different to our modelling approach as we handle every compound slightly differently, integrating all the knowledge available at the time.

As an example, we describe the case of a solubility-limited absorption in a FIH study with a dose range of 100–1,300 mg. The compound characteristics were as follows:

HCl salt, high MW (>740 g/mol), log P around 4.4, pK_a (base) of 6.9, solubility at pH 2 and 4 around 0.04 and 0.02 mg/mL, respectively, and solubility in simulated gastric fluid; pH 1.2 around 0.02–0.06 mg/mL, a FaSSIF solubility of 0.005 mg/mL, and an FeSSIF solubility of 0.12 mg/mL. In the FIH study, the drug was given as film-coated tablets (X90, about 16 µm, e.g. micronized API). In vitro dissolution at pH 1, 4.5, and 6.8 showed incomplete dissolution with 10–40% of dose dissolved. The permeability of the compound in the Caco-2 permeability model was high (P_{app} of 5×10^{-6} cm/s) with saturated efflux at high concentration (>20 µM) and thus no impact due to efflux transporters expected for doses of 50 mg or higher. High hepatic clearance estimated from intrinsic clearance data in human microsomes and hepatocytes predicted a rather low bioavailability in man. Based on PKPD predictions and competitor data in patients, a very high efficacious dose was predicted (and the need of a sevenfold increase of the exposure observed at 400 mg). Modelling enabled to predict that increasing the dose will reduce F_a (solubility limited) and PSA showed that no improvement was anticipated with the exception of further particle size reduction. Predictions were confirmed using a dynamic in vitro gut system (TIM1 model at TNO, The Hague, Netherlands) showing a marginal increase of F_a with twofold increase in dose. Addition of food was tested in the clinic, showing a significant increase in drug exposure (two- to fourfold), as expected from modelling (three- to fivefold) based on drug solubility in FeSSIF and FaSSIF. Switch to a solid dispersion together with food was further tested in vitro (TNO model [40, 41]) and in vivo in the dog, which were quite predictive of the human situation, but ultimately the predicted high exposure needed for efficacy was not reached at highest dose tested.

Solubility-limited absorption is considered difficult to predict in pre-clinical species (rat, dog, monkey) with input of in vitro data and low-dose pharmacokinetics alone. From our observations, it is likely that the in vitro solubility values underestimate the true in vivo solubility in the gut and we are missing possible supersaturation effects. Solubility-limited absorption in human SAD study is considered predictable with confidence by absorption/PK modelling on the basis of a correct description of low- and high-dose experiments in two different pre-clinical species. These predictions remain rather qualitative like anticipating a non-linear absorption in a planned FIH dose-range finding and predicting the likelihood to reach efficacious targeted exposure. However, it is our experience that it is not really possible to predict with enough confidence at which dose level the non-linearity will start to kick-in and the steepness of this non-linearity.

Poorly soluble molecules should be intensively studied in vitro (solubility, dissolution) and in vivo (exposure vs dose in tox species). These data should be integrated in PBPK models to assess the risk of not reaching the predicted efficacious concentrations and study mitigation options using PSA.

4.5.1 Non-oral routes

For non-oral routes of drug administration, such as parenteral, nasal, topical, and ocular, solubilization has an equally significant impact on drug penetration, distribution, and ultimately on the therapeutic effect. Formulations for these routes of administration must take into account solubility, or drug loading, in the formulation, as well as miscibility, which can have an impact on stability. Many parenteral formulations are lyophilized powders, which are reconstituted prior to administration. Lyophilization is generally employed to control stability, and re-constitution is performed within a few hours of administration to the patient. When a powder is reconstituted, in general it must occur very rapidly, thus solubility in the formulation and wetting of the lyophilized cake is a key consideration. In general, lyophilized products are co-lyophilized with excipients, which are used to facilitate rapid wetting in an aqueous-based media (e.g. saline, dextrose). Additionally, other excipients can be co-lyophilized which solely function to improve solubilization when the drug is reconstituted. An example of this is the use of cyclodextrins to improve intrinsic solubility through complexation. The drug may be dissolved into the cyclodextrin, lyophilized, and then reconstituted with saline prior to administration to the patient. It is not uncommon for a cyclodextrin to improve solubility in aqueous media by more than 10-fold, compared to solubility of the drug in aqueous media alone. Cyclodextrins have additionally been utilized successfully in nasal sprays and topical ocular solutions [42].

4.6 Conclusion and outlook

A lot of knowledge has accumulated in the last two decades and our understanding of how solubility impacts bioavailability and affects compound attrition and development timelines improved tremendously. Solubility only marginally affects oral exposure in the low-dose rodent PK screen and the challenge is more the early identification of molecules for which it will be difficult to get efficacious exposure and/or exposure multiples for safety assessment. Our ability to predict the impact of solubility on exposure for molecules that are within the rule of five property space is usually good, although cross-species prediction is not always accurate. Failure to correctly predict in vivo exposure can be divided into two main categories. First, we may not describe the properties of the molecule correctly. For example, the impact of the common ion effect on in vivo solubility may not be sufficiently well captured. Larger molecules are also more likely to form soluble aggregates than traditional low-molecular-weight compounds, and the impact on absorption is not yet well understood. Second, the in vitro media used to simulate intestinal fluids may not completely describe the in vivo solubility and neglect the dynamic aspect of the system.

The accurate prediction of more complex, larger molecules is more challenging because traditional assay protocols may not work, and semi-empirical rules derived from past experience may not necessarily apply. For example, clog P becomes less meaningful for compounds with MW > 1,000 and in this space, it is not uncommon to get quite soluble molecules with clog P values >8. Another challenge is the prediction of very potent compounds that could be efficacious at a relatively modest exposure and would tolerate a relatively low oral bioavailability. Solubility remains a key attribute that influences successful compound progression as illustrated by the analysis of the GlaxoSmithKline portfolio by Bayliss et al. [43]. On average, the time from development candidate selection to either termination or successful progression is 2 years longer for poorly soluble molecules. Interestingly, the attrition of low soluble compounds is higher in phase I to phase II transition compared to preclinical to phase I clinical study. The amount of risk taken with the selection of a low soluble development candidate needs to be put in perspective with the competitive landscape and the severity of the unmet medical need.

References

[1] Oh, D-M, Curl, R.L, & Amidon, G.L. Estimating the fraction dose absorbed from suspensions of poorly soluble compounds in humans: a mathematical model. Pharm. Res. 1993, 10, 264–270.

[2] Curatolo, W. Physical chemical properties of oral drug candidates in the discovery and exploratory development settings. PSTT 1998, 1, 387–392.

[3] Hilgers, A R, D P, Smith, J J, Biermacher et al. Predicting oral absorption of drugs: a case study with a novel class of antimicrobial agents. Pharm. Res. 2003, 20, 1149–1155.

[4] Wuehling, P W, Kwong, E, & Higgins, J. Identification of suitable formulations for high dose oral studies in rats using in vitro solubility measurements, the maximum absorbable dose model, and historical data sets. Mol. Pharm. 2012, 9, 1163–1174.

[5] Wuehling Peter, W, Daublain, P, Kesisoglou, F, Templeton, A, & McGregor, C. Preclinical dose number and its application in understanding drug absorption risk and formulation design for preclinical species. Mol. Pharm. 2015, 12, 1031–1039.

[6] Morrison, J.S, Michelle, J, & Haskell, R.J. A Combination turbidity and supernatant microplate assay to rank-order the supersaturation limits of early drug candidates. J. Pharm. Sci. 2014, 103, 3022–3032.

[7] Skolnik, S M, Geraci, G M, & Dodd, S. Automated supersaturation stability assay to differentiate poorly soluble compounds in drug discovery. J. Pharm. Sci. 2018, 107(1), 84–93.
[8] Brouwers, J, M E, Brewster, & Augustijns, P. Supersaturating drug delivery systems: the answer to solubility-limited oral bioavailability?. J. Pharm. Sci. 2009, 98(8), 2549–2572.
[9] Johnson, K. C, & Swindell, A. C. Guidance in the setting of drug particle size specifications to minimize variability in absorption. Pharm. Res. 1996, 13, 1795–1798.
[10] Hatton, GB, Yadav, V, Basit, AW, & Merchant, HA. Animal farm: considerations in animal gastrointestinal physiology and relevance to drug delivery in humans. J Pharm Sci. 2015, 104(9), 2747–2776.
[11] DeSesso, JM, & Jacobson, CF. Anatomical and physiological parameters affecting gastrointestinal absorption in humans and rats. Food Chem Toxicol. 2001, 39(3), 209–228.
[12] Kararli, TT. Comparison of the gastrointestinal anatomy, physiology, and biochemistry of humans and commonly used laboratory animals. Biopharm Drug Dispos. 1995(5), 351–380.
[13] Merchant, HA, McConnell, EL, Liu, F, Ramaswamy, C, Kulkarni, RP, Basit, AW, & Murdan, S. Assessment of gastrointestinal pH, fluid and lymphoid tissue in the guinea pig, rabbit and pig, and implications for their use in drug development. Eur. J. Pharm Sci. 2011, 42(1–2), 3–10.
[14] Davies, B, & Morris, T. Physiological parameters in laboratory animals and humans. Pharm. Res. 1993, 10(7), 1093–1095.
[15] Sjogren, E, Abrahamsson, B, Augustijns, P, Becker, D, Bolger, MB, Brewster, M et al. In vivo methods for drug absorption - comparative physiologies, model selection, correlations with in vitro methods (IVIVC), and applications for formulation/API/excipient characterization including food effects. Eur. J. Pharm. Sci. 2014, 57, 99–151.
[16] Hogben, CAM, Schanker, LS, Tocco, DJ, & Brodie, BB. Absorption of drugs from the stomach. II. The human. J. Pharm. Exp. Ther. 1957, 120(4), 540–545.
[17] Schanker, LS, Shore, PA, Brodie, BB, & Hogben, CAM. Absorption of drugs from the stomach I the rat. J. Pharm. Exp. Ther. 1957, 120(4), 528–539.
[18] Rubbens, J, Mols, R, Brouwers, J, & Augustijns, P. Exploring gastric drug absorption in fasted and fed state rats. Int. J. Pharm. 2018, 548(1), 636–641.
[19] Vertzoni, M, Dressman, J, Butler, J, Hempenstall, J, & Reppas, C. Simulation of fasting gastric conditions and its importance for the in vivo dissolution of lipophilic compounds. Eur. J Pharm. Biopharm. 2005, 60(3), 413–417.
[20] Kalantzi, L, Goumas, K, Kalioras, V, Abrahamsson, B, Dressman, JB, & Reppas, C. Characterization of the human upper gastrointestinal contents under conditions simulating bioavailability/bioequivalence studies. Pharm. res. 2006, 23(1), 165–176.
[21] Schneider, F, Grimm, M, Koziolek, M, Modeß, C, Dokter, A, Roustom, T et al. Resolving the physiological conditions in bioavailability and bioequivalence studies: Comparison of fasted and fed state. Eur. J. Pharm. Biopharm. 2016, 108, 214–219.
[22] Bevernage, J, Hens, B, Brouwers, J, Tack, J, Annaert, P, & Augustijns, P. Supersaturation in human gastric fluids. Eur. J. Pharm. Biopharm. 2012, 81(1), 184–189.
[23] Tanaka, Y, Kawakami, A, Nanimatsu, A, Horio, M, Matsuoka, J, Wada, T et al. In vivo evaluation of supersaturation/precipitation/re-dissolution behavior of cinnarizine, a lipophilic weak base, in the gastrointestinal tract: The key process of oral absorption. Eur. J. Pharm. Sci. 2017, 96, 464–471.
[24] Lindahl, A, Ungell, A-L, Knutson, L, & Lennernäs, H. Characterization of fluids from the stomach and proximal jejunum in men and women. Pharm. res. 1997, 14(4), 497–502.
[25] Kalantzi, L, Persson, E, Polentarutti, B, Abrahamsson, B, Goumas, K, Dressman, JB et al. Canine intestinal contents vs. simulated media for the assessment of solubility of two weak bases in the human small intestinal contents. Pharm. res. 2006, 23(6), 1373–1381.

[26] Jantratid, E, Janssen, N, Reppas, C, & Dressman, JB. Dissolution media simulating conditions in the proximal human gastrointestinal tract: An update. Pharm. res. 2008, 25(7), 1663–1676.
[27] Fadda, HM. Drug solubility in luminal fluids from different regions of the small and large intestine of humans. Mol Pharm. 2010, 7(5), 1527–1532.
[28] Mithani, SD, Bakatselou, V, TenHoor, CN, & Dressman, J. Estimation of the increase in solubility of drugs as a function of bile salt concentration. Pharm. res. 1996, 13(1), 163–167.
[29] Berghausen, J, Seiler, FH, Gobeau, N, & Faller, B. Simulated rat intestinal fluid improves oral exposure prediction for poorly soluble compounds over a wide dose range. Admet & DMPK 2016, 4(1), 35.
[30] Augustijns, P, Wuyts, B, Hens, B, Annaert, P, Butler, J, & Brouwers, J. A review of drug solubility in human intestinal fluids: Implications for the prediction of oral absorption. Eur. J. Pharm. Sci. 2014, 57, 322–332.
[31] Schiller, C, C-P, Fröhlich, Giessmann, T, Siegmund, W, Mönnikes, H, Hosten, N, & Weitschies, W. Intestinal fluid volumes and transit of dosage forms as assessed by magnetic resonance imaging. Aliment. Pharma. & Ther. 2005, 22(10), 971–979.
[32] Daublain, P, Feng, KI, Altman, MD, Martin, I, Mukherjee, S, Nofsinger, R, Northrup, AB, Tschirret-Guth, R, Cartwright, M, & McGregor, C. Analyzing the potential root causes of variability of pharmacokinetics in preclinical species. Mol. Pharm. 2017, 14(5), 1634–1645.
[33] Klein, S. The use of biorelevant dissolution media to forecast the In vivo performance of a drug. AAPS J. 2010, 12(3), 397–406.
[34] Gad, SC, Spainhour, CB, Shoemake, C, Pallman, DR, Stricker-Krongrad, A, Downing, PA, Seals, RE, Eagle, LA, Polhamus, K, & Daly, J. Tolerable levels of nonclinical vehicles and formulations used in studies by multiple routes in multiple species with notes on methods to improve utility. Int. J. Toxicol. 2016, 35(2), 95–178.
[35] Diehl, KH, Hull, R, Morton, D, Pfister, R, Rabemampianina, Y, Smith, D, Vidal, JM, & van de Vorstenbosch, C European federation of pharmaceutical industries association and european centre for the validation of alternative methods. A good practice guide to the administration of substances and removal of blood, including routes and volumes. J. Appl. Toxicol. 2001, 21(1), 15–23.
[36] Jamei, M, Turner, D, Yang, J et al. Population-based mechanistic prediction of oral drug absorption. AAPS J. 2009, 11(2), 225–237.
[37] Yu, LX, & Amidon, GL. A compartmental absorption and transit model for estimating oral drug absorption. Int. J. Pharm. 1999, 186(2), 119–125.
[38] Davies, B Morris T Physiological parameters in laboratory animals and humans. Pharma. Res. 1993, 10, 1093–1095.
[39] Margolskee, A, Darwich, AS, Pepin, X et al. IMI – Oral biopharmaceutics tools project – Evaluation of bottom-up PBPK prediction success part 2: An introduction to the simulation exercise and overview of results. Eur. J. Pharm. Sci. 2017, 96, 610–625.
[40] Barker, R, Abrahamsson, B, & Kruusmägi, M. Application and validation of an advanced gastrointestinal in vitro model for the evaluation of drug product performance in pharmaceutical development. J. Pharm. Sci. 2014, 103(11), 3704–3712.
[41] Minekus M. In: Verhoeckx K, Cotter P, López-Expósito I, Kleiveland C, Lea T, Mackie A, Requena T, Swiatecka D, Wichers H editors The Impact of Food Bioactives on Health: in vitro and ex vivo models, Springer, Cham (CH), 2015, Chapter 5.
[42] Muankaew, C, & Loftsson, T Cyclodextrin-based formulations: a non-invasive platform for targeted drug delivery. Basic Clin. Pharm. Toxicol. 2018, 122, 46–55.
[43] M K, Bayliss, Butler, J, Feldman, P L. et al. Quality guidelines for oral drug candidates: Dose, solubility and lipophilicity. DDT 2016, 21, 1719–1727.

Christos Reppas and Maria Vertzoni

5 Estimation of intraluminal drug solubility

5.1 Intraluminal solubility and drug absorption

Oral dosing continues to be the most popular route of administration; 33 out of 59 total novel drug approvals by the FDA in 2018 were for oral administration (www.fda.gov/Drugs/DevelopmentApprovalProcess/DrugInnovation/ucm592464.htm). After oral administration, the drug is released from the dosage form and gets molecularly dispersed in the luminal contents in a timely and efficient manner to be available for local action or for transport through the gastrointestinal (GI) epithelium [1, 2]. For systemically acting drugs, intraluminal concentrations influence the rate of appearance in portal blood and determine the total amount reaching the general circulation [3].

Over the past decade, luminal drug concentrations have been measured after oral administration of drugs in humans using intubation techniques. The added value of gastrointestinal concentration profiling to gain in-depth knowledge of intraluminal drug and formulation behaviour and to identify the key processes of drug absorption have been demonstrated [4]. Examples include a better understanding of intestinal precipitation of weakly basic drugs (see chapter 2) clarifying inter-individual or food-induced variability in absorption, and an improved insight into the solubility–permeability interplay. Drawbacks with direct measurements of intra-gastric or intra-intestinal drug concentrations are the ethical issues in terms of exposing humans to the intubation procedure and drugs without any direct therapeutic benefit to the subject [3]. As a result, data are usually collected from a limited number of subjects and information is typically limited to one active pharmaceutical ingredient (API) per clinical study. One way to improve the cost:benefit ratio of the experiments is to collect human aspirates – without prior administration of the drug – and use them for measuring the parameters of interest.

Thermodynamic solubility of an API is the concentration in equilibrium with a given medium. Solubility in the GI contents is a key parameter that affects luminal dissolution and maximum apparent absorption rates of the API, especially if the API is passively absorbed [1, 2]. Solubility measurements of drugs in GI contents are performed ex vivo. Values are typically different than in water and can vary dramatically depending on the region of the GI lumen and the dosing conditions (Table 5.1). As a result, the use of aqueous solubility as an indicator of luminal solubility can lead to very pronounced underestimation of intraluminal values especially for lipophilic APIs.

Christos Reppas, Maria Vertzoni, Department of Pharmacy, National and Kapodistrian University of Athens, Athens, Greece.

https://doi.org/10.1515/9783110559835-005

Table 5.1: Aqueous thermodynamic solubility (μg/ml, 37 °C) of danazol (non-ionizable, logP 4.2) and felodipine (non-ionizable, logP 4.5) in comparison with values in aspirates collected at various times after administration of a glass of water to fasted adults (fasted) and after administration of a liquid meal to fasted adults (fed) [5, 6].

	Danazol	**Felodipine**
Water	0.42	1
HGF_{fasted}	1.61 ± 0.05 (5 adults)	0.410 ± 0.077 (20–40 min, pooled sample from 12 adults)
HIF_{fasted}	2.04 ± 1.45 (10 adults) 4.9 ± 1.1 (pooled sample from12 adults) 16 ± 21 (30 min, 4 adults) 18 ± 18 (60 min, 4 adults) 13 ± 7 (90 min, 3 adults) 15 ± 18 (120 min, 3 adults)	14.3 ± 0.03 (pooled sample from 12 adults)
HCF_{fasted}	7.7 ± 4.5 (5h, 8 adults)	11.6 ± 5.8 (5 h, 8 adults)
HGF_{fed}	Not available	77 ± 35 (90 min, pooled sample from 12 adults) 43 ± 21 (150 min, pooled sample from 12 adults) 57 ± 4 (180 min, pooled sample from 12 adults)
HIF_{fed}	101 ± 38 (pooled sample from 6 adults) 37 ± 11 (30 min, 5 adults) 27 ± 18 (60 min, 5 adults) 25 ± 21 (90 min, 5 adults) 14 ± 12 (120 min, 5 adults) 18 ± 26 (150 min, 5 adults) 12 ± 10 (180 min, 5 adults) 9 ± 5 (210 min, 5 adults) 14 ± 5 (240 min, 5 adults) 12 ± 8 (270 min, 5 adults) 6 ± 5 (300 min, 5 adults)	413 ± 52 (pooled sample from 6 adults)
HCF_{fed}	6.1 ± 3.7 (6 h, 8 adults)	12.2 ± 7.0 (6 h, 8 adults)

*Data are mean ± SD. HGF, human gastric fluid; HIF, human intestinal fluid (collected from the upper small intestine); HCF, human colonic fluid. Aspirations were performed at specific times after water or liquid meal that are indicated in parenthesis unless they are not known.

It should be acknowledged that although thermodynamic solubility designates the highest apparent drug concentration which can be achieved in equilibrium, apparent luminal drug concentration may temporarily reach higher values since supersaturation of contents may occur. Indeed, due to high-throughput drug discovery methodologies and the selection of targets requiring more lipophilic compounds, an increasing number of drug candidates have limited solubility and drug delivery systems are designed to generate supersaturation [7]. Also, for lipophilic weak bases, solubility is expected to be higher in the stomach in the fasted state than in any other region of the GI lumen. This can result in supersaturation as the dissolved drug moves out from the acidic gastric environment to the almost neutral intestinal environment and precipitation may occur in the duodenum [8]. Drug delivery systems designed to generate supersaturation include lipid-based formulations and amorphous drug solid dispersions (See Chapter 8).

This chapter aims, first, to present the procedures necessary for obtaining reliable results for solubility measurements in human aspirates and, secondly, to describe situations where drug solubility in a specific region of the GI lumen would be useful to know.

5.2 Procedures for aspirating gastrointestinal contents

The methodology for estimating drug solubility in the GI tract involves aspiration of contents from the human GI tract and measurement of solubility of the APIs of interest ex vivo. When the aim is to collect gastric contents or contents of the upper small intestine, aspirations can be performed after naso- or oro-gastro intestinal intubation. If the aim is to measure solubility in contents of the lower intestine (distal ileum or colon), then contents could be collected during colonoscopy. In all cases, aspiration methodologies require highly trained and experienced staff. In addition, well-defined protocols for the aspiration study and handling of samples after aspiration are important to generate reliable data. As a result, collection of human aspirates is a costly procedure.

5.2.1 Aspiration protocols

As in oral drug absorption studies pertaining to measurement of drug concentration in plasma, healthy adults are enrolled with the main exclusion criteria being the existence of a major health problem (cardiovascular, pancreatic, hepatic, thyroid, etc.), the existence of any condition requiring prescription drug therapy, recent history of gastrointestinal symptom regardless of the severity (e.g. heartburn, constipation, haemorrhoids, etc.), the intake of an investigational agent (new or generic) during the past 30 days prior to the initiation of study, the presence of antibodies indicating active acute or chronic HIV, HBV, or HCV infection, the use of medication which may affect GI function (including antibiotics) during the past 30 days prior to the study, or irregular bowel habits.

5.2.1.1 Aspiration from the stomach and the upper intestine

Aspirations can be performed after nasal or oral intubation. The nasal route considered more practical by gastroenterologists since the tube reaches the tracheal–oesophageal junction at an angle which is more favourable for entry into the oesophagus [3].

After spraying the back of the mouth with lidocaine hydrochloride, two sterile tubes (one positioned into the stomach and one positioned in the duodenum) or one sterile two-lumen duodenal tube (Figure 5.1) is introduced through the mouth or nose of a human volunteer. Typically, the two-lumen duodenal tube is 150 cm long with external diameters of ~5 mm for the gastric and ~3 mm for the jejunal area, respectively, and a plastic tip at its distal end. A series of holes, 55-65 cm proximal to the tip is used to access the antrum of the stomach. A further series of holes, 13.5-23.5 cm proximal to the tip was used to aspirate samples from the ligament of Treitz [9, 10]. Insertion of the tube(s) is assisted by a guiding wire, and the position is usually monitored fluoroscopically. After reaching the final position and removing the wire, the subject is allowed to lay semi supine and any nasal, oesophageal, and/or gastric secretions in response to the intubation process should be aspirated out of the stomach.

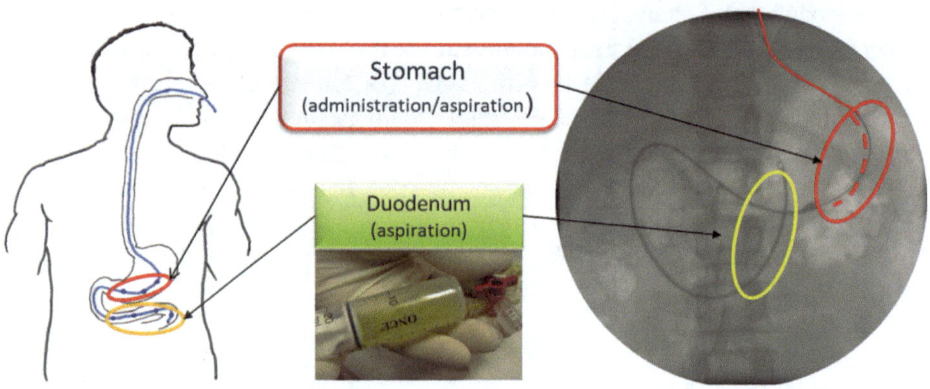

Figure 5.1: Schematic overview of intraluminal sampling technique. Left: A two lumen tube is introduced through nose and positioned into the stomach and the intestine. Right: X-ray images of the tube positioned inside the stomach (dotted line) and the intestine.

Although intubation studies are invasive and might disturb the GI physiology, it has been concluded that transpyloric tubes with slightly longer than 3 mm external diameter had only minor or non-significant effects on gastric emptying, gastric secretions, and duodenogastric reflux [11].

In the fasted state, contents from the stomach and/or the upper intestine are collected after the administration of a glass of water, that is, mimicking conditions to which drugs are administered during bioavailability/bioequivalence (BA/BE) studies in healthy adults [12].

In the fed state, to date, only liquid meals have been used to estimate solubility in gastric and intestinal contents as the administration of the high-fat, high-calorie meal proposed by regulatory agencies (reference meal [12]) could create problems with sample aspiration, due to potential clogging of the aspiration port [3, 13]. Liquid meals (e.g. NuTRIflex, Ensure Plus, Scandishake Mix) with similar composition, origin of calories, calorie content, and/or volume to that of the high-fat, high-calorie meal proposed by regulatory agencies [14–17] had instead been employed. It should be noted that, based on the few studies published to date on the impact of homogenization of solid-liquid meals, [18, 19] the extent to which the luminal environment after administration of liquid meals is similar to that after administration of the reference (solid-liquid) meal is not clear [20].

Once the water or the meal is administered, samples of contents can be manually aspirated over a period of 1 h (fasted state) or up to 4 h (fed state). It is important to aspirate as little as possible for the sampling to have minimum impact on the physiological process of transit of contents from the stomach to subsequent regions of GI lumen.

At the end and before removing the tube and discharging the subject, the final position of the tube should be confirmed fluoroscopically.

5.2.1.2 Aspiration from the lower intestine

Collection of contents from the lower intestine (distal ileum and ascending colon) is performed via colonoscopy. Colonoscopy requires prior cleaning of the distal colon. In clinical practice, colon preparation procedures involve the ingestion of varying volumes of polyethyleneglycol (PEG 3350) solution in non-carbonated water. In order for the environment of lower intestine to reflect the environment during a BA/BE study, any effect(s) of colon preparation procedure must be reversed before the sampling.

During the past decade, a protocol for direct sampling from the lower intestine with minimal effects on its physiology has been validated [21] and applied repeatedly [22, 23]. Based on this protocol, subject work-up involves the administration of 10 mg of bisacodyl 50 h prior to colonoscopy and 10 mg of bisacodyl, 44 h prior to each colonoscopy. From the time of first bisacodyl administration until the night prior to each colonoscopy day, subjects eat only liquified food (e.g. fish/chicken/rice soups, fruit juices) and white bread ad libitum. From the night prior to colonoscopy day until arrival to the clinic, that is, for at least 12 h, subjects remain fasted (Figure 5.2). With this regimen, bisacodyl effects on the intracolonic environment at the time of colonoscopy was shown to be non-significant [21] and did not have major effect on stools consistency [22].

Samples from distal ileum, caecum, and ascending colon are collected 5 h after the administration of a glass of water or after the consumption of the high-fat, high-calorie meal as proposed by regulatory agencies (reference meal [12]), that is, about the time drugs administered as conventional products or modified release products

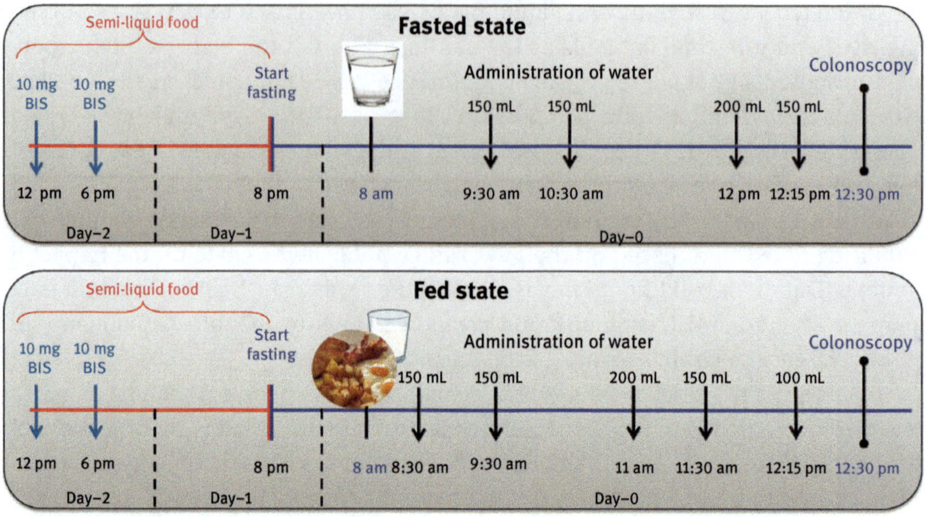

Figure 5.2: Subject work-up prior to colonoscopy and administration of water and/or food during the colonoscopy day. BIS: bisacodyl.

are expected to reach the lower intestine after oral administration. In order to control water consumption on colonoscopy day, non-carbonated water is administered to subjects according to the schedule described in Figure 5.2. Colonoscopy can last up to 1 h and collection of contents from the distal ileum, the cecum, and the ascending colon is performed under anaerobic conditions.

5.2.1.3 Handling of aspirated samples

Immediately upon collection, precautions should be taken to avoid any post-sampling alteration in the aspirated contents [4].

In order to maintain the composition of the aspirated sample until the solubility experiment is performed, it is necessary to deactivate enzymes immediately using a method that itself only minimally (or preferably, not at all) affects the composition of the sample, before storing the sample under deep-freeze conditions (–20 °C or lower). Especially in the fed state, enzymatic digestion should be inhibited by adding enzyme inhibitors.

In order to inhibit ex vivo lipolysis, the lipase inhibitor Orlistat dissolved in ethanol was used [16]. Orlistat was immediately added to the collected duodenal aspirates in a final concentration of 1 μM such that the ethanol concentration in the samples were 0.1% v/v. Hernell et al proposed the addition a cocktail of lipase/protease inhibitors (2% v:v) consisting of 50 mM diisopropylfluorophosphate, 50 mM diethyl (p-nitrophenyl)phosphate, 50 mM acetophenone, and 250 mM phenylboronic acid [9, 24].

Immediately upon collection, care should also be taken to limit exposure to the atmospheric environment, since this may affect the pH and alter the buffer capacity of the aspirates [4, 25]. If the aim is to correlate the solubility values with parameters obtained from physicochemical characterization of the contents, the pH and buffer capacity should be measured immediately upon collection.

To avoid second freeze-thawing cycle of samples [25], samples could be divided in sub-samples before storage [7]. It has been shown that application of one freeze-thaw cycle with lipase-containing media prior to using them, lowered the observed solubility values for dipyridamole and ketoconazole by 18%–34% and 13%–45%, respectively [26].

In the case of contents collected from the upper small intestine in the fed state, the micellar phase of the contents can be obtained using ultracentrifugation at 410,174 g for 2 h at 37 °C [9, 24]. After ultracentrifugation, four phases are obtained, that is, triglyceride phase, interphase, primarily micellar phase, and pellet.

In the case of contents collected from the lower intestine, the sample is ultracentrifuged (30,000 g, 15 min, 25 °C) under anaerobic conditions and the aqueous content is separated and used for solubility measurements. These ultracentrifugation conditions were proved appropriate for eliminating both solids and bacteria without affecting the structure of bacteria, that is, without liberating intracellular components which could contribute to degradation in the supernatant [26].

Individual or pooled aspirates can be used for solubility estimations. The use of individual samples enables correlations with levels of specific components in each sample and thus to determine the most important factors affecting luminal solubility of the tested compound.

A pooled sample from a large number of volunteers offers greater volumes and solubility measurements of many number of compounds in contents with the same composition and enables comparisons across a set of compounds. If samples are to be pooled, the volume taken from each individual sample should be held constant to ensure that the pooled sample is equally representative of all subjects [3].

5.3 Methodology for estimating solubility in gastrointestinal contents

Typically, solubility is measured with the shake flask method at 37 °C [13]. An excess amount of the API is added to glass vials containing small volumes of aspirated samples (2–3 ml). The excess amount should be 2-3 times higher than the expected amount needed to saturate the medium [27]. The equilibration time should be as short as possible to minimize composition changes in the aspirated samples but long enough to allow the system to reach equilibrium.

Since aspirates cannot be easily filtered through 0.45 um filters, samples are frequently centrifuged at 37 °C in order to separate the undissolved API. Adequacy of centrifugation could be partly evaluated from the variability of estimated values. After measuring the pH, the supernatant has to be immediately diluted with the appropriate solvent in order to avoid any precipitation due to the lower room temperature.

One problem of the shake-flask method when measuring solubility of ionized compounds is the possible alteration of the final pH of the medium at equilibrium. However, for highly dosed compounds, similar alterations may also occur intraluminally, especially in the fasted small intestine where buffering capacity is relatively low [3].

In the case of solubility measurements in aspirates collected in the fed sate, experiments are usually performed in the total luminal contents [14, 16, 17]. One issue with this approach is the difficulty to find the appropriate conditions (speed of centrifugation) in order to separate the excess solid drug from the rest of the sample prior to analysis [28]. Another issue is that the concentrations achieved in the phase driving drug absorption (likely the micellar phase) may be over or underestimated if only the total concentration is reported, leading to false estimates of the expected impact on absorption [3]. Assuming that it is the micellar phase concentration which drives absorption, it would be preferable to report solubilities determined in this phase of the postprandial luminal contents to predict food effects [3, 9].

5.4 Estimating drug solubility in the gastrointestinal contents

The physicochemical characteristics of human GI fluids change with the location, the dosing conditions, and the time after water or food ingestion. Table 5.2 presents an overview of the variable composition after the administration of a glass of water to healthy adults from different studies [20]. Table 5.3 summarizes the complex composition of luminal contents at various times after administration of a meal to fasted adults [20]. Based on the different composition of luminal contents along the GI tract, it is expected that solubility can vary dramatically with the region of the GI lumen and the dosing conditions.

5.4.1 Estimation of drug solubility in the stomach

In the fasted state, intragastric solubility is of interest, primarily for compounds that are administered in liquid form from which precipitation in stomach may occur (e.g. lipid dosage forms) and for formulations or compounds that are highly soluble and rapidly dissolving in the stomach but that have limited solubility in the small

Table 5.2: Physicochemical characteristics and composition of gastric and intestinal contents at various times after administration of a glass of water to fasted adults [20].*

	Stomach		Duodenum/Jejunum	Distal Ileum	Caecum	Asc. Colon
	10–20 min	**30–40 min**	**30 min**	**5 h**	**5 h**	**5 h**
pH	1.7–3.3	1.6–2.7	6.1–7.0	7.7–8.1	7.4	7.8
Buffer capacity (mmol/L/ΔpH)	4.7–21.3 (NaoH titration)	18–27.6 (NaoH titration)	6.9–9.0 (HCl titration)	8.9(3.6) (HCl titration)	19.2(10.2) (HCl titration)	21.4 (HCl titration)
Osmolality (mOsm/kg)	44.9–103.6	117–178	115–206	60(50)	144(65)	81 (102)
Surface tension (mN/m)	43.2	43.0	32.7–35.3	na	na	42.7
Bile salts (mM)	0.014–0.032	0.013–0.147	3.66–7.74	0.071(0.151)	0.183(0.221)	0.115(0.119)
Phospholipids (mM)	na	na	0.320–0.910	0.073(0.041)	0.166(0.110)	0.362(0.210)
Cholesterol (mM)	na	na	0.08–0.44	0.413(0.309)	1.004(1.072)	1.703(1.764)
Free fatty acids (mM)	na	na	0.95–1.55	0.063(0.047)	0.143 (0.118)	0.119
Monoglycerides (mM)	na	na	0.36–0.39	<LOQ**	<LOQ**	<LOQ**
Total protein content (mg/mL)	na	na	1.1–2.8	5.1(3.3)	10.2(2.2)	9.7(4.6)
Total carbohydrate content (mg/mL)	na	na	na	1.55(0.99)	2.3(1.0)	8.1(8.6)
Total SCFAs (mM)	na	na	traces	8.6(6.6)	32.2 (17.6)	30.9

*Range of median values for pH, range of mean values for all other parameters, based on various published relevant studies; na: not available.
**LOQ: 5 μM

Table 5.3: Physicochemical characteristics and composition of gastric and intestinal contents at various times after administration of a meal to fasted adults [20].[1]

	Stomach[2]					Duodenum/Jejunum[3]				Distal Ileum[4]	Cecum[4]	Asc. Colon[4]
	0.5 h	1 h	2 h	3 h	4 h	0.5 h	1 h	2 h	3 h	5 h	5 h	5 h
pH	3.6–4.1	2.7–3.3	2.0–2.3	1.5–2.2	0.7–1.6	6.2–6.6	6.3–6.5	5.3–6.1	5.6–5.8	8.1	6.4	6.0
Buffer capacity[5] (mmol/L/ΔpH)	25	23	23.2	29.8	na	28	22–27.4	18–23.3	12–25.6	15.2(8.4)	33.6(13.1)	37.7
Osmolality (mOsm/kg)	531	474	442	321	na	291–391	360–402	274–423	215–364	252(245)	267(197)	224(125)
Surface tension (mN/m)	31.2	30.3	30.6	30.7	na	31.3–35.1	30.3–32.9	30.4–34.0	30.2–34.0	na	na	39.2
Bile salts (mM)			<LOQ[6]			10.1–14.0	5–18.2	3.9–7.7	3.7–7.3	0.182(0.132)	0.280(0.305)	0.587 (0.413)
Phospholipids (mM)	na	2.9	1.9	0.9	0.4	3.9–6.0	2.87–7.1	1.5–5.6	1.4–4.3	0.040(0.051)	0.082(0.077)	0.539 (0.393)
Cholesterol (mM)	na	1.2	1.2	0.7	0.4	0.75–1.50	0.68–3.12	0.40–1.2	0.30–1.4	0.317(0.426)	0.640(0.771)	1.8(1.3)
Free fatty acids (mM)	na	9.4	14.0	15.3	7.3	30.2–52.0	21.7–54	42–46	34.7–56.9	0.0640(0.072)	0.150 (0.141)	0.225
Monoglycerides (mM)	na	2.5	3.2	4.6	1.1	5.9–9	7.08–11	5.21–9.6	4.20–18.4	<LOQ[6]	< LOQ[6]	< LOQ[6]

Diglycerides (mM)	na	8.2	17.7	13.5	9.1	1.1–6.5	1–10.7	4.20–12.6	2.6–33.7	<LOQ[6]	< LOQ[6]	< LOQ[6]
Triglycerides (mM)	na	157.1	150.4	154.0	42.7	1.16–4.7	0.76–60.7	1.90–44.7	0.60–63.3	<LOQ[6]	< LOQ[6]	< LOQ[6]
Total protein content (mg/mL)	na	na	na	na	na	6.1–13.7	5.5–15.1	4.3–11.0	2.9–11.4	3.39(0.74)	6.2(3.2)	6.9(2.3)
Total carbohydrate content (mg/mL)	na	na	na	na	na	62.7	72.0	74.3	70.1	12.7(5.3)	9.8(7.0)	14.0(7.4)
Total SCFAs (mM)	na	na	na	na	na	traces				5.8 (4.7)	29.3 (15.4)	48.1

[1]Range of median values for pH, range of mean values for all other parameters, based on various published relevant studies; na: not available
[2]Data have been collected after administration of liquid or homogenized meals. Only pH values were measured after the administration of solid-liquid meals.
[3]Data have been collected after administration of liquid or homogenized meals
[4]Only data after administration of reference meal have been published
[5]HCl titration
[6]LOQ: (Bile salts: 500 μM, Monoglycerides: 5 μM, Diglycerides: 5 μM, Triglycerides 12 μM)

intestine [29]. The latter applies to supersaturating drug delivery systems and to lipophilic weak bases and their salts. Lipophilic weak bases can induce supersaturation as the dissolved drug moves out from the stomach to the duodenum and precipitation in the duodenum may occur [8].

In the fed state, gastric residence times of dosage forms are increased substantially, so dissolution of solid particles in the stomach can be more extensive than in the fasted state. For lipophilic compounds, solubility in the contents of the fed stomach is expected to increase, due to the increased presence of various components that promote solubility and the extent of such an increase will vary with time after meal administration (Table 5.3). Intragastric solubility in the fed state should, therefore, be of interest for BCS Class II and Class IV compounds. Meal components can have an adverse effect on solubility if an insoluble complex is formed with the drug. A relevant example is complexation with calcium, which precipitates bisphosphonates and tetracyclines, rendering them insoluble and thus unavailable for absorption [3].

5.4.2 Estimation of drug solubility in the upper intestine

As the upper intestine is the primary site of absorption for the majority of drugs, the emphasis few years ago has been given to the measurement of solubility in human intestinal aspirates. Solubility values covering 59 different APIs was summarized (102 solubility values in the fasted state and 37 solubility values in the fed state) [13]. In the upper intestine, the presence of bile salts, phosphatidylcholine, and cholesterol create mixed micelles that can solubilize lipophilic APIs. The presence of mixed micelles is much higher after meal ingestion (Table 5.3) as the gall bladder contracts in response to meal and empties its content into the duodenum.

5.4.3 Estimation of drug solubility in lower intestine

The impact of absorption from the lower intestine, that is, from distal ileum and proximal colon, on plasma pharmacokinetic profile after oral administration of a drug product is of particular interest when considering the development of modified release products [30]. Conditions in the lower intestine are of interest when the product targets the API to the colonic region for local action. It could also be useful for understanding the pharmacokinetic performance of poorly soluble active pharmaceutical ingredients (APIs), BCS Class II and Class IV APIs, which are administered in immediate release products, when drug absorption is incomplete in the upper intestine. Studies have shown that solubility of lipophilic APIs in the lumen of the ascending colon can be different than in the contents of stomach and upper intestine during fasting state (Table 5.1, [5]) and such differences together with subsequent difference in luminal dissolution rates may be crucial for the prediction of intraluminal behaviour of highly dosed APIs [31].

5.5 Concluding remarks

The apparent thermodynamic solubility of orally administered active pharmaceutical ingredients in the gastrointestinal lumen is a useful parameter at various stages of the discovery and development programs in pharmaceutical industry. Today, several issues relating with aspiration procedures and methodology for measuring solubility in aspirates have been resolved and collection of high-quality data is possible. However, the knowledge on physicochemical characteristics of luminal contents accumulated during the past two decades in conjunction with ethical issues and costs associated with the collection of human aspirates make the use of in vitro surrogate media for estimating luminal drug solubility an attractive alternative, at least in certain situations.

References

[1] Buckley, ST., Frank, KJ., Fricker, G., & Brandl, M. Biopharmaceutical classification of poorly soluble drugs with respect to "enabling formulations". Eur. J. Pharm. Sci. 2013, 50(1), 8–16.

[2] Frank, KJ., Rosenblatt, KM., Westedt, U., Hölig, P., Rosenberg, J., Mägerlein, M., Fricker, G., & Brandl, M. Amorphous solid dispersion enhances permeation of poorly soluble ABT-102: True supersaturation vs. apparent solubility enhancement. Int. J. Pharm. 2012, 437(1–2), 288–93.

[3] Dressman, J.B., Vertzoni, M., Goumas, K., & Reppas, C. Estimating drug solubility in the gastrointestinal tract. Adv. Drug Del. Rev. 2007, 59, 591–602.

[4] Brouwers, J., & Augustijns, P. Resolving intraluminal drug and formulation behavior: Gastrointestinal concentration profiling in humans. Eur. J. Pharm. Sci. 2014, 61, 2–10.

[5] Vertzoni, M., Diakidou, A., Chatzilias, M., Söderlind, E., Abrahamsson, B., Dressman, J.B., & Reppas, C. Biorelevant media to simulate fluids in the ascending colon of humans and their usefulness in predicting intracolonic drug solubility. Pharm. Res. 2010, 27, 2187–2196.

[6] Reppas, C., & Vertzoni, M. Biorelevant *in-vitro* performance testing of orally administered dosage forms. J. Pharm. Pharmacol. 2012, 64, 919–930 (2012).

[7] Bevernage, J., Brouwers, J., Brewster, ME., & Augustijns, P. Evaluation of gastrointestinal drug supersaturation and precipitation: strategies and issues. Int J Pharm. 2013, 453(1), 25–35.

[8] Psachoulias, D., Vertzoni, M., Goumas, K., Kalioras, V., Beato, S., Butler, J., & Reppas, C. Precipitation in and supersaturation of contents of the upper small intestine after administration of two weak bases to fasted adults. Pharm. Res. 2011, 28, 3145–3158.

[9] Vertzoni, M., Markopoulos, C., Symillides, M., Goumas, C., Imanidis, G., & Reppas, C. Luminal lipid phases after administration of a triglyceride solution of danazol in the fed state and their contribution to the flux of danazol across caco-2 cell monolayers. Mol. Pharm. 2012, 9, 1189–1198.

[10] Petrakis, O., Vertzoni, M., Angelou, A., Kesisoglou, F., Bentz, K., Goumas, K., & Reppas, C. Identification of key factors affecting the oral absorption of salts of lipophilic weak acids: A case example. J. Pharm. Pharmacol. 2015, 67, 56–67.
[11] Longstreth, GF., Malagelada, JR., & Go, VL. The gastric response to a transpyloric duodenal tube. Gut. 1975, 16, 777–80.
[12] FDA, Food-Effect Bioavailability and Fed Bioequivalence Studies. Guidance for Industry 2002
[13] Augustijns, P., Wuyts, B., Hens, B., Annaert, P., Butler, J., & Brouwers, J. A review of drug solubility in human intestinal fluids: Implications for the prediction of oral absorption. Eur. J. Pharm. Sci. 2014, 57, 322–32.
[14] Clarysse, S., Brouwers, J., Tack, J., Annaert, P., & Augustijns, P. Intestinal drug solubility estimation based on simulated intestinal fluids: Comparison with solubility in human intestinal fluids. Eur. J. Pharm. Sci. 2011, 43(4), 260–9.
[15] Kalantzi, L., Goumas, K., Kalioras, V., Abrahamsson, B., Dressman, JB., & Reppas, C. Characterization of the human upper gastrointestinal contents under conditions simulating bioavailability/bioequivalence studies. Pharm. Res. 2006, 23(1), 165–76.
[16] Clarysse, S., Psachoulias, D., Brouwers, J., Tack, J., Annaert, P., Duchateau, G., Reppas, C., & Augustijns, P. Postprandial changes in solubilizing capacity of human intestinal fluids for BCS class II drugs. Pharm. Res. 2009, 26(6), 1456–66.
[17] Persson, EM., Gustafsson, AS., Carlsson, AS., Nilsson, RG., Knutson, L., Forsell, P., Hanisch, G., Lennernäs, H., & Abrahamsson, B. The effects of food on the dissolution of poorly soluble drugs in human and in model small intestinal fluids. Pharm. Res. 2005, 22(12), 2141–51.
[18] Grimm, M., Scholz, E., Koziolek, M., Kühn, JP., & Weitschies, W. Gastric water emptying under fed state clinical trial conditions is as fast as under fasted conditions. Mol. Pharm. 2017, 14(12), 4262–4271.
[19] Malagelada, JR., Go, VL., & Summerskill, WH. Different gastric, pancreatic, and biliary responses to solid–Liquid or homogenized meals. Dig. Dis. Sci. 1979, 24, 101–110.
[20] Pentafragka, C., Symillides, M., McAllister, M., Dressman, J., Vertzoni, M., & Reppas, C. The impact of food intake on the luminal environment and performance of oral drug products with a view to in vitro and in silico simulations: A PEARRL review. J. Pharm. Pharmacol. 2019, 71, 557–580.
[21] Diakidou, A., Vertzoni, M., Dressman, J., & Reppas, C. Estimation of intragastric drug solubility in the fed state: Comparison of various media with data in aspirates. Biopharm. Drug Dispos. 2009, 30, 318–325.
[22] Markopoulos, C., Andreas, C., Vertzoni, M., Dressman, J., & Reppas, C. In-vitro simulation of luminal conditions for evaluation of performance of oral drug products: Choosing the appropriate test media. Eur. J. Pharm. Biopharm. 2015, 93, 173–182.
[23] Vertzoni, M., Goumas, K., Söderlind, E., Abrahamsson, B., Dressman, J., Poulou, A., & Reppas, C. Characterization of the ascending colon fluids in ulcerative colitis. Pharm. Res. 2010, 27, 1620–1626.
[24] Hernell, O., Staggers, J. E., & Carey, M. C. Physical-Chemical behaviour of dietary and biliary lipids during intestinal digestion and absorption. 2. Phase analysis and aggregation states of luminal lipids during duodenal fat digestion in healthy adult human beings. Biochemistry 1990, 29, 2041–2056.
[25] Litou, C., Vertzoni, M., Goumas, C., Vasdekis, V., Xu, W., Kesisoglou, F.,& Reppas, C. Characteristics of the human upper gastrointestinal contents in the fasted state under hypo- and a-chlorhydric gastric conditions under conditions of typical drug – drug interaction studies. Pharm. Res. 2016, 33, 1399-1412.

[26] Diakidou, A., Vertzoni, M., Goumas, K., Söderlind, E., Abrahamsson, B., Dressman, J., & Reppas, C. Characterization of the contents of ascending colon to which drugs are exposed after oral administration to healthy adults. Pharm. Res. 2009, 26, 2141–2151.
[27] Kawakami, K., Miyoshi, K., & Ida, Y. Impact of the amount of excess solids on appararent solubility. Pharm. Res. 2005, 22, 1537–1543.
[28] Riethorst, D., Mitra, A., Kesisoglou, F., Xu, W., Tack, J., Brouwers, J., & Augustijns, P. Human intestinal fluid layer separation: The effect on colloidal structures & solubility of lipophilic compounds. Eur. J. Pharm. Biopharm. 2018, 129, 104–110.
[29] Reppas, C., & Augustijns, P. Drug solubility in the Gastrointestinal Tract, in Oral Drug Absorption, 2nd ed., Dressman J., and Reppas C., editors, Informa Healthcare, ISBN 1-4200-7733-3, New York, USA, 2010.
[30] Tannergren, C., Bergendal, A., Lennernäs, H., & Abrahamsson, B. Toward an increased understanding of the barriers to colonic drug absorption in humans: Implications for early controlled release candidate assessment. Mol. Pharm. 2009, 6, 60–73.
[31] Georgaka, D., Butler, J., Kesisoglou, F., Reppas, C., & Vertzoni, M. Evaluation of dissolution in the lower intestine and its impact on the absorption process of high dose low solubility drugs. Mol. Pharm. 2017, 14, 4181–4191.

Jennifer Dressman

6 Biorelevant media

6.1 A short history of biorelevant media

The media used in the various pharmacopoeia to represent the conditions in the gastrointestinal (GI) tract were designed to reflect primarily the difference in pH between the stomach and the small intestine in order to test whether different oral dosage forms could release the active pharmaceutical ingredient (API) as intended. For example, an immediate release dosage form should release the drug in the stomach or latest in the proximal small intestine, whereas an enteric-coated product should prevent release in the stomach but provide for adequate release under intestinal conditions. Initially, the pH of the medium representing the stomach was set at pH 1.2, whereas the pH for the intestine was set at pH 7.4 [1]. Other aspects of GI physiology such as the effects of meal intake on fluid composition, fluid volumes, motility, and transit times were largely ignored in favour of attaining sink conditions for dissolution and a short test period, both of which are favoured for quality control purposes. As long as the API in question had good solubility, the standard quality control tests were often satisfactory for predicting product performance.

For less soluble drugs, however, problems were encountered with respect to sink conditions, and at first, solvents were added to the dissolution medium or larger volumes were used (up to 4 L vessels were available). Gradually, it was realized that these approaches were neither representative of GI physiology nor very practical, and many companies switched to using surfactants in the media to induce sink conditions, often employing very high levels (2% and more) [2].

In the meantime, scientists became aware of the influence of the naturally occurring surfactants in the GI tract and started to investigate their influence on drug solubility and dissolution [3–5]. It became clear that not only could bile salts solubilize and improve the dissolution rate of steroidal drugs, but that addition of lecithin (which is also secreted with the bile) enhanced these effects considerably. Simultaneously, clinical studies were being conducted to determine how the pH profile in the GI tract might be affected by meal ingestion [6], by ageing [7], and by disease (e.g. achlorhydria [7] and cystic fibrosis [8]). These studies were further supplemented by investigations of gastric emptying, focusing on the interaction between particle size and viscosity on gastric emptying behaviour of non-digestible dosage form surrogates [9, 10].

This emphasis on understanding the physiology in the GI tract and how it might interact with dosage forms to affect drug absorption attracted the interest of

Jennifer Dressman, Goethe University, Frankfurt

https://doi.org/10.1515/9783110559835-006

the FDA and a project was funded in collaboration with Dr Vinod Shah at FDA to propose a set of biorelevant media which could be used to help understand dosage form performance after oral administration [11, 12].

Initially, four media were proposed: Simulated gastric fluid (SGF) was used to simulate gastric fluid in the fasted state and milk was used to simulate the contents of the stomach in the fed state, while new media were introduced to simulate conditions in the intestine: fasted state simulating intestinal fluid (FaSSIF) and fed state simulating intestinal fluid (FeSSIF). Since then, research on the biorelevant media has continued uninterrupted and the timeline in Figure 6.1 describes the evolution of biorelevant media over the last 20 years.

– FaSSIF/FeSSIF	[12]
– $\text{FaSSGF}_{\text{achlorhydria}}$	[13]
– FaSSGF	[14]
– Canine FaSSIF/FaSSGF	[15]
– FaSSIF V2/FeSSIF V2/FeSSGF	[16]
– FaSSCoF/FeSSCoF	[17]
– FaSSGF V2	[18]
– FaSSIF V2 Plus	[19]
– Canine FaSSIF	[20]
– FeSSGEm (now FeSSGF)	[21]
– FaSSIF V3	[22]
– Set of variations on FaSSIF	[23]
– $\text{FaSSIF}_{\text{ileum}}$	[24]

Figure 6.1: Timeline for evolution of biorelevant media from 1998 to 2018.

6.1.1 Fasted gastric media

Although FaSSIF and FeSSIF had been proposed by Dressman et al. [12], they at first relied on the compendial medium SGF to represent fasted state conditions in the stomach. With a pH of 1.2, achieved with hydrochloric acid, together with sodium chloride to adjust osmolarity and the option to add pepsin to hydrolyse proteins (e.g. the gelatine in capsules), this medium already addressed several aspects of gastric physiology. A slight drawback was the very acidic pH of the SGF, which lies at the lower end of pH values that have been recorded in healthy subjects [5, 6]. Through a collaboration with Boehringer Ingelheim, it was also apparent that patients taking H_2-receptor antagonists would have an elevated gastric pH due to the lack of acid secretion into the stomach, and that this could severely affect the bioavailability of poorly soluble, weakly basic drugs. To address this gap, Martin Wunderlich developed and proposed an "achlorhydric" fasted state gastric fluid which had a pH of 5 and consisted of a very weak acetate buffer (0.005 M) [13]. The weak buffer capacity reflected the fact that the elevation in pH is not due to addition of a buffer (as is the case after administration of some antacids), but rather due to the absence of gastric acid.

The next development was made at the National and Kapodistrian University of Athens, where after discussion with the group at the University of Frankfurt, it was deemed necessary to lower the surface tension and increase the pH of the fluid in order to more closely resemble the characteristics of human gastric fluid aspirates. The resulting medium, FaSSGF, replaced the use of SGF as a surrogate for gastric fluid of healthy volunteers in the fasted state [14]. Later, a revised version, FaSSGF V2, was proposed by Erceg et al. [18].

6.1.2 Fed gastric media

After a detailed analysis of the pH profile in the stomach of healthy adults during and after eating [5, 6], it was concluded that the best way of modelling the composition of the gastric contents after ingestion of a meal was to divide the profile into three parts: early (the first half-hour after meal ingestion), middle (30-90 min after eating), and late (90 min and longer after eating). Three media were designed to reflect the differences among these three phases, with pH and fat content declining from "early" to "middle" to "late" [16]. A "global" medium was also proposed by the authors for those researchers only interested in an overall assessment of how the solubility of a drug in the stomach fluids might change from the fasted to the fed state. The compositions of each of these media were based on milk, which had been the first medium used to represent conditions in the stomach after eating [11].

With time, researchers using milk as the basis for the dissolution media reported difficulties in separating the dissolved drug from the medium in preparation for analysis. The separation procedure involved precipitation of the milk proteins and centrifugation, which added considerably to the variability in results. Murat Kilic [21] set about determining whether a simpler medium, free of proteins, could also serve as a surrogate for the gastric chyme in the fed state. Instead of using milk, he switched to Lipofundin®, a parenteral nutrition product marketed by

B. Braun, containing soybean oil and medium-chain triglycerides as the fat component as well as glycerol, egg-yolk phospholipids, and sodium oleate, as the basis for the fed state media. As results for solubility and dissolution for several drugs and formulations were comparable in the milk and Lipofundin®-based media, it was concluded that the Lipofundin®-based media would be suitable for predicting drug behaviour in the fed state stomach.

6.1.3 Further versions of intestinal media

Starting with Jantratid et al. [16], several different versions of media representing the fasted and fed conditions in the small intestine have been proposed.

The main difference between the original FaSSIF (henceforth "FaSSIF V1") and the second version (FaSSIF V2) was the change in the ratio of bile salts to lecithin in the composition. Whereas the ratio in FaSSIF V1 is 4:1, the ratio in FaSSIF V2 is 15:1 [16]. This results in some differences in solubility behaviour, as addressed below. While the ratio of 4:1 addresses average relative concentrations of the two bile components in bile from healthy humans, the 15:1 ratio reflects a value closer to the upper bound and few sets of data in the literature have reported mean ratios this high. In 2012, Psachoulis et al. [19] put forward another FaSSIF version, FaSSIF Plus. This medium was based on FaSSIF V2 but included sodium oleate and cholesterol, which are also found in human intestinal aspirates collected in the fasted state. Cholesterol is also secreted in bile at low concentrations, whereas free fatty acids such as oleate are produced in the intestinal lumen by lysis of lecithin. It was thought that both of these components could have an effect on solubility and drug dissolution.

Finally, after a thorough review of all available data in the literature about the composition of intestinal aspirates collected from healthy humans in the fasted state, Fuchs et al. [22, 25] proposed another version of FaSSIF, FaSSIF V3. Additional to the Psachoulis medium, this one also took into account the rapid lysis of lecithin in the intestine by using lysolecithin as the phospholipid component.

Jantratid et al. [16] also updated the medium reflecting fed state conditions in the intestine, proposing a FeSSIF V2. To date, no further versions of this medium have been reported in the literature.

6.1.4 Media for the lower intestine

Since there are many formulations on the market, which release the drug over a long time or at a position lower in the GI tract, it is also of interest to know the solubility of the drug in this region. Using colonoscopy techniques, researchers in Athens were able to recover samples of colonic fluid from the upper ascending colon and analyse them for physicochemical characteristics as well as chemical composition [26], as described in chapter 5 of this book. Based on these data, compositions were proposed to represent the colonic conditions, FaSSCoF and FeSSCoF, for the fasted and fed state, respectively [17]. More recently, adopting a similar approach, luminal fluid samples were also recovered from the ileum, and a biorelevant medium representing this region has also been published [24].

Therefore, as of this writing, a complete set of biorelevant media covering the stomach, small intestine, and ascending colon in humans has been developed.

6.1.5 Canine media

Another area in which scientists in the early 2000s were interested was the extrapolation of bioavailability results from dogs to humans. At that time, most major pharmaceutical companies relied heavily on testing in dogs to determine which formulation (s) of a new drug would be most likely to perform well in humans. For this reason, it was also of interest to be able to predict a drug's solubility in the canine GI tract. The first proposals for fluids to simulate canine gastric and intestinal fluids were made by the Athens group in 2006 [15]. Later, the Frankfurt team elaborated on this work in collaboration with F. Hofmann-LaRoche to propose further canine media [20]. In order to reflect the essentially bimodal gastric pH in dogs, two media were proposed: since some dogs have low gastric pH, similar to the great majority of heathy adult humans, whereas others have consistently elevated gastric pH. Thus one medium has a pH of 6.5 to address the dogs with high gastric pH, while the second medium has a pH of 1.5 to address the pH in dogs with "normal" gastric pH. A revised canine intestinal medium was also proposed by these authors.

6.2 Current versions of biorelevant media

The currently recommended media compositions of fasted and fed state versions of the biorelevant media are shown in Table 6.1 for the fasted state, and Tables 6.2 and 6.3 for the fed state [27]. They are all referred to in the table as "Level II" media since they contain bile components and, if appropriate, fat digestion products as well as adjusting for pH, buffer capacity and osmolarity. Level I media lack the bile components/fat digestion products but are adjusted for pH, buffer capacity, and osmolarity, whereas level 0 media are simply adjusted to the relevant pH (Figure 6.2).

Table 6.1: Biorelevant media for simulating the environment in the stomach, upper/middle/lower small intestine, and ascending colon in the fasted state in experiments with poorly soluble drugs[1].

Ingredients	Level II FaSSGF	Level II FaSSIF-V1	Level II FaSSIF-V2	Level II $FaSSIF_{midgut}$	Level II SIF_{ileum}	Level II FaSSCoF
Sodium taurocholate (mM)	0.08	3	3	1.5	0.8	–
Sodium cholate (mM)	–	–	–	–	–	0.15
Lecithin (mM)	0.02	0.75	0.2	0.1	0.2	0.3
Sodium oleate (mM)	–	–	–	–	–	0.1
Tris (mM)	–	–	–	–	–	45.4
Maleic acid (mM)	–	–	19.1	19.3	52.8	75.8
Potassium dihydrogen phosphate (mM)	–	28.7	–	–	–	–
NaOH (mM)	–	10.5	34.8	36.5	105	120
HCl	qs pH 1.6	–	–	–	–	–
Sodium chloride (mM)	34.2	105.85	68.6	76.1	30.1	–
Parameters						
Osmolality (mOsm/kg)	121	270	180	190	190	196
Buffer capacity (HCl) [(mmol/L)/ΔpH]	n.a.	12	10	10	10	16
pH	1.6	6.5	6.5	6.8	7.5	7.8

[1]Osmolality, pH, and buffer capacity are measured values; osmolality was adjusted with NaCl if needed; n.a., not applicable.

Table 6.2: Biorelevant media used to simulate the environment in the stomach in the fed state in experiments with poorly soluble drugs[1].

Ingredients	Level II $FeSSGF_{early}$	Level II $FeSSGF_{middle}$	Level II $FeSSGF_{late}$
Maleic acid (mM)	47.0	–	–
Acetic acid (mM)	–	18.31	–
Sodium acetate (mM)	–	32.98	–
Ortho-phosphoric acid (mM) (mM)	–	–	5.5
Sodium dihydrogen phosphate (mM)	–	–	32
Lipofundin®:buffer ratio	17.5:82.5	8.75:91.3	4.375:95.6

Table 6.2 (continued)

Ingredients	Level II $FeSSGF_{early}$	Level II $FeSSGF_{middle}$	Level II $FeSSGF_{late}$
HCl/NaOH	*qs* pH 6.4	*qs* pH 5	*qs* pH 3
Sodium chloride (mM)	270.1	181.7	127.5
Parameters			
Osmolality (mOsm/kg)	559	400	300
Buffer capacity (HCl) [(mmol/L)/ΔpH]	21	25	25
pH	6.4	5	3

[1]Osmolality, pH, and buffer capacity are measured values; osmolality was adjusted with NaCl.

Table 6.3: Biorelevant media for simulating the fed state environment in the upper/middle/lower small intestine and ascending colon in experiments with poorly soluble drugs[1].

Ingredients	Level II FeSSIF	Level II FeSSIF-V2	Level II $FeSSIF_{midgut}$	Level II SIF_{ileum}	Level II FeSSCoF
Sodium taurocholate (mM)	15	10	5	0.8	–
Sodium cholate (mM)	–	–	–	–	0.6
Lecithin (mM)	3.75	2	1	0.2	0.5
Glyceryl monooleate (mM)	–	5	2.5	–	–
Sodium oleate (mM)	–	0.8	0.4	–	0.2
Glucose (mg/mL)	–	–	–	–	14
Tris (mM)	–	–	–	–	30.5
Maleic acid (mM)	–	71.9	46.5	52.8	30.15
Acetic acid (mM)	144	–	–	–	–
NaOH (mM)	101	102.4	83	105	16.5
Sodium chloride (mM)	–	125.5	102.6	30.1	34
Potassium chloride (mM)	204	–	–	–	–
Parameters					
Osmolality (mOsm/kg)	635	390	300	190	207
Buffer capacity (HCl) [(mmol/L)/ΔpH]	76	25	25	10	15
pH	5.0	5.8	6.5	7.5	6.0

[1]Explanation of abbreviations is provided in the text; osmolality, pH, and buffer capacity are measured and are not nominal values; osmolality was adjusted with KCl or NaCl.

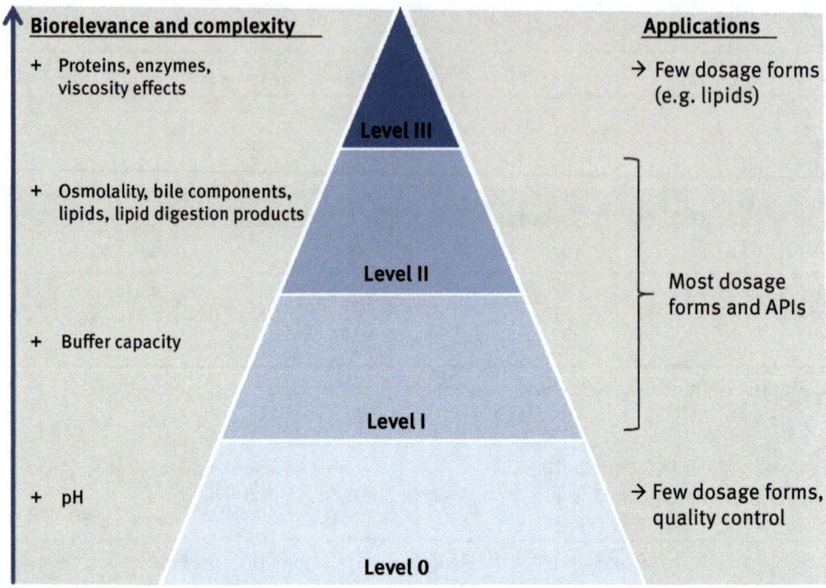

Figure 6.2: Levels of biorelevant dissolution media (adapted from [27]).

6.3 Selection of appropriate media for solubility and dissolution studies

6.3.1 Solubility studies

In early development, whenever the drug candidate is being considered for an oral formulation, a key question is the solubility behaviour in the GI tract, since the established tenet is that a drug must be in solution in the gut lumen before it can be quantitatively absorbed into the systemic circulation.

The usual standard applied is the solubility under intestinal conditions, as this is where the great majority of drug absorption takes place. Several studies have been published showing the solubility of drugs in aspirates recovered from the intestinal lumen of healthy volunteers (human intestinal fluid, HIF) (see Chapter 5), and solubility in HIF is usually the benchmark by which media that attempt to reproduce the conditions in the human intestine are judged. Although there are some issues with solubility values in HIF around questions such as intersubject variability, pooling of aspirates, and their handling prior to solubility measurements [25], which for certain drugs have led to a wide range of reported values, this is the most reliable benchmark currently available. That being said, use of HIF as a medium for solubility assays is not a feasible approach for routine screening and optimization of drug candidates. Biorelevant media are suitable surrogates that have been proven to be advantageous to understand solubility behaviour of drug candidates, especially those that are poorly soluble, in the GI milieu.

A comparison of results in three versions of FaSSIF with HIF solubility has now been published by Fuchs et al. [22] and by Klumpp et al. [28]. The results from Fuchs et al. are shown in Table 6.4.

The results show that all three of the biorelevant media are able to give a good approximation of the average solubility of the drug in the human fasted intestine. By contrast, for the poorly soluble drugs studied, using the corresponding simple buffer solutions resulted in an underestimation of the HIF solubility, while the surfactant-containing media typically used in quality control dissolution of such drugs (0.5% SDS) resulted in an overestimation of the HIF solubility.

The bile salt composition and pH appear to have an impact on the solubility, as do the differences in lecithin (lecithin/lysolecithin) and further components of the media. These data, as well as the variations in the solubilities measured in HIF, illustrate the intersubject variability of the luminal contents and hence the drug's ability to get into solution and be absorbed in a given subject.

Recent studies by Khadra and Halbert at the University of Strathclyde [23] have aimed at developing a set of biorelevant media to capture the potential variability in drug solubility in vivo using a statistical approach. Data on the various parameters that can be important to intestinal solubility were identified, and the typical range of values employed in a statistical design to tease out combinations which together would cover a broad range of solubility results. This work is ongoing, with the aim of not only capturing the parameters themselves, but also covariate effects, in a set of media that is small enough to be practical, but large enough to predict the in vivo variability for a given drug.

Following on from this research, the Strathclyde group has also published a similar set for fed state intestinal conditions [29]. Going forward, it would also be extremely helpful to extend such studies to investigation of variability in gastric solubility, as dissolution in the stomach often provides a key first driving force for dissolution and hence absorption from the intestine (see Chapter 2).

6.3.2 Dissolution studies

The designation "Level II" comes from a seminal paper by Markopoulous et al. [27], which recognized that since some drugs and formulations are more sensitive to GI conditions than others, it would be prudent to choose media accordingly. Following this logic, if a compound that is highly soluble over a wide range of pH is housed in an immediate release dosage form, release from the formulation can be tested in a simple buffer medium. This principle is essentially followed by the BCS-based Biowaiver concept [30] to cut down on the regulatory burden associated with applying for continued

Table 6.4: Solubility of 10 model compounds in two earlier versions of FaSSIF and prototype FaSSIF V3 biorelevant media, in blank maleate buffer, and in a 0.5% SDS solution.

	Neutral compounds							Weak acid	Weak bases	
API	**Carbamazepine**	**Cyclosporine**	**Danazol**	**Fenofibrate**	**Celecoxib**	**Felodipine**	**Nifidepine**	**Glibenclamide**	**Dipyridamole**	**Ketoconazole**
pK_a					**10.7[a]**	**5.4[b*]**	**5.4[b*]**	**4.3[b]**	**6.6[b]**	**6.8[b]**
Medium	**Solubility (µg/mL)**									
FaSSIF-V3-blank buffer	*244.9 ± 3.8*	*4.2 ± 0.1*	*0.3 ± 0.0*	*0.3 ± 0.0*	*1.7 ± 0.1*	*1.1 ± 0.1*	*11.8 ± 0.9*	*2.7 ± 0.2*	*5.1 ± 0.1*	*8.3 ± 0.3*
FaSSIF-V3-TC	304.3 ± 2.8	15.9 ± 0.7	3.1 ± 0.2	1.2 ± 0.1	8.6 ± 0.1	14.6 ± 0.6	17.5 ± 0.6	4.8 ± 0.1	14.3 ± 0.3	22.9 ± 0.7
FaSSIF-V3-GC /TC	299.6 ± 5.1	16.2 ± 0.6	3.3 ± 0.1	1.4 ± 0.1	6.7 ± 0.4	16.1 ± 0.5	18.3 ± 0.1	4.6 ± 0.0	12.9 ± 0.3	22.0 ± 0.5
FaSSIF-V3-GC	294.2 ± 5.8	17.3 ± 0.2	2.8 ± 0.1	1.1 ± 0.0	6.9 ± 0.3	14.7 ± 0.3	17.6 ± 0.6	4.7 ± 0.1	15.2 ± 0.4	22.7 ± 0.9
FaSSIF-V3-TC-$^1/_2$ PL	303.8 ± 3.7	10.5 ± 0.3	1.7 ± 0.1	0.4 ± 0.0	4.1 ± 0.3	9.0 ± 0.2	16.0 ± 0.3	3.6 ± 0.3	11.6 ± 0.3	15.6 ± 0.1
FaSSIF-V3-T-Urso	303.6 ± 1.8	23.8 ± 0.5	3.6 ± 0.1	1.7 ± 0.1	6.9 ± 0.3	12.3 ± 0.3	17.9 ± 0.1	4.4 ± 0.2	16.0 ± 0.4	30.4 ± 0.8
FaSSIF-V3-T-Cheno	307.9 ± 3.9	21.0 ± 0.5	4.1 ± 0.1	2.2 ± 01	9.7 ± 0.4	15.9 ± 0.1	19.7 ± 0.3	4.3 ± 0.1	22.3 ± 0.4	33.0 ± 0.7
FaSSIF-V3-GC /TC_Chol	311.7 ± 2.5	12.5 ± 0.9	3.0 ± 0.1	1.4 ± 0.0	7.4 ± 0.2	11.6 ± 0.4	16.2 ± 0.1	5.7 ± 0.1	27.3 ± 1.4	19.1 ± 0.6
FaSSIF-V3-GCDC	295.6 ± 10.0	17.9 ± 0.3	3.1 ± 0.2	2.0 ± 0.2	10.6 ± 0.1	15.2 ± 0.1	18.9 ± 0.6	6.0 ± 0.1	15.9 ± 0.5	26.0

FaSSIF (original)	298.1 ± 15.9	31.9 ± 0.2	10.1 ± 0.2	11.0 ± 0.4	41.5 ± 1.0	46.4 ± 0.6	24.2 ± 0.1	3.2 ± 0.3	29.9 ± 1.8	21.7 ± 0.8
FaSSIF-V2	295.3 ± 16.8	12.9 ± 0.8	2.0 ± 0.0	3.3 ± 0.2	12.0 ± 0.2	12.8 ± 0.3	15.5 ± 0.1	3.0 ± 0.2	16.3 ± 0.6	15.3 ± 0.3
FaSSIF-V3-buffer_0.5% SDS	*1388.0 ± 5.3*	*2488.6 ± 75.9*	*209.5 ± 5.6*	*165.4 ± 6.1*	*249.8 ± 1.4*	*606.7 ± 5.1*	*162.0 ± 2.2*	*36.7 ± 0.1*	*831.0 ± 10.1*	*1787.1 ± 34.6*
fasted state HIF (values from different literature sources)	305.8 ± 68.0 c	13.0 ± NA g	6.7 ± 1.4 e	11.9 ± NA j		14.0 ± NA d	45.0 ± NA k	15.4 ± 3.8 e	22.5 ± 2.0 m	111.3 ± 49.0 e
	283 ± 5.7d	3.5 ± 0.4 d	8.8 ± 5.2 c	19.7 ± 25.9 c		14.0 ± 0.0 g	23.1 ± 6.3 c	9.2 ± 11.0 c	29.0 ± 0.4 d	102.6 ± 76.5 c
	336.0 ± 17.9 e		13.2 ± 14.1 h				40.5 ± 7.5 e		20.0 ± NA k	326.4 ± 366.0 h
	170 ± NA f		2.04 ± 1.5 i				38.6 ± 24.0 h			28.8 ± 3.0 m
			9.0 ± NA g				10.0 ± NA l			56.0 ± 4.3 d
			4.9 ± 1.1 d							
			8.25 ± NA j							

*Even though a weak base pK_a for felodipine has been reported as 5.07 [31] and for nifedipine 5.33 [32], these compounds have been shown in former investigations to have similar solubility at acidic and neutral pH (e.g. felodipine [33] and nifedipine [20]).

marketing authorization after changes in the formulation or manufacture for existing products and for approval of generic versions. In this case, a level 0 medium should be appropriate for dissolution testing (see Figure 6.2). On the other hand, if the solubility of the drug or a formulation component is highly sensitive to pH, bile components, and/or the presence of fats in the gut lumen, these factors need to be adequately addressed in the choice of dissolution media. In this case, level II media (as described in Tables 6.1–6.3) will be more appropriate to gauge how the formulation will release the drug in the GI tract.

Further examples of the selection of the level of biorelevant media, based on the characteristics of both the API and the formulation, are addressed in [27].

6.4 Some considerations in solubility measurement with a view to predicting in vivo solubility

The in vivo solubility of an API is an important parameter for estimating whether or not the API can be fully absorbed at the dose administered. However, it is not possible to measure the thermodynamic solubility of an API in situ in the human GI tract as the equilibration time (usually at least several hours are required to attain equilibrium) is severely limited by GI motility. As a surrogate, samples of GI fluids are often aspirated and the solubility is measured in the laboratory (see Chapter 5). The solubility value achieved can be influenced by diverse factors that affect the composition of the aspirated samples, including the duration of the aspiration, whether the subject was fed prior to sample collection, how long before sampling water was ingested, whether the subject was perfused during sampling, how the sample was handled after aspiration (sample pooling, time on bench before storage), and during storage (duration and temperature of storage, repeated freezing and thawing). These factors and their influence on the composition of aspirates have been thoroughly reviewed by Fuchs and Dressman [25]. Not surprisingly, large variations in reported solubilities in HIF have been reported in the literature [22] for several APIs.

Due to the difficulty in obtaining human aspirates and the uncertainty as to their composition and hence the reliability of extrapolating solubility data obtained with them to the general patient population, the in vivo solubility is usually estimated using biorelevant media. The following points should be considered when determining solubilities in biorelevant media with a view to obtaining a useful estimate of the value in vivo.

6.4.1 Which is more relevant – thermodynamic or kinetic solubility?

The thermodynamic solubility of an API expresses how much drug is in solution at equilibrium in a given solvent and at a given temperature. As such, it is a useful parameter with which to judge the possibility of the API being completely dissolved in the GI tract and thus available for absorption. However, the GI tract is a dynamic environment. With changes in the composition of the fluid as the API moves through the GI tract, the concentration of drug in solution may fall well below the thermodynamic solubility or temporarily exceed it. An example of the former case is during the release of API from an extended release formulation, while examples of the latter case are the release of API from an "enabling formulation" (i.e. one which aspires to produce a supersaturation of drug in the gut lumen) or when a basic drug dissolves in the gastric fluids and then moves into the small intestine. In these cases, the thermodynamic solubility is likely to be a poor predictor of the concentration actually achieved in the GI tract and thus the driving force for absorption.

Many scientists now view the so-called kinetic solubility as being a more useful parameter with which the driving force for absorption can be estimated, especially for drug/formulation combinations where a supersaturation in the GI fluids may be expected. For example, for "enabling" formulations, concentrations far in excess of the thermodynamic solubility can be achieved over a considerable period of time in in vitro dissolution experiments and these concentrations appear to yield a more accurate estimate of the plasma concentration profile when coupled with physiologically based pharmacokinetic (PBPK) models like Simcyp® or STELLA than when a thermodynamic solubility is used [34, 35].

6.4.2 When can the "solubility" change during measurement?

The classic case in which supersaturation can be achieved under GI conditions, but then result in precipitation is the conversion of a salt to a free acid or base. Salts of ionizable substances are often used to achieve higher concentrations than would be possible with the protonated acid or unprotonated base form, respectively. Although the Henderson–Hasselbalch equation tells us that at a given pH value, the solubility will revert to the thermodynamic solubility at that pH over time, the kinetic solubility (i.e. the observed highest concentration) often exceeds the thermodynamic solubility for long enough to make a considerable difference in absorption and hence the plasma profile of the API. A typical case, using various salts of ibuprofen, has been discussed in the literature by Cristofoletti et al. [36].

Other approaches to enhancing the absorption of (especially) poorly soluble APIs include the conversion of a crystalline substance to an amorphous form, for example by spray drying, creating a solid solution of the API in a hydrophilic polymer using hot melt extrusion, preparing a lipid-based solution of the API or adsorbing the API onto mesoporous silica. In all of these cases, the kinetic profile of the API in solution is more likely than the thermodynamic solubility to yield a realistic estimate of the plasma profile of the drug when combined with PBPK modelling.

6.4.3 Precautions in measuring solubility in biorelevant media

When using biorelevant media to measure solubility, there are some specific precautions that should be taken. The first is the preparation of the medium itself. Although the media can be made from the individual components, this is a rather arduous procedure, such that most scientists now use the instant powder versions available from Biorelevant.com (www.biorelevant.com). Although the availability of these powders has significantly streamlined the preparation procedure, an equilibration time prior to starting the solubility experiment is still required for some compositions [28]. After this period, the medium can be stored for up to several days prior to use, adding flexibility to scheduling of experiments. Reproducibility of results with these powders has also been investigated and shown to be reliable both within and across laboratories [37]. Therefore, in contrast to measuring solubility in HIFs, the biorelevant media offer standardization and reproducibility on an intra- and inter-laboratory basis.

As the buffer capacity of both HIFs and biorelevant media reflecting the fasted state is quite low compared to standard compendial media, a change in pH during the solubility determination can occur, especially if the weak base/acid has a significant solubility in the medium. Thus, the pH of the medium should always be checked both before and after the solubility experiment when an ionizable API is being tested and, if the pH is changing, it should be adjusted with HCl or NaOH as needed during the experiment.

A third consideration is that the API may degrade during the solubility determination. Thus, the concentration should always be analysed using a stability indicating method. Plöger et al. [38] discussed this issue in detail and made recommendations about the appropriate duration of the experiment under such circumstances, concluding that the duration of the experiment should be adjusted to the passage time through the GI region of interest and thus supporting the determination of a kinetic solubility.

6.4.4 Miniaturization of the solubility determination

The standard method for determining solubility is the shake flask method, whereby aliquots of drug substance are sequentially added to a volume of the solvent in an Erlenmeyer flask and shaken at a predetermined temperature (usually

20 °C) for long enough to determine whether the entire aliquot has been dissolved. The process is repeated until by addition of a further aliquot, a sediment is formed. The respective setup is depicted in Figure 6.3. Although this method is generally suitable for classifying the solubility of an API according to the pharmacopoeial categories (e.g. “freely soluble” or “sparingly soluble”), there are three issues with this approach when determining the solubility of an API in biorelevant media. The first is that for a new API, there may only be a few milligrams available for testing. In this case, the shake flask method needs to be miniaturized to avoid using large quantities of API. The second is the temperature, as the temperature of interest for biorelevant media is 37 °C (body temperature) rather than 20 °C. Depending on the heat capacity of solution of the API, this can make a considerable difference in the solubility. Third, the solubility in biorelevant media needs to be precisely measured and requires a stability indicating assay, preferably HPLC or similar.

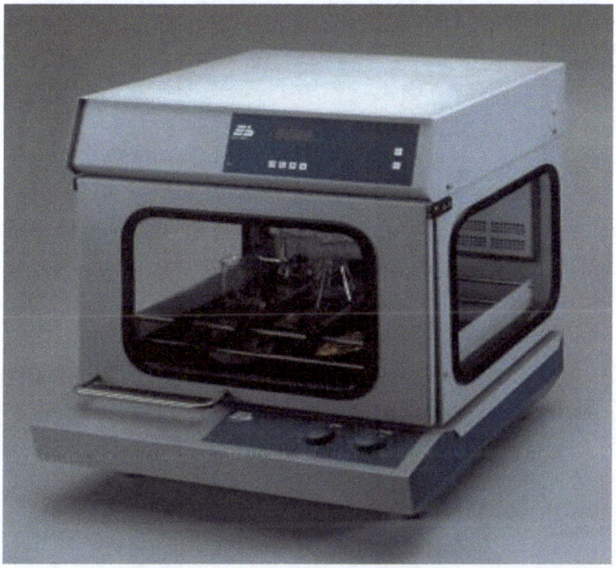

Figure 6.3: Set-up of a standard shake flask method for determining solubility.

For all these reasons, a miniaturized version of the shake flask method using Uniprep filters was devised by Glomme et al. [39]. In order to minimize the amount of API required for the solubility determination, an approximate solubility can be calculated using algorithms such as the Yalkowsky equations, WSKOWWIN, ABSOLV, and HYBOT. An excess amount of API corresponding to approximately two times the calculated solubility is weighed and filled into a Whatman UniPrep filter chamber as illustrated in Figure 6.4. Two millilitres of the medium is added,

Figure 6.4: Set-up of the UniPrep filter device.

and the chamber is closed with the plunger. The plunger has a filter membrane on one end and a pre-attached cap on the other end. The sample is then shaken for a pre-determined time on an orbital shaker at 37 °C. The pH value (except for neutral compounds) and the presence of a sediment (undissolved API) should be checked after several hours. If dissolution is complete, more sample can be added at this time. The pH should additionally be checked at the end of the solubility determination. After the equilibration time, the plunger is inserted into the sample-containing chamber and pushed through the medium, forcing the filtrate into the reservoir of the plunger. A distinct advantage of this set-up is that there is no contact between the sample and the filter prior to lowering the plunger.

To obtain the thermodynamic solubility of the API at 37 °C in the biorelevant media, the experiment should be set up to take samples at 2, 4, 8, and 24 h. Occasionally, an API may take longer to reach a constant concentration, in which case the experiment should be extended, while in other cases equilibrium may be reached within a couple of hours. In yet other cases, it may be of more physiological relevance to measure the solubility over the usual passage time through the GI segment of interest, for example, for solubility in FaSSGF, determination of the concentration dissolved after 1–2 h would be appropriate. Assuming that a stability indicating assay is used, this alignment with the physiology will also indicate whether substantial decomposition of the API is likely during its residence in the given GI segment (see also [38]).

6.4.5 Single or multiple media?

Since biorelevant media have been developed for the various segments in the GI tract, estimates of the API solubility in vivo corresponding to residence in each of these segments can be obtained and potential problems for absorption, for example, decomposition in the stomach or very poor solubility at intestinal pH, can easily be identified.

Another consideration is the variability in solubility among individuals. As there is a wide variance in parameters such as pH and bile salt concentration across the population, it is of great interest, especially for APIs that have high lipophilicity (and therefore a high dependency of solubility on bile concentration) and/or that are ionizable (and therefore have a high dependency of solubility on pH), to not only determine the solubility under average conditions – as achieved with biorelevant media – but also to study the solubility in variations of the biorelevant media to predict how variable the in vivo solubility might be. Already there has been good progress in this area, with the Strathclyde group recommending a set of variations on FaSSIF to fully explore the range of solubilities an API might have in the small intestine in the fasted state [23] and a set of variations on FeSSF for the fed state small intestine [29]. Since drugs are often taken on an empty stomach or with food, and since the small intestine is the key region of the GI tract for drug absorption, the variability in the concentration that can be achieved in the small intestine in the fasted and fed states is extremely valuable information for the formulation scientist.

6.5 Outlook

Over the last 20 years, enormous progress has been made in understanding and estimating solubility of drugs in the GI tract. But the biorelevant media journey is far from over. Although we now have biorelevant media for all sections of the GI tract and, for the small intestine, a way forward for estimating the range of solubilities that might occur within the population, all of these media are directed towards healthy adults. Still to be evolved are corresponding media for different age groups, for drug–drug interactions such as proton pump inhibitor-mediated hypochlorhydria and for disease states that affect the composition of GI fluids.

Additionally, biorelevant solubilities are being increasingly used in physiologically based pharmacokinetic modelling, enabling more accurate predictions of the in vivo release from the dosage form and thus of the resulting plasma profile. Finally, biorelevant media are an important component of regulatory endeavours, for example, by the US-FDA to set clinically relevant specifications for oral drug products.

References

[1] USP 24, 1995 The United States Pharmacopeia: USP 24: The National Formulary: NF 19, 1995, Rockville MD United States Pharmacopoeia Convention

[2] USP 25, 2000 The United States Pharmacopeia: USP 25: The National Formulary: NF 20, 2002 Rockville MD United States Pharmacopoeia Convention

[3] Naylor, L.J., Bakatselou, V., Rodriguez-Hornedo, N., Weiner, N.D., & Dressman, J.B. Dissolution behavior of steroids in bile salt solutions is modified by the presence of lecithin. Eur. J. Pharm. Biopharm. 1995, 41, 346–353.

[4] Bakatselou, V., Oppenheim, RC., & Dressman., JB. Solubilization and wetting effects of bile salts on the dissolution of steroids. Pharm. Res. 1991, 8, 1461–1469.

[5] Naylor, L.J., Bakatselou, V., & Dressman, J.B. Comparison of the mechanism of dissolution of hydrocortisone in simple and mixed micelle systems. Pharm. Res. 1993, 10, 865–870.

[6] Dressman, JB., Berardi, RR., Dermentzoglou, LC., Russell, TL., Schmaltz, SP., Barnett, JL., & Jarvenpaa., K. Upper gastrointestinal pH in young, healthy men and women. Pharm. Res. 1990, 7, 756–761.

[7] Russell, T.L., Barnett, J.L., Berardi, R.R., Dermentzoglou, L.C., Jarvenpaa, K.M., Schmaltz, S.S., & Dressman., J.B. Upper gastrointestinal pH in seventy-nine healthy, North American men and women ≥ 65 years. Pharm. Res. 1993, 10, 187–196.

[8] Youngberg, C.A., Berardi, R.R., Howatt, W.F., Hyneck, M.L., Amidon, G.L., Meyer, J.H., & Dressman., J.B. Comparison of gastrointestinal pH in cystic fibrosis and healthy subjects. Dig. Dis. Sci. 1987, 32, 472–480.

[9] Meyer, J.H., Elashoff, J., Porter-Fink, V., Dressman, J.B., & Amidon., G.L. Human postprandial gastric emptying of 1-3 millimeter spheres. Gastroenterology 1988, 94, 1315–1325.

[10] Meyer, J.H., Gu, Y., Elashoff, J., Reedy, T., Dressman, J.B., & Amidon., G.L. Effects of viscosity and fluid outflow on postcibal gastric emptying of solids. Am. J. Physiol. 1986, 193, G161–164.

[11] Dressman, JB., Amidon, GL., Reppas, C., & Shah., VP. Dissolution testing as a prognostic tool for oral drug absorption: Immediate release dosage forms. Pharm. Res. 1998, 15, 11–22.

[12] Galia, E., Nicolaides, E., Hörter, D., Löbenberg, R., Reppas, C., & Dressman., J.B. Evaluation of various dissolution media for predicting in vivo performance of Class I and II drugs. Pharm. Res. 1998, 15, 698–705.

[13] Wunderlich, Martin. (Doctorate, Goethe University, Germany). Biorelevante in vitro Methoden zur Vorhersage des in vivo Verhaltens von schlecht wasserlöslichen, schwach basischen Arzneistoffen (2004)

[14] Vertzoni, M., Dressman, J., Butler, J., Hempenstall, J., & Reppas., C. Simulation of fasting gastric conditions and its importance for the *in vivo* dissolution of lipophilic compounds. Eur. J. Pharm. Biopharm. 2005, 60, 413–417.

[15] Kalantzi, L., Persson, E., Polentarutti, B., Abrahamsson, B., Goumas, K., Dressman, J., & Reppas., C. Canine intestinal contents vs. Biorelevant media for the assessment of intestinal solubility of two weakly basic drugs. Pharm. Res. 2006, 23, 1373–1381.

[16] Jantratid, E., Janßen, N., Reppas, C., & Dressman., J. Dissolution media simulating conditions in the proximal human gastrointestinal tract: An update. Pharm. Res. 2008, 25, 1663–1676.

[17] Vertzoni, M., Diadikou, A., Chatzilias, M., Söderlind, E., Abrahamsson, B., Dressman, J., & Reppas., C. Biorelevant Media to simulate fluids in the ascending colon of humans and their usefulness in prediction of intracolonic drug solubility. Pharm. Res. 2010, 27, 2187–2196.

[18] Erceg, M., Vertzoni, M., Ceric, H., Dumic, M., Cetina-Cizmek, B., & Reppas, C. In vitro vs. canine data for assessing early exposure of doxazosin base and its mesylate salt. EJPB 2012, 80, 402–409.

[19] Psachoulias, D., Vertzoni, M., Butler, J., Busby, D., Symillides, M., Dressman, J., & Reppas., C. An In vitro methodology for forecasting luminal concentrations and precipitation of highly permeable lipophilic weak bases in the fasted upper small intestine. Pharm. Res. 2012, 29, 3486–3498.
[20] Arndt, M., Chokski, H., Tang, K., Reppas, C., & Dressman., J. Dissolution media simulating the proximal canine gastrointestinal tract in the fasted state. Eur. J. Pharm. Biopharm. 2013, 84, 633–641.
[21] Kilic, Murat. (Doctorate; Goethe University, Germany). Die Lipidtoolbox für Arzneimittel-ein systematischer Ansatz für die Entwicklung von oralen Lipidformulierungen schwerlöslicher Arzneistoffe (2013).
[22] Fuchs, A., Leigh, M., Kloefer, B., & Dressman., J. Advances in the design of fasted state simulating intestinal fluids: FaSSIF-V3. EJPB 2015, 90, 229–240.
[23] Khadra I, I., Zhou, Z., Dunn, C., Wilson, C.G., & Halbert., G. Statistical investigation of simulated intestinal fluid composition on the equilibrium solubility of biopharmaceutics classification system class II drugs. Eur. J. Pharm. Sci. 2015, 67, 65–75.
[24] Georgaka, D., Butler, J., Kesisoglou, F., Reppas, C., & Vertzoni., M. Evaluation of dissolution in the lower intestine and its impact on the absorption process of high dose, low solubility drugs. Mol. Pharm. 2017, 14, 4181–4191.
[25] Fuchs, A., & Dressman., J. Composition and physical chemical properties of fasted state human duodenal and jejunal fluids: A critical evaluation of the available data. J. Pharm. Sci. 2014, 103, 3398–3411.
[26] Diakidou, Amalia., Vertzoni, Maria., Goumas, Konstantinos., Söderlind, Erik., Abrahamsson, Bertil., Dressman, Jennifer., & Reppas, Christos. Characterization of the contents of ascending colon to which drugs are exposed after oral administration to healthy adults pharm. Res. 2009, 26, 2141–51.
[27] Markopoulos, C., Andreas, C., Dressman, J., Vertzoni, M., & Reppas., C. In-vitro simulation of luminal conditions for evaluation of performance of oral drug products: Choosing the appropriate test media. EJPB 2015, 93, 173–182.
[28] Klumpp, Lukas., Nagasekar, Kalpa., McCullough, Orla., Seybert, Anja., Ashtikar, Mukul., & Dressman., Jennifer. Stability of Biorelevant media under various storage conditions. Dissolution Technologies 2019, 26(2), 6–18.
[29] Zhou, Z., Dunn, C., Khadra, I., Wilson, CG., & Halbert, GW. Influence of physiological gastrointestinal surfactant ratio on the equilibrium solubility of bcs class ii drugs investigated using a four component mixture design. Mol. Pharm. 2017, 14(12), 4132–4144.
[30] ICH M9 draft guidance,2018: ICH harmonised guideline, biopharmaceutics classification system-based biowaivers M9 Draft version. https://www.ich.org/fileadmin/Public_Web_Site/ICH_Products/Guidelines/Multidisciplinary/M9/M9EWG_DraftGuideline_Step2_2018_0606.pdf.
[31] Pandey, M., Jaipal, A., Kumar, A., Malak, R., & Chandra, S. 2013. Determination of pK(a) of felodipine using UV-Visible spectroscopy. Spectrochim Acta A Mol. Biomol. Spectrosc. 2013, Nov, 115, 887–90.
[32] Drug Bank. Entry for nifedipine. Accessed on 28.04.2019 at https://www.drugbank.ca/drugs/DB01115
[33] Wagner, C., Thelen, K., Willmann, S., Selen, A., & Dressman., J. Utilizing *in vitro* and PBPK tools to link ADME characteristics to plasma profiles: Case example nifedipine immediate release formulation. J. Pharm. Sci. 2013, 102, 3205–3219.
[34] Fei, Y., Kostewicz, ES., Sheu, MT., & Dressman, JB. Analysis of the enhanced oral bioavailability of fenofibrate lipid formulations in fasted humans using an in vitro-in silico-in vivo approach. Eur. J. Pharm. Biopharm. 2013, 85, 1274–84.

[35] Hansmann, S., Miyaji, Y., & Dressman, J. An in silico approach to determine challenges in the bioavailability of ciprofloxacin, a poorly soluble weak base with borderline solubility and permeability characteristics. Eur J Pharm Biopharm. 2018, 122, 186–196. 2018.
[36] Cristofoletti, R., Dressman, J. Dissolution Methods to Increasing Discriminatory Power of In Vitro Dissolution Testing for Ibuprofen Free Acid and Its Salts Journal of Pharmaceutical Sciences 2017, 106, 92–99.
[37] Mann, J., Dressman, J., Rosenblatt, K., Ashworth, L., Muenster, U., Frank, K., Hutchins, P., Williams, J., Klumpp, L., Wielockx, K., Berben, P., Augustijns, P., Holm, R., Hofmann, M., Patel, S., Beato, S., Ojala, K., Tomaszewska, I., Bruel, J.-L., & Butler, J. Validation of Dissolution Testing with Biorelevant Media: An OrBiTo Study Molecular Pharmaceutics 2017, 14, 4192–4201.
[38] Plöger, GF., Hofsäss, MA., & Dressman, JB. Solubility determination of active pharmaceutical ingredients which have been recently added to the list of essential medicines in the context of the biopharmaceutics classification system-biowaiver. J. Pharm. Sci. 2018, 107(6), 1478–1488.
[39] Glomme, A., März, J., & Dressman, JB. Comparison of a miniaturized shake-flask solubility method with automated potentiometric acid/base titrations and calculated solubilities. J. Pharm. Sci. 2005, 94, 1–16.

Christoph Saal and Klara Valko

7 The role of solubility to optimize drug substances – a medicinal chemistry perspective

7.1 Introduction

Solubility is a term that is very common in everyday use. We consider a compound to be soluble, when it appears to form a clear solution when a solid form is put into a solvent. Solubility depends on several experimental conditions, such as, what is the solvent, what is the temperature, and what is the solid-state form of the material? From a medicinal chemistry perspective, it is very common to be concerned about low solubility of compounds early in the drug discovery process as it can cause problems in preparing solutions for biological testing. Low solubility can affect the results of a screening assay as used by the medicinal chemist, which relies on a known concentration of the molecule in solution under the assay conditions. Low solubility also affects oral absorption when the compound is administered in a solid form. The compound must be in solution form before it can be absorbed through the gastrointestinal tract. It follows that an ideal drug molecule should be soluble enough to be absorbed and circulate in solution in the blood stream and reach the site of action. Additionally, solubility matters for medicinal chemists for synthesis. In contrast to the other parameters, which have been mentioned, here solubility in organic solvents is the key. This topic will be addressed in chapter 10 of this book. Pharmacopoeias usually describe the solubility of a compound by stating how many units of solvent are needed to dissolve a unit weight of a solid compound in a defined solvent. For example, 1 g of caffeine can be dissolved in 60 g of water. The simplest way to measure solubility is to weigh a certain amount of compound and add known volume of solvents until it forms clear solutions. Some compounds can take longer to dissolve and need more stirring. Therefore, the definition and precise measurement of solubility are complex. It is even more difficult to define good or bad solubility because it depends on the purpose and condition for which we need to achieve the desired solubility.

The thermodynamic definition of solubility is the concentration of the saturated solution of a substance that is in equilibrium with a defined solid-state form of the compound, as discussed more in detail in Chapters 1 and 9. This concentration must be cited together with the solvent. It is also necessary to define the temperature at which the equilibrium between the solid and dissolved form was achieved and when the solvent is water, to state the pH and ionic strength. It is very

Christoph Saal, Merck KGaA, Frankfurter Straße 250, 64293 Darmstadt, Germany
Klara Valko, School of Pharmacy, University College, London, UK

https://doi.org/10.1515/9783110559835-007

important to describe the solid-state form of the compound (amorphous or crystalline including polymorph) as the solubility of different solid-state forms of the same compound may differ by orders of magnitude. To assess when equilibrium has been reached and a saturated solution is obtained, it is necessary to measure the concentration of the compound at various time points, and to check that the solid-state form of the compound has not changed during the measurement.

In drug discovery it may be advantageous to approach solubility from a physiological point of view. The drug molecules must be dissolved in the gastrointestinal tract to be able to be absorbed. The pH ranges from about 2 to 8 during the drug transit through the gastrointestinal tract to reach absorption. Therefore, it is essential to know the solubility – pH profile. The FDA defines the solubility as good when the highest dose strength is dissolved in 250 mL of water in a pH range of 1–7.5 [1]. This is because it can be expected that the patient will drink a glass of water when taking the drug orally, and there is a wide pH range from the stomach to the small intestine in which the drug should stay in solution and not precipitate. The dissolved drug molecule then needs to be able to cross the intestinal wall that requires it to have suitable permeability and lipophilicity. Therefore, the solubility together with the partition coefficients of drugs are important factors in influencing the oral absorption and in vivo distribution [2, 3]. The initial rate of dissolution of a drug molecule from the formulation in aqueous media is a direct function of the aqueous solubility [4], but it can be significantly altered by formulation as discussed in chapter 8 of this book. Therefore, the solubility of putative drug molecules is an important property to assess their oral absorption. The determination of the aqueous solubility of a drug molecule requires the measurement of the concentration of the substance in an aqueous medium that is in equilibrium with an excess of the solid-state form of the substance.

It is very difficult, however, to develop a solubility screening method for drug discovery to assess the FDA requirement for good solubility. The dose is not known at this stage and it would also require solubility measurements for a range of pHs, which is not feasible for large numbers of compounds. The solubility, and especially the dissolution rate, can be enhanced by formulation, and pharmaceutical technology has been developed that can get poorly soluble compounds absorbed using, for example, micronization [5] or nanoparticles [6, 7]. But before this is done, the medicinal chemist should try to design molecules with good inherent solubility.

When a drug molecule fulfils the requirement for good solubility and good permeability, good absorption and bioavailability may be expected. Therefore, authorities may waive expensive bio-equivalence studies in the clinic and this can save a significant part of the cost of drug development and registration. This aspect is discussed in more detail in Chapter 11.

In early drug discovery, potency is expressed in molar concentration units and therefore it is good to know the solubility in molar terms. It is easy to calculate the molar concentration of the saturated solution by knowing the molecular weight. However, when the aim is to design more soluble compounds, the solubility must be related to the structure, and it is important to express the solubility as the molar concentration of the saturated solution. A compound can be considered more soluble than another compound if higher numbers of molecules can be dissolved. If the solubility is expressed in mg/mL for two molecules that differ in molecular weight, the comparison is misleading. For example, molecule A has a molecular weight of 200 g/mol, while molecule B has a molecular weight of 600 g/mol. The two molecules have the same solubility in molar terms. The bigger molecule shows higher solubility when expressed in mg/mL unit. Wrong conclusions can be drawn by using different units as a measure of solubility, for example, bigger molecules are more soluble, which is not true if we consider the number of molecules in solution. Therefore, in drug discovery, to design more soluble molecules and set up structure solubility relationships, the solubility should be expressed as a molar concentration of the saturated solution. Care must be taken to avoid confusion, when molecules move from discovery to development, as during discovery medicinal chemists and biologists usually express solubility as mol/L, while during development pharmacists usually express solubility as mg/mL.

In conclusion, the medicinal chemist should aim to design drug molecules that have a good aqueous solubility for several reasons. First to have the compounds in solution during the initial potency screenings that are carried out using a stock solution of the putative drug molecule in dimethylsulfoxide (DMSO) diluted down by buffer solution of the target enzyme or medium for a cellular assay, typically at pH 7.4. The second important reason to design soluble compounds is to help absorption in the gastrointestinal transit. It is better to build the inherent good solubility into the molecule than at later stages spend expensive resources on formulations and risk the developability of the molecule to a drug. When fragments are screened for, this is usually carried out at a higher concentration level, which requires higher solubility of the fragment molecules to avoid precipitation. On the other hand, fragments have a lower molecular weight and accordingly higher solubility. Based on the above-mentioned considerations, there is a need for solubility measurements at the early stages of drug discovery, in a high throughput manner, to be able to have information about the aqueous solubility of compounds. It is important to consider that at the early stages of drug discovery there are usually only small quantities of the substances available, the solid-state form is not characterized, and the samples may not be as pure as in the development stage. These factors may influence the measured solubility significantly.

7.2 Molecular interactions with water

Investigating the process of dissolution of a compound from solid into water from a physicochemical point of view is important for the design of water-soluble molecules.

To get a compound into solution, first the intermolecular interactions holding the crystal or amorphous solid together must be broken. This process requires energy. Every crystal has crystal lattice energy, which is required to disrupt the crystal packing. For an amorphous material, in general, we need to invest less energy to break the interactions between the molecules in the solid form than in a purely crystalline form. Once removed from the crystal lattice, the molecules must create a cavity in the water by disrupting the hydrogen bonds. This dissolution process requires further energy. However, if the solute and solvent molecules can form new interactions (e.g., dipole–dipole and H-bond donor–acceptor interactions), then some of the energy may be regained during the dissolution process. Lipophilic molecules, however, cannot form strong interactions with polar water molecules, and so energy cannot be retrieved from these interactions, and they have poor solubility in water.

The most popular equation for the estimation of solubility of organic molecules is the Yalkowsky equation [8] (eq. (7.1)), which has been already introduced in Chapter 1. This equation describes the logarithmic value of the water solubility of an organic compound (log *S*) in relation to its lipophilicity expressed by the octanol/water partition coefficient (log $P_{\text{oct:water}}$) and its melting point (MP).

$$\log S_{\text{solid in water}} = 0.5 - 0.01(\text{MP} - 25) - \log P_{\text{oct:water}} \tag{7.1}$$

Note that the melting point is set to zero for solutes that are liquid at room temperature and that is why 25 °C is subtracted from the melting point (MP-25). Equation (7.1) relates solubility to lipophilicity, and the melting point is taken as proportional to the energy of the crystal lattice. The octanol/water partition coefficient represents the lipophilicity of the compound that summarizes the energy of all the interactions between the solute molecule and water. It is also important to note that the log *S* refers to the logarithmic value of the equilibrium solubility of the compound expressed in molar concentration units. Equilibrium with the particle in the respective solid-state form must be achieved, and other conditions such as particle size and stirring should not unduly influence the result. From the coefficient of the melting point in the equation, an increase of 100 °C in melting point will reduce the intrinsic solubility by a factor of 10, that is, a unit lower in log *S* terms.

Abraham and Le [9] applied the solvation equation model to predict the solubility of over 400 diverse compounds for which the molecular descriptors were available. The solvation process is depicted in Figure 7.1.

Abraham and Le [9] constructed a solvation equation that contained five molecular descriptors (where *E* is the excess molar refraction, *S* is dipolarity/polarizability, *V* is the molecular size (McGowan volume), A and B account for H-bond acidity and basicity, respectively, and MP is the melting point) as given by eq. (7.2).

$$\log S_w = 0.58 - 0.58E + 0.98S + 1.23A + 3.39B - 4.08V - 0.01(\text{MP} - 25) \tag{7.2}$$

$n = 411$, $r^2 = 0.915$, $\text{sd} = 0.564$, $F = 724$

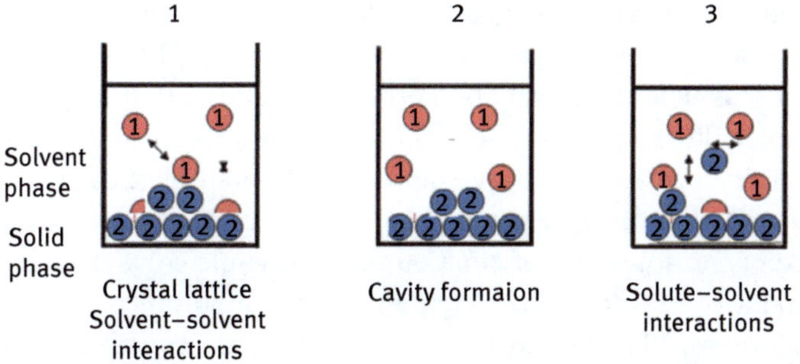

Figure 7.1: The solubility process by solvation the solute (2) by the solvent molecules (1).

where n is the number of compounds, r^2 is the correlation coefficient, sd is the standard error of the estimate, and F is the Fisher-test value.

Note that the signs of the coefficients are the opposite of the signs of the Abraham equation obtained for lipophilicity as there is an inverse correlation between lipophilicity and solubility. The coefficients are of similar magnitude to the coefficients in the octanol/water partition [10]. However, the H-bond acidity term has a positive value and is not zero, meaning that the H-bond acidity enhances solubility. The solvation equation for the octanol/water partition process describes the concentration ratio of the compound in the two solvents. The solubility process can be considered as the partition equilibrium between the solid-state form and the solute in the water. As the properties of the solid-state forms vary for different types of molecules and still today crystal forms cannot be predicted with enough reliability, it is very difficult to derive a precise solvation equation for solubility. Considering this variability, the solvation equation for solubility is statistically significant for many compounds. Abraham and Le [9] replaced the melting point with molecular descriptors using the product term of H-bond acidity and H-bond basicity, which resulted in eq. (7.3).

$$\log S_W = 0.52 - 1.00E + 0.77S + 2.168A + 4.24B - 3.36A^{*}B - 3.99V \qquad (7.3)$$
$$n = 659,\ r^2 = 0.920,\ \text{sd} = 0.56,\ F = 1256$$

where n is the number of compounds, r^2 is the correlation coefficient, sd is the standard error of the estimate and F is the Fisher-test value.

The Abraham solvation equation for describing the logarithmic value of aqueous solubility expressed in molar concentration units as a solute property demonstrates that larger compounds that have both H-bond donor and acceptor groups with low polarity and polarizability will have low solubility. The effect of H-bond donors and acceptors can be explained by the fact that molecules that have both H-bond donor and acceptor functionality may form strong intermolecular H-bonding in the crystal lattice.

The parameters mentioned in the last paragraph result in increasing crystal lattice energy, a higher melting point, and consequently a reduction in solubility. Another thermodynamic approach has been published by Perlovich et al. [11] that helps in understanding of the molecular solvation mechanism and the crystal lattice energy. They proposed that crystal lattice energy could be estimated by measuring the temperature dependence of the vapour pressure of saturated solutions and the temperature dependence of the solubility of compounds in organic solvents such as octanol and hexane relative to water. They highlighted the importance of lipophilicity and other polar forces, like H-bond formation both in the crystal and in water. For the estimation of the effect of H-bond formation on solubility, an "inert" solvent such as *n*-hexane is used.

All the above theoretical considerations for the description of the dissolution process and for the equilibrium solubility refer to a pure compound forming a perfect crystalline solid-state form. Compounds never have 100% purity. It is commonly observed in pharmaceutical industry that the solubility of a compound gradually decreases as it moves forward in the drug discovery and development process. The simple explanation for this is that the compound is synthesized in an increasingly pure form. Impurities cause disruptions in the crystal lattice, thus disrupting the forces between the molecules, and so reducing the crystal lattice energy and melting point. Additionally, impurities might enhance solubility as they can act as solubilizers such as surfactants.

Although there are many pharmaceutical formulation approaches that might help to overcome the problem of a poorly soluble drug molecule, the scope and the aim at early drug discovery lead optimization is to synthesize and progress the potent compounds that have good or acceptable physicochemical and biopharmaceutical properties for developability, which includes solubility and oral absorption. To be able to help drug design and lead optimization, to progress the best compounds, solubility measurements must be developed that can be carried out on the required number of compounds in a desired time frame and so provide results adequate for problem solving. Typical methods for solubility measurements at the early stages of the drug discovery and lead optimization process are described in the following section.

7.3 Various solubility measurements that can be applied during the early drug discovery process

The solubility obtained for a compound significantly depends on the experimental conditions used in the experiment to measure solubility. The exact measurement of the thermodynamic solubility of a pure solid substance may take a week and may require sensitive analytical methods. Even when using a high level of automation, it is practically impossible to characterize thousands of compounds per month.

However, if we understand the factors that influence the solubility, and we would like to answer a question about the solubility of the compound that is relevant to the drug discovery stage, we can carry out an adequate number of measurements with a fit for purpose quality of results. Thus, first we need to clarify by the measurements the questions that we would like to have the answers to. These are as follows:

A. Is the compound in solution under the biological screening condition?
B. Will the required dose dissolve in the gastrointestinal tract?

The first question arises early in the drug discovery process and a high-throughput precipitative solubility measurement can give a satisfactory answer. It can reveal whether the measured biological activity is compromised by low solubility. In this case the best results can be achieved by applying the same amount of DMSO and the same buffer (pH, ionic strength, and other constituents) as it is used for screening. The effect of the DMSO concentration and the type of buffer on the measured apparent precipitative solubility must be investigated.

The need for assessing rapidly the solubility of a large number of compounds emerged with the introduction of combinatorial chemistry to obtain compound libraries used for high-throughput potency screening. As the molecules became larger and more lipophilic to increase potency, the reproducibility of the potency screens decreased. Compounds synthesized by combinatorial chemistry are usually dissolved and stored in DMSO. Solutions are then typically dispensed using liquid dispensing robots for various screens. DMSO is a universally good solvent for both polar and non-polar compounds and has a relatively high boiling point (189 °C), which means it does not easily evaporate when dispensed in micro-litre (μL) quantities onto well plates. DMSO is compatible with proteins and cells and accordingly can be used in biochemical and cellular assay formats. Storing the compounds in liquid form is made possible by the accurate dispensing of large numbers of compounds with robots.

Another advantage of DMSO is that it is readily miscible with water. In a typical screening procedure, the compounds are dispensed as 0.1–10 mM DMSO stock solutions and are diluted with aqueous buffer solution before adding to the appropriate target protein or enzyme. Although the compound might be in solution in DMSO, when it is diluted by aqueous buffer the compound might precipitate out. Typically, 10–100 times dilution is applied, resulting in only 10–1% of DMSO in the final solution. As the dilution is carried out by robotic systems, the precipitation occurring during the dilution often remains unobserved and unrecorded. Therefore, there is a need to find an automatic procedure that detects the precipitation of compounds during dilution of the DMSO solution with aqueous buffer. Lipinski et al. [3] published a method that detects the scattering of light due to solid particulates when precipitation occurs. This turbidity measurement is carried out by instruments for measuring the optical characteristics of a solution and can operate in 96- and 384-well plate formats. They observed that compounds that precipitate out of solution as crystals gave a sudden strong signal. For these compounds, the dilution factor was more reproducible than for compounds that precipitated as amorphous material. The latter compounds produced a gentle steadily increasing signal [3], and the solubility results were not very reproducible. The turbidity measurement does not give an absolute value of the solubility, which is unlikely to be the same as an equilibrium solubility obtained from solid. However, it provides a warning signal when the compound precipitates out of solution. Bevan and Lloyd [12] used a nephelometer to detect precipitation using a laser beam. The instrument could detect the laser light-scattered signals for a 96-well plate within a couple of minutes. However, it was also found that when compounds precipitated out in large crystals the signal was weaker. Some compounds with surface-active properties caused a change of the background signal due to the meniscus change in the solution in the well, the so-called lensing effect. Nevertheless, these technologies are used frequently in screening laboratories not for providing an accurate solubility measurement but for detecting precipitation during the biological screening processes.

7.3.1 Precipitative solubility measurements by diluting DMSO solutions

Several pharmaceutical companies [13–16] have developed their high-throughput solubility screens to be applied at the early drug discovery stage using DMSO solutions of the compounds that are diluted down with aqueous buffer. A typical measurement process is to dilute down 5 μL of 10 mM DMSO stock solution with 95 μL phosphate-buffered saline solution. It results in a 500 μM total concentration of the compound containing 5% DMSO. The wells can be screened by measuring the light scattering. For these measurements, a simple UV plate reader can be used. Precipitation should result in a distinct distortion in the compound's UV spectrum. If no precipitation is detected, the compound can be declared soluble as equal to or greater than 500 μM.

Other screens use 2–5 μL of 10 mM DMSO stock solution of the compound that is dispensed onto a 96-well plate format by a robotic liquid dispenser. Duplicate plates are prepared, each containing 2–5 μL of a 10 mM DMSO solution. For the so-called standard plate, the DMSO solution is diluted with known amounts of solvent (methanol or DMSO) possibly dissolving the compounds. The wells on the so-called sample plate are diluted with the same known amounts of aqueous buffer as used in the biochemical or cellular assay. Compounds having poor aqueous solubility will eventually precipitate out in the wells of the sample plates. The precipitation and the equilibration between the saturated solution and the solid can be promoted by applying sonication. Before HPLC analysis of the wells of the standard and sample plates, the contents of the "sample plate" are filtered with a vacuum filtration device or centrifuge using 96-well plate format filters. Alternatively, on-line filtration can be used before injecting the sample solutions (for non-soluble compounds in suspension) into the HPLC system. Applying the generic fast reversed-phase gradient chromatographic method, two chromatograms are collected for one compound: one from the "standard plate" and one from the filtrated "sample plate". Nowadays "ultra high-pressure liquid chromatography" (UPLC) represents a good alternative to HPLC, which can even accelerate analytics. The proportions of the peak areas reveal whether the compound stayed in solution or precipitated out. From the ratio of the peak areas and the molar concentration of the compound, the sample solution can be estimated. If the sample peak is the same size as the standard peak, it means that the compound is soluble at the concentration level given by the standard solution. If no peak is detected in the sample solution, then the solubility of the compound is less than the detection limit. As the amount of sample is limited in the standard solution, this screen provides rather a "yes" or "no" answer to the solubility. The use of a larger amount of DMSO solution – that means a larger amount of compound in the standard solution – could provide a solubility parameter, which is very far from the true aqueous solubility of the compound, as the co-solvent concentration increases.

7.3.1.1 Conditions that affect solubility

7.3.1.1.1 Effect of the co-solvent concentration

The effect of DMSO in the buffer during the precipitative solubility measurements must be taken into consideration. The above-described buffer solutions contain 2–5% DMSO, which enhances aqueous solubility. Rubino and Yalkowsky [17] investigated the solubility enhancement of co-solvents in aqueous systems for poorly soluble compounds (diazepam, benzocaine, and phenytoin). Equation (7.4) describes the solubility enhancement and Figure 7.2 shows the increase of aqueous solubility by increasing the concentration of DMSO solution:

$$ {}_{\text{total}}S^{\text{cos}} = S + S{*}10^{\sigma\text{u}*\text{fc}} \tag{7.4}$$

where ${}_{\text{total}}S^{\text{cos}}$ is the solubility of a neutral compound with the co-solvent, σu is the solubilizing factor of the co-solvent, and fc is the fraction of co-solvent used. The effect of pH on solubility of the ionized compounds will be discussed later in more detail, and so the pH effect has been omitted at this stage, from eq. (7.4).

Figure 7.2 shows that 1–5% of DMSO does not cause a significant enhancement in aqueous solubility. There are publications for high-throughput measurements of solubility that apply 2% or 5% DMSO solutions [15, 16]. Using eq. (7.5) we can calculate the approximate differences in the measured solubility under these two conditions as follows:

$$ S_{\text{increase}} = 1{*}10^{5*0.05}/1{*}10^{5*0.02} = 1.4 \tag{7.5}$$

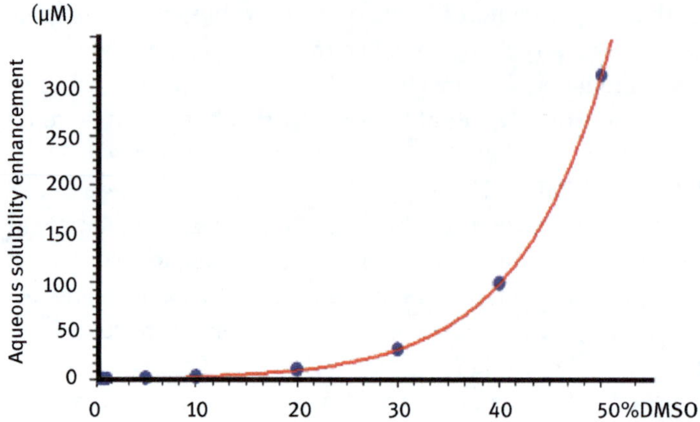

Figure 7.2: Effect of the DMSO concentration on the enhancement of aqueous solubility based on eq. (7.4). (The figure is courtesy of Chris Bevan, GSK.).

This means that we can expect 1.4 times increase in solubility when the DMSO concentration is raised to 5% from 2%. When the DMSO stock solution concentration is fixed (typically 10 mM), which is the case in most pharmaceutical companies, applying the lower percentage DMSO solution during the solubility measurement also means that the maximum amount of compound is less (e.g., instead of 500 μM, the total concentration of the compound is only 200 μM). This means that the upper limit of the solubility measurement is lower. As the detection limits of the analytical methods used are similar, the solubility range of the measurement will be also smaller. Using the DMSO precipitative method to assess the solubility of the compound and the nephelometry or turbidity measurements as detection, a "yes" or "no" answer is obtained to the question: is the compound still in solution under the assay condition? To obtain a quantitative measure of solubility, repeated measurements using a range of dilutions of the sample are needed. The highest concentration when the precipitation is detected provides an estimate for solubility. When more sophisticated detection methods are used that measure the concentration of the compound in solution, it is enough to ensure that precipitation has occurred. Hence, the measured concentration is a result of the equilibrium between the dissolved and solid material.

The upper limit of the solubility measurement depends on the amount of the compound that is added to the buffer. If no precipitation occurs, the dissolved compound is not in equilibrium with the solid. In this case, the measured concentration is not equal to the solubility. It can be considered that the solubility is equal to or above the maximum concentration applied in the buffer. The range in which the solubility can be quantified is between the detection limit (typically about 1 μM when HPLC-UV detection is used) and 200 or 500 μM depending on the amount of compound (DMSO solution) used.

7.3.1.1.2 The effect of equilibration time

Ideally equilibration time should be enough to allow equilibrium to develop between the precipitated solid material and the saturated buffer solution of the compound. This can depend on the compound. When the equilibration time is too short, some compounds might not have enough time to precipitate. Some compounds tend to form super-saturated solutions, which means that the concentration in the buffer maybe temporarily higher than the compound solubility. A little movement of the plate or a tiny dust particle can initiate precipitation and terminate the energetically unstable super-saturation. On the other hand, too long equilibration time (more than 24/48 h) may cause decomposition of the compound, or the precipitated solid material may transform to a less soluble solid-state form.

Figure 7.3 shows a comparison of the measured solubility using different equilibration times, but otherwise under exactly the same conditions [18]. It can be seen from Figure 7.3 that some compounds showed lower solubility after 3 h than after 1 h, indicating that the precipitation process had not reached equilibrium within 1 h.

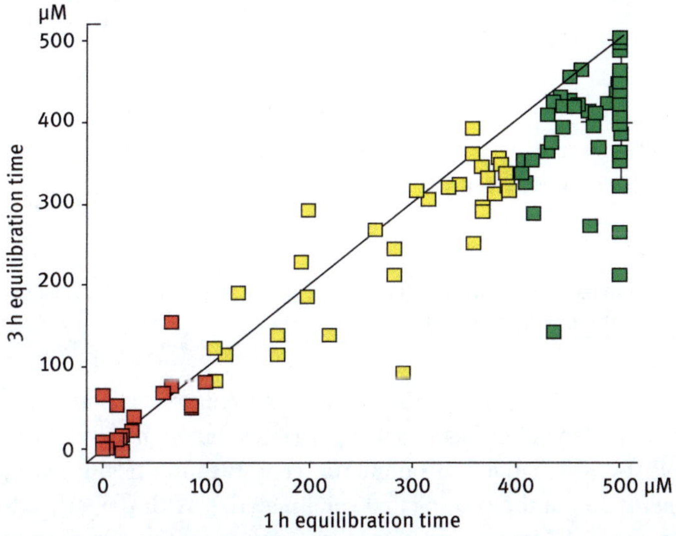

Figure 7.3: Comparison of the measured DMSO solubility of drug discovery compounds using 1 and 3 h equilibration time. (From reference [18]).

7.3.1.1.3 Effect of filtration on the measured precipitative solubility

The nature of the filtration process is also very important. Inert filters need to be used, as adsorption can occur onto the filter (or the 96-well plate wall) that reduces the apparent concentration of the compound in solution. The pore size of the filter may also be important when the precipitation results in very fine particles that can penetrate certain types of filters. Figure 7.4 shows the plot of solubility results obtained with the same set of compounds, applying the same equilibration time and buffer, but altering the filter type [18, 19].

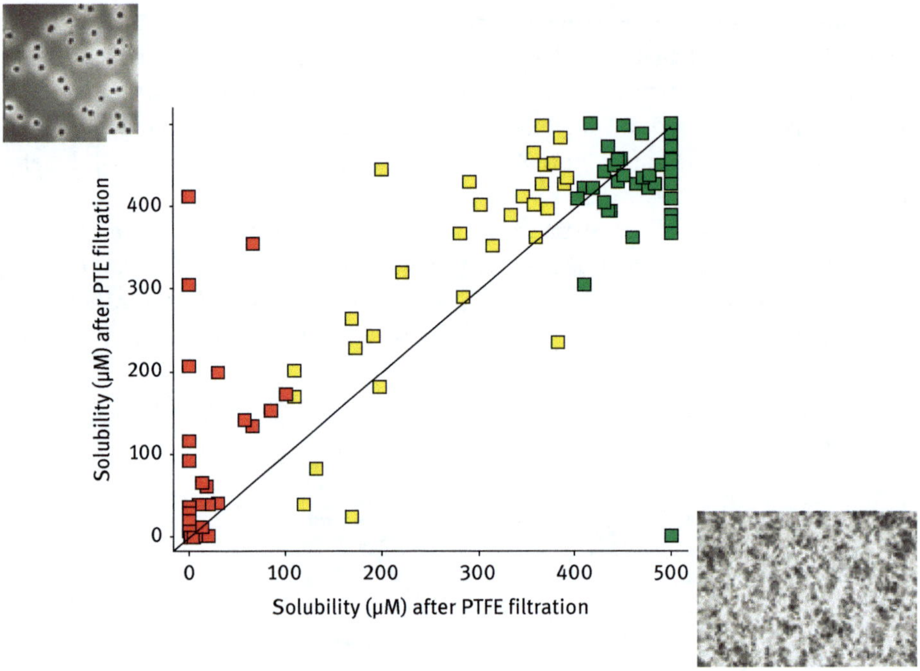

Figure 7.4: Comparison of the obtained precipitative solubility of drug discovery compounds using two different types of filter plates (from reference [18]).

The polycarbonate track-etched (PTE) filter has a precisely controlled pore size, whereas the polytetrafluoroethylene filter has variable pore sizes, but is more resistant to acids, bases, and solvents. Alternatively, to filtration, centrifugation of the plates can result in rapid sedimentation of the precipitated solid material. With properly adjusted needle positioning, the HPLC instrument can inject solely from the supernatant liquid. However, this methodology is difficult to apply in high-throughput situations as some research compounds may have low density, resulting in floating precipitates that will not sediment. Without manual inspection such samples may block the HPLC instrument's injector.

Phase separation using filtration tends to underestimate solubility, as compounds might be adsorbed to the filter. Adsorption might be an issue with filters. Phase separation using centrifugation might overestimate solubility as small undissolved particles might not be separated and dissolve in the eluent of the chromatographic system. The latter introduced a pronounced error especially for low-soluble compounds [20].

7.3.1.1.4 Effect of buffer and ionic strength

The effect of the ionic strength and type of the buffer on solubility has also been investigated on diverse drug discovery compounds from several programs [18]. It was found that the addition of saline to the phosphate buffer (0.9% (w/v) NaCl solutions to match the ionic strength of blood and plasma) altered the solubility of several compounds (Figure 7.5). The explanation for this could be that the solubility of protonated bases is different when a phosphate or a chloride salt is formed.

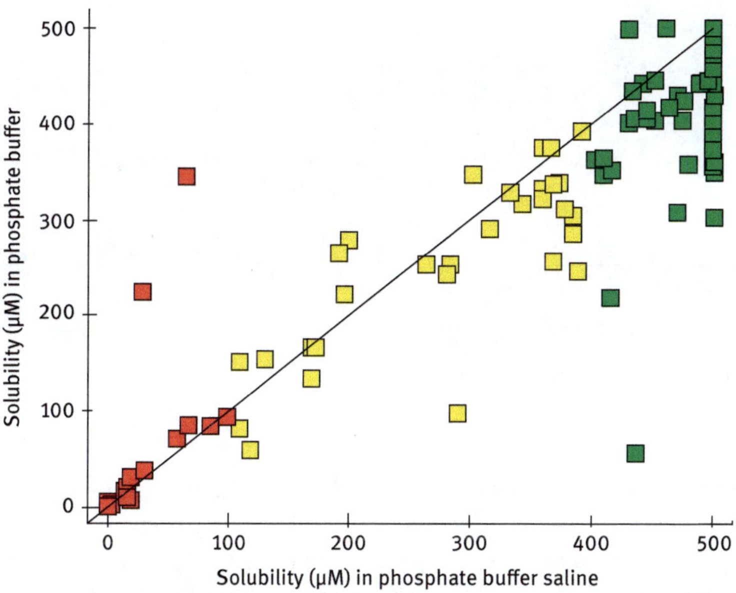

Figure 7.5: Plot of the solubility of drug discovery compounds in phosphate buffer vs phosphate-buffered saline. (From reference [18]).

7.3.1.1.5 Effect of buffer pH on the solubility; solubility–pH profile

The Henderson–Hasselbach equation can be used to calculate the percentage of the ionized form of a compound at a particular pH. Similar to the lipophilicity pH profile, the pH dependence of solubility can be described by the solubility of the neutral form and the solubility of the ionized from as given by eq. (7.6).

$$\log S_{\mathrm{pH}} = \log S_{\mathrm{intrinsic}} + \log S_{\mathrm{ionized}}\left(1 + 10^{\mathrm{ch}(\mathrm{pH} - \mathrm{p}K_\mathrm{a})}\right) \tag{7.6}$$

S_{pH} is the molar solubility of the compound at a particular pH, $S_{\mathrm{intrinsic}}$ and S_{ionized} are the molar solubility of the compound in neutral and fully ionized form, ch is the charge (+1 for monoprotic acids and −1 for monoprotic bases), pH is the negative logarithm of the proton concentration, and $\mathrm{p}K_\mathrm{a}$ is the acid dissociation constant of the compound (the pH where 50% of the ionized and 50% of the non-ionized forms are present). Note that while ionization decreases the lipophilicity of compounds, it increases their aqueous solubility. Figure 7.6 shows a typical solubility–pH profile of a weakly basic compound.

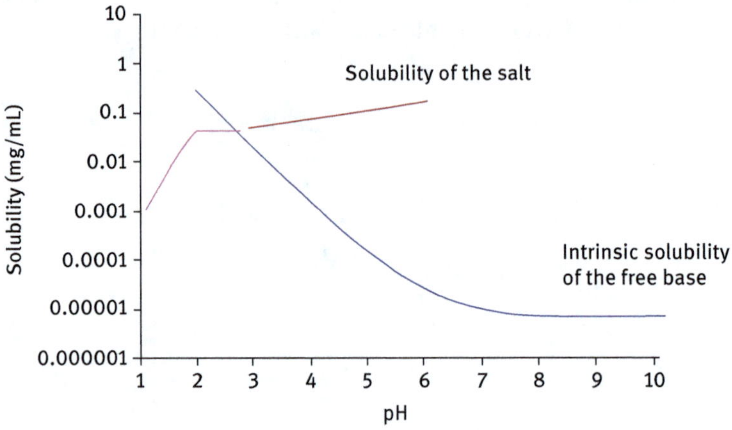

Figure 7.6: pH dependence of the solubility of a basic compound.

The solubility of the fully ionized species depends on the type of buffer, especially on the type of counter-ion in the buffer. The increase in solubility of a base with decreasing pH leads to a point of maximum solubility called pH_{max}. Above this pH the free base represents the solid-state form, which is in equilibrium with the solution. Below this pH the solid-state form changes to a salt of the compound. It is possible that, for example, a phosphate salt of a base can have a very low solubility, while the HCl salt will be very soluble. This would alter the pH-dependent solubility at the lower pH range. For acids a mirror-like behaviour is observed: solubility increases with increasing pH up to pH_{max}. At the early stages of drug discovery, high-throughput screening of solubility is usually carried out only at neutral pH, but it is important to know the pH dependence of solubility for estimating absorption as the solubility must be acceptable within the pH range of the gastrointestinal tract (between pH 1 and 8).

The solubility–pH profile for a compound can be obtained by measuring the solubility at various pHs between 1 and 8, or by measuring the intrinsic solubility and the acid dissociation constant (pK_a) to estimate the lowest solubility value within the pH range of 1 and 8.

The pH dependence of solubility of ionized compounds has been utilized by the recently introduced "chasing equilibrium method" (CheqSol) [21]. This method has the advantage that it is not necessary to wait until equilibrium is reached, as the precipitation is forced by changing the pH. Ionizable compounds turn to less soluble neutral forms. A known amount of solid compound should be accurately weighed and a known quantity of acid (HCl) for basic compounds or base (KOH) for acidic compounds is added until the solid compound is fully dissolved due to salt formation. Titrating the acidic or alkaline solution then started with a known amount and strength of base or acid until precipitation is detected due to the loss of ionization of the compound. The precipitation is detected using a spectroscopic dip probe. From the amount of the alkaline and acid solutions, one can calculate the exact pH when precipitation occurs. The kinetic solubility of the compound is derived from the pH at the point of precipitation. To calculate the equilibrium solubility of the compound, the sample is back titrated with an acid or base, until the compound starts to dissolve. Then additional aliquots of titrant are added to produce additional solid, after which the pH change is measured. The direction of the pH gradient upon precipitation gives an indication of the saturation level of the solution. For example, an unionized base precipitating from a super-saturated solution produces a negative pH gradient, contrary to an unionized base dissolving from a sub-saturated solution that produces a positive pH gradient. These changes of the sign of the pH gradient are used to determine the exact pH at which the solid is in equilibrium with the saturated solution. The concentration of the saturated solution, for example, the intrinsic solubility is calculated from the pH at equilibrium by applying the principles of mass and charge balance [22]. The advantage of this method is that it is not required to wait for 18–48 h to reach equilibrium between the solid and the saturated solution. There is no need to separate the solid material and use an analytical method to determine the concentration of the saturated solution. As the amount of solid used to start the experiment is exactly known, one knows precisely how many moles of acid or base are used to dissolve the compound and precipitate it out again. One can calculate how many moles are in solution at the time of precipitation. Another interesting observation has been made by using the CheqSol method: some compounds form super-saturated solutions very easily ("chasers"), while other compounds instantly precipitate out without super-saturation when the equilibrium has been achieved ("non-chasers") [23]. Box et al. [24] observed that around 5% of acids and 25% of bases are "non-chasers". For example, "non-chasers" are chlorpromazine, imipramine, and nortriptyline. They have relatively large rigid aromatic systems with attached flexible side chain containing hydrophilic basic nitrogen. The "chaser's" kinetic solubility is much higher (it can be four times higher) than their equilibrium solubility. These types of experiments highlight the important differences between the kinetic and the equilibrium solubility and reveal that the kinetic solubility is always higher or equal to the equilibrium solubility. The difference shows the ability of the compound to form super-saturated solutions.

7.3.1.1.6 Effect of the method of quantification on the measured solubility

Various analytical techniques can be applied to quantify the amount of compound in the buffer solution relative to the standard solution. When the UV-plate reader technology is used, the reader will acquire the UV spectrum of the compound in the standard solution and compare this with the absorbance in the sample solution. The "standard" plate should be prepared by diluting the DMSO stock solution with a solvent that will not precipitate out the compound. It should be also less UV active than the, DMSO which has strong UV absorbance around 230 nm. When comparing UV spectra of the same compound from the standard and from the sample solution, which have different solvents, there must be awareness of the possible solvatochromic effect of the solvent, which may shift the UV maximum, or increase or decrease the molar extinction coefficient of the compounds. Avdeef [25] developed a "universal" solvent mixture that dissolves most compounds at 500 μM level and does not have a significant solvatochromic effect on the UV spectra.

Applying HPLC/UPLC with UV quantification has several advantages over the plate reader method. Only 5–10 μL solutions are required from both the "standard" and the "sample" plate for injection. The solvent in the standard will elute at the beginning of the chromatogram. It appears only as a solvent peak separated from the retained compound. UV-active solvents can also be used without influencing the quantification of the compounds. The separation power of the HPLC/UPLC method also makes it possible to analyse compounds that are impure, because the impurities are separated from the compound of interest. With a generic reversed phase gradient method there is no need to develop a specific method for each compound on the plate. By injecting the compound from the standard plate, one can expect to see a peak on the chromatogram with a retention time that is characteristic for the compound. The filtered "sample" is injected after the injection from the "standard" plate. The peaks of the solute appear at the same retention time in the chromatogram of the filtrate "sample" and the "standard" plate. When a peak appears with the same retention time and approximately the same size (peak area), the compound was fully soluble in the buffer. When the peak area is smaller than that of the peak obtained from the "standard" plate, we can calculate the solubility from eq. (7.7).

$$\text{Solubility}(\mu\text{M}) = \frac{\text{Peak area}(\text{Sample})}{\text{Peak area}(\text{Standard})} {}^{*}\text{Conc standard}(\mu\text{M}) \tag{7.7}$$

When there is no peak to be found in the "sample" chromatogram that has the same retention time as the peak in the "standard" sample, the solubility of the compound is less than the detection limit. It is common to observe other peaks in the "sample" chromatogram that are not visible in the standard chromatogram. The explanation for this is that the impurities may be more soluble in the buffer than the compound itself or that the compound might degrade. The relative concentration of the impurities may have increased in the filtered buffer solution.

The HPLC-UV detection method has another advantage: namely when a reversed phase gradient with acetonitrile is used and the mobile phase pH is adjusted to 7.4, the chromatographic hydrophobicity index of the compound can also be obtained from the retention time at the same time as when the solubility result is obtained. This means that the solubility and lipophilicity characteristics of a compound can be obtained from a single measurement. The limitation of the HPLC-UV quantification is that in principle it is based on a "one-point" calibration to calculate the concentration of the "sample" solution. There is also an assumption that the original DMSO stock solution was accurately prepared as a 10 mM solution. In practice, the concentration of the stock solution may deviate from the original value significantly. When the stock solutions go through several freeze-thaw cycles, they can absorb moisture from the air due to the hygroscopic nature of the DMSO. The wet DMSO solution may decrease the compound solubility, resulting in precipitation in the stock solution. When the robot dispenses from the suspension, it either can pick up a crystal, resulting in a higher concentration of the compound in the DMSO stock solution, or just dispense from the moist DMSO solution containing less than the 10 mM of compound.

There are techniques that can determine absolute concentration of the DMSO stock solution, such as the chemiluminescence nitrogen detector (CLND). This measures the nitrogen content of molecules and can be calibrated by any nitrogen-containing compound. The CLND signal is proportional to the number of nitrogen atoms in a molecule. The compound goes through pyrolysis at a high temperature (over 1,000 °C), which oxidizes the nitrogen to nitric oxide. Nitric oxide reacts with ozone to form electronically excited nitrogen dioxide. The excited nitrogen dioxide emits light in the red and near-infrared region of the spectrum (600–3,200 nm) as it relaxes back to its ground state. The light emitted is directly proportional to the amount of nitrogen in the sample. When the detector is calibrated with a known amount of nitrogen-containing compound, the calibration curve obtained can be used to quantify any other nitrogen containing compound. For example, various concentrations of caffeine solutions can be used to calibrate the CLND instrument signal. The CLND methodology can be used for analysing a sample directly without a chromatographic separation process. However, without separation, it will measure the total quantity of nitrogen in the sample, which may come from impurities or nitrogen-containing solvent residues, such as acetonitrile. When the CLND is connected to an HPLC system, it is essential to ensure that the mobile phase does not contain nitrogen (especially do not use ammonium acetate as buffer, or acetonitrile as organic modifier). Methanol, ethanol, and propanol can be used as organic solvent additives in the mobile phase. The application of the CLND technology for solubility determination has been discussed by Bhattachar et al. in more detail [26]. When the compound contains adjacent nitrogen atoms, the additivity of the signal decreases as the nitric oxide mole for mole proportional formation from both nitrogen atoms cannot be assured. The CLND method could also be very useful to measure the absolute concentration of a pure standard solution made from solid samples without accurate weighing. The weighing procedure is still carried out manually even in large pharmaceutical companies as the accurate handling of solid samples has not yet been reliably automated. Because of the high sensitivity of the CLND detector for nitrogen, even small amounts of ammonia in the air or other nitrogen containing contaminants can distort the results. Special care and handling should always be applied. Fortunately, the nitrogen in the air does not react to give a signal.

In conclusion, the precipitative solubility measurements in early drug discovery are useful for detecting compound precipitation under screening conditions, and may provide an approximate value of solubility, albeit in a relatively narrow range. The amount of DMSO, the equilibration time, the buffer composition, the filtration or centrifugation step, and other experimental conditions can alter the results. The data cannot be regarded as a precise thermodynamic constant for solubility as it refers only to the solubility at a particular pH, at a particular time, without any description of the precipitated solid form.

7.3.2 Solubility measurements from solid

Wenlock et al. [27] described an excellent approach to automate the aqueous solubility measurements from solid using the shake-flask equilibrium principle. They used a liquid handling robot equipped with constituent parts for transportation, decapping and recapping and centrifugation of the sample tubes. The principle of the method is very similar to that which has been described for the solubility determination from DMSO stock solution. The robot prepares several dilutions from 1 mg solid material with DMSO that serve as standard solutions. Another 1 mg solid compound under test is dissolved in phosphate buffer. The equilibration time used is a minimum of 18 h. The tubes then are centrifuged, and several dilutions are prepared by the robot automatically from the supernatant. Each dilution is analysed by HPLC-UV with identification of the compound by mass spectrometry (MS). The solubility range by this method is up to 1.33 mg/mL, which is equivalent to 1–2,000 μM depending on the compound's molecular weight.

Another semi-automated procedure has been described by Alsenz and Kansy [28]. The solid material is dissolved in hexane, a non-polar solvent. Then a robotic system dispenses the hexane solutions to a 96-well plate. After evaporating the hexane, a known amount of solid is present in each well. As the next step, the buffer solutions can be added to the solid material. After and appropriate equilibration time, the undissolved solid particles are filtered from the solution. The concentration of the compound that remained in solution in the buffer is then determined by HPLC.

It is interesting to investigate the differences between the solubility values obtained from solid material and those obtained from DMSO stock solution, when pH, equilibration time and other significant conditions, are constant. As expected the DMSO precipitative solubility should be higher as the DMSO content in the buffer enhances the solubility. There could be a significant difference in the solid material's crystallinity when freshly precipitated from solution or prepared as solid material. The DMSO precipitates are more likely to be amorphous. This would also make it more likely to measure higher solubility for the same compound when it is measured from DMSO stock solutions than from solid. Figure 7.7 shows the comparison of the solubility data for 160 compounds obtained from solid and from DMSO stock solutions using the same pH 7.4 phosphate buffers [18]. The compound set contained a diverse set of known drugs and research compounds. For the precipitative solubility measurements 10 mM DMSO stock solution was used and diluted down with phosphate buffer saline (pH 7.4) to obtain 5% DMSO in the final solution. As expected, the precipitative solubility was equal or higher than the solubility obtained from solid for the same compounds using the same filters, buffers and analytical method (HPLC-UV) for the concentration determination.

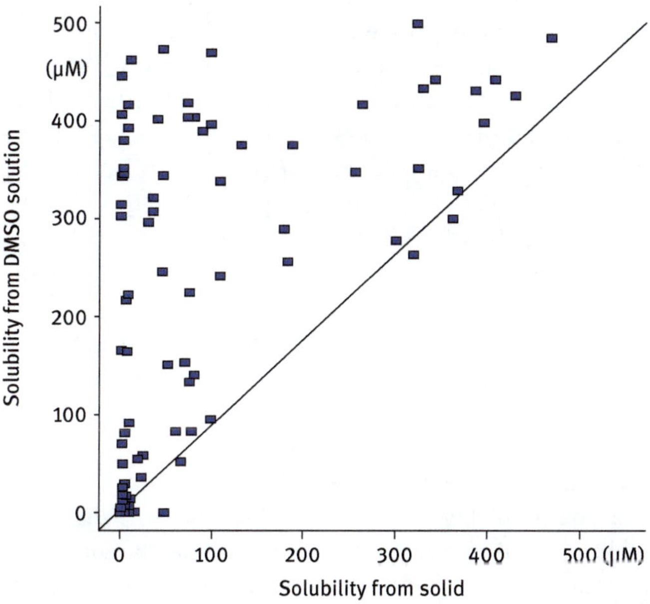

Figure 7.7: Comparison of the measured solubility values of a diverse set of 160 compounds obtained by the DMSO precipitative method and by using a solid form of the compounds.

A proper thermodynamic aqueous intrinsic solubility (S_0) measurement should be carried out with a buffer where the pH is adjusted to ensure the compound is completely unionized. Sufficient amounts of well-characterized solid material should be added to the buffer and a sufficient amount of time then given to progress to equilibrium. The concentration of the saturated solution can be checked every 3–6 h. When it is constant, equilibrium had been established. An appropriate amount of clear supernatant solution – preferably to avoid any application of filtration – can be diluted and subjected to some form of quantitative analysis, using a UV spectrophotometer or an HPLC-UV/MS system. The pH of the saturated solution should be checked at the end of the equilibration process to ensure the compound is still in its fully unionized state. The proper characterization of the solid form in equilibrium with the saturated solution should be carried out to ensure that the solubility value relates to a well-defined solid-state form of the compound. The temperature should be controlled for the whole equilibration process. Applying smaller particle size solid material and stirring may accelerate the equilibrium process but should not affect the results [29].

Variables such as temperature, equilibration time, type of buffer and phase separation techniques may alter the solubility results, while the excess amount of solid material should not cause any deviations. Baka et al. [29] suggested a minimum of 24 h equilibration time that consisted of 6 h stirring followed by 18 h of sedimentation. They claimed that when these conditions are strictly controlled, the experimental error of the solubility measurement should be less than ±4%.

It is much more difficult to design an experiment for solubility measurements to answer the second question, namely, whether the required amount of drug will dissolve in the gastrointestinal tract within a certain time frame. At early stages of the drug discovery process, the required dose is not known, although it is usual for the program team to have a target dose and a target potency before they start lead optimization. For example, let us say a 100 mg dose of the compound is required to be dissolved in the gastrointestinal tract. This 100 mg of compound will have to be dissolved in 250 mL of aqueous medium based on the definition of good solubility in the Biopharmaceutical Classification System (BCS), as also discussed in Chapter 11 [1]. The pH change in the gastrointestinal system also has to be taken into account, as it changes from acidic (pH 1–2) in the stomach to basic (pH 8) in the intestine. Solubility of ionizable compounds will vary depending on the pH. Therefore, the pH dependence of solubility should also be measured or at least understood.

To answer the question of whether the solubility can be enhanced at a later stage by formulation, the causes of poor solubility should first be investigated. When poor solubility is due to the lipophilicity of the compound, there is a hope that the compound will dissolve in bio-relevant media as it can partition into phospholipid and taurocholate micelles having a reasonably large lipophilic environment. When poor solubility is due to high crystal lattice energy, only expensive formulation studies may help. Micronization, including polymer coating of the drastically reduced size of the crystals, may improve the dissolution rate and solubility just enough to get the formulated drug absorbed. These aspects have been discussed in Chapters 2 and 8 of this book.

7.4 Structure–solubility relationships; designing soluble compounds

As seen from the general solubility equation (GSE) as introduced by Yalkowsky and coworkers [30] and shown by eq. (7.1), decreasing the lipophilicity and the melting point of compounds will enhance the aqueous solubility. The melting point is related to the crystal lattice energy. A lower melting point generally is a consequence of reduced lattice energy, making it easier for solvents to break down the crystal lattice, which leads to higher solubility. The two ways to increase solubility represent increased interaction of the molecule with water in the solution or decreased interaction with itself in the crystal lattice. Sometimes other components also come into play, such as in pharmaceutical salts, hydrates, solvates and co-crystals. In these cases, the crystal lattice energy is not only defined by interactions of the molecule with other molecules of the same species but also with other components like counterions in pharmaceutical salts, water in hydrates, solvents in solvates and co-crystal formers in co-crystals. In the same way, interaction between the solute and solvent can relate not only to water but also to other components in solution if co-solvents or solvent mixtures are used. Respective solid-state forms and their effect on solubility are discussed in Chapter 9.

The melting point is supposed to account for the solid-state properties of the compounds. Before the molecules are solvated from the solid form by the solvent molecules, the intermolecular bonds in the crystal must be broken. Therefore, the solubility not only depends on the chemical structure of the molecules and its interactions with the solvent molecules but also on the crystal lattice energy in the solid-state form. The validity of the GSE equation has been tested on large sets of simple organic molecules, herbicides and pesticides [31, 32]. This semi-empirical equation works very well, but it requires the measurement of melting points. While in the past the melting point was used as an indicator of purity and identity of a newly synthesized compound, it is not now measured routinely by medicinal chemists as it would decrease the throughput of the discovery chemistry efforts. Discovery compounds are synthesized in small quantities, often purified by HPLC, UPLC or SFC and recovered from the mobile phase in an amorphous form, which does not allow the determination of the melting point. The estimation of melting point or the enthalpy required for melting is a difficult task [33, 34]. Wassvik et al. [35] have investigated the contribution of the solid-state properties to the aqueous solubility on a set of carefully selected drug molecules. They have tried to develop a solubility prediction model for drug molecules by considering the properties of the solid-state form, such as the melting point (T_m) and the enthalpy (ΔH) and entropy (ΔS) contributions of melting. The intrinsic solubility (S_0) of the selected drug molecules showed a very poor inverse correlation with their lipophilicity ($\log P$). Therefore, it was an ideal set to use to investigate the contribution of the solid-state properties in addition to the lipophilicity of the drugs. The data analysis revealed that the solid-state form of the compounds has significantly influenced the intrinsic solubility and explained the majority of the deviations from the solubility–lipophilicity relationship. The authors suggested to include an entropy term in the GSE as shown by eq. (7.8), which is closely related to eq. (7.1):

$$\log S_0 = 0.5 - \frac{\Delta S_m}{5705.85}(T_m - 25) - \log P \tag{7.8}$$

Equation (7.8) shows an inverse correlation between log P and solubility with a slope of −1 and an intercept of 0.5 that provides an estimate of the maximum solubility expected from a compound without considering the effect of the crystal lattice energy.

To design soluble drugs, medicinal chemists must rely on a few empirical observations of structural motifs that might contribute to higher crystal lattice energy besides the lipophilicity of the molecules. One of these empirical rules can be derived from the general solvation equation to describe intrinsic aqueous solubility derived by Abraham and Le [9] as shown in eq. (7.3). They added the product term of H-bond acidity and H-bond basicity to the solubility equation for the prediction of possible intermolecular H-bond formation in the crystals. The equation suggests that the presence of H-bond donor and acceptor groups in the molecule potentially decreases the solubility by three log units, as the coefficient of the product term is −3.36.

The influence of simultaneous presence of the H-bond donor and acceptor groups in molecules on the melting points has been demonstrated by the matched molecular pair analysis by Schultes et al. [36]. An algorithm described by Hussain and Rea [37] can be used to assess the effect of substituents on solubility. The matched molecular pair analysis can be carried out on large numbers of molecules with measured data. Their computer program pulls pairs of molecules and their measured data together, which differ only by a specified functional group. The average effect of a substituent on solubility is calculated. This powerful technique can be used to identify and quantify the effect of molecular changes on any measured property when large amounts of measured data are available. The changes are usually averaged, and the calculation is carried out to test whether the substituent significantly contributes to the change in properties. This technique provides a quantitative assignment of the group contributions to a certain property. The work by Schultes et al. [36] analysed the functional group contributions to the measured melting points of 746 molecules of free bases. Pharmaceutical salts were excluded from the analysis as different pharmaceutical salts of the same compound usually show different melting points. Simple calculated properties were assigned for the structural changes such as number of H-bond donors, number of H-bond acceptors, total polar surface area, the number of rotatable bonds, clog P, and the number of halogen atoms. H-bond donor and H-bond acceptor groups increased the melting point on average by 48 °C and 36 °C, respectively. This is probably due to the increased crystal lattice energy caused by the possible intermolecular H-bond interactions between the donor and acceptor groups. Increasing the number of rotatable bonds showed a significant decrease in the melting point. This can be explained by the hypothesis of Dannenfelser and Yalkowsky [33] that the rotatable bonds provide a higher flexibility of the molecule resulting in a higher melting entropy, as discussed in the next section of this chapter. The matched molecular pair analysis also revealed that halogen atoms significantly influence the melting point of compounds. Bromine and iodine increase the melting point, while the presence of fluorine and chlorine causes a slight decrease of the melting point. The effect of the hydroxyl group on solubility is quite complex. In general, the addition of a hydroxyl group reduces the lipophilicity by around a log unit, and so the solubility increases. On the other hand, the hydroxyl group might form hydrogen bonds in the crystal that reduces solubility. Zhang et al. [38] observed in a study of intrinsic solubility of 2,974 compounds that an addition of a hydroxyl group in general increased the solubility.

Another interesting observation that can help medicinal chemists to design soluble potent compounds is the effect of aromatic rings on the general ADME properties of molecules. Lovering et al. [39] noticed that saturation of aromatic rings generally improved the clinical success of drug candidates. They also observed that saturation and the introduction of a chiral centre in general provided higher probability of a molecule becoming a drug, but that the number of aromatic carbons significantly reduces the solubility. In addition, for these aspects specific examples are provided later in this chapter. Ritchie and Macdonald [40] have analysed the effect of the number of aromatic rings on solubility among other developability characteristics: more than three aromatic rings in the molecule usually result in low solubility and developability. Solubility data in pH 7.4 phosphate buffer were measured for over 30,000 research molecules in GSK. Most of the molecules had 3–4 aromatic rings and while the average solubility of compounds with only two aromatic rings was around 100 µg/mL, the average solubility of compounds with four aromatic rings dropped to 25 µg/mL. It is very likely that if they had expressed the solubility in molar term, the drop would have been even more pronounced. Later Ritchie et al. [41] investigated the effect of hetero-atoms in aromatic rings and fused aromatic rings: the hetero-aromatic rings showed less of a detrimental effect on solubility. Table 7.1 lists the ranking of various hetero-aromatic rings on solubility, with the higher number representing a greater enhancement of solubility.

Hill and Young [42] reported an interesting observation by analysing the measured solubility data of over 100,000 research compounds at GSK. They found that in addition to the decrease of solubility due to the lipophilicity of the compounds, the number of aromatic rings caused a further decrease in the measured solubility. Compounds with the same calculated log *D* (referring to the octanol/water distribution coefficient at pH 7.4) were less likely to be soluble if they had more aromatic rings. They introduced "solubility forecast index", abbreviated as SFI, which is the sum of the log *D* value and the number of aromatic rings in the molecules. When the SFI value was around 6, 50% of the compounds had less than 30 µM solubility. Every unit increase in SFI decreased the probability of good solubility (>30 µM) by approximately 10%. Thus, over 90% of the investigated compounds with an SFI value of over 12 had lower than 30 µM solubility, while over 90% of compounds had solubility above 30 µM when their SFI value was below 4. The explanation of the observation is that multiple aromatic rings in the molecule provide a planar structure that can "stack" in the crystals, increasing their melting points and crystal lattice energy. Based on the empirical observations described earlier, the number of aromatic rings and the presence of both H-bond donor and acceptor groups can decrease the solubility above the contribution of lipophilicity. Specific examples will be discussed in the next part of this chapter.

Table 7.1: The ranking of the hetero-aromatic rings based on their contribution to solubility [41]. The higher the score, the better is the ring in terms of increasing solubility.

Hetero-aromatic ring	Prevalence in the set	Solubility score
Pyrazine	1.90%	22
Pyridine	24.40%	21
1,3,4-Oxadiazole	1.30%	20
Tetrazole	0.80%	19
Pyrazole	10.80%	18
Pyridazine	1.80%	17
Furan	2.50%	16
1,2,4-Triazole	2.20%	15
Imidazole	8.10%	14
Oxazole	2.00%	11
1,2,4-Oxadiazole	1.50%	10
Pyrrole	7.90%	9
1,2,3-Thiadiazole	0.20%	8
Isoxazole	1.90%	7
1,2,3-Triazole	0.90%	6
1,3,4-Thiadiazole	2.10%	5
Thiophene	5.60%	4
Pyrimidine	15%	3
Thiazole	6.90%	2
1,3,5-Triazine	2.20%	1

The structure–solubility relationships discussed above are related to the intrinsic aqueous solubility of compounds. Ionization has a significant impact on solubility. Compounds that have ionizable groups can have increased solubility at pH where the compounds are charged. The increase in solubility can be 3–4 orders of magnitude higher at a pH where the compound is fully ionized. It is difficult to predict the solubility of the fully ionized form because the counter ion may affect the crystal lattice energy and may reduce the solubility. The effect of ionization is considered both in the calculated lipophilicity and the solubility measurements. One of the most popular approaches to increase solubility is the introduction of ionizable groups in the molecule. This makes the molecule partially ionized in the pH range of the gastrointestinal tract. The introduction of a positive or negative charge may significantly increase the solubility, but there are other serious consequences from the charge. Negatively charged compounds usually have impaired permeability. The presence of charge also significantly influences the in vivo distribution of the compounds between the plasma and tissue compartments. As the pH in the gastrointestinal tract ranges from 2 to 8, there is a possibility of precipitation during the transit from the stomach to the duodenum. The safest way to increase solubility of compounds is to increase the intrinsic aqueous solubility, i.e., the solubility of the neutral form. There are some favoured groups that chemists introduce for solubility enhancement, such as the morpholine ring or the recently introduced oxetane fragment [43].

There are several approaches published for the prediction of solubility from chemical structure [44–48], but in general solubility is more difficult to predict than other properties of molecules as it is not only influenced by the molecular structure but also by the crystal structure, which is still difficult to predict. When the solubility prediction provided acceptable results in terms of statistics, it was usually carried out on large sets of compounds covering a 6–10 order of magnitude solubility range. However, in the range of solubility of importance in pharmaceutical research and medicinal chemistry, normally between 10 and 1,000 μM, the 1–2 log unit error of prediction is too large to be able to rank compounds and to make decisions about their solubility for acceptable absorption. There have also been concerns about the quality of the measured solubility data, which makes model development difficult. This is mainly due to the lack of characterization of the solid-state form or changes in the conditions of the solubility measurements. Many compounds exist as amorphous material or in various crystal forms, of which the solubility may differ by several orders of magnitude. In 2008 Llinàs et al. [49] published a manuscript inviting scientists to a "solubility challenge". The intrinsic solubility data of 100 compounds was determined by the CheqSol method, which was discussed earlier [21]. The ionic strength, the temperature, and the composition of the buffer were provided. All the compounds' solubility was measured under identical conditions. Modellers were invited to predict the solubility of 32 molecules for which data was withheld by the authors. Hewitt et al. [50] then published the results of the solubility predictions for the challenge, comparing various methods, such as multiple regression analysis using various molecular descriptors and artificial neural networks. The root mean square errors from the prediction by several commercially available in silico programs (ChemSilico, Optibrium, Pharma Algorithms, Simulation Plus) [50] (and references therein) were compared. The errors ranged from 0.87 to 1.56 logarithmic molar solubility units. Authors suggested dividing the training set of compounds based on structural similarity and developed local models for the subsets that worked slightly better. It was concluded that a simple linear regression approach often was superior to more complex modelling methods. A more in-depth discussion of the modelling approaches used in the "solubility challenge" is given in Chapter 3.

Chemists intuition in designing compounds with better aqueous solubility is also important and may be reliable and even more reliable when supported by experience.

7.5 Examples to increase solubility by structural modification

Methods to improve solubility based on the molecular structure include the addition of solubilizing, polar groups to a molecule, increasing the fraction of atoms with sp^3-hybridization, disrupting planarity, and hence decreasing crystal lattice energy. In the following part, some examples are shown where the solubility in research projects has been increased by one or more of these approaches. As a first example described by Huang et al. [51] the solubility of non-nucleoside reverse transcriptase inhibitors was increased starting from the approved drug Etavirine (**1**) by increasing the fraction of sp^3-hybridized atoms and breaking planarity. Respective structures are depicted in Figure 7.8. Additionally, basic nitrogen has been introduced in the structure. Even if the molar mass of the compound increases from 435 g/mol for Etavirine (1) to 521 g/mol for (**2**), the solubility increases by a factor of more than 10. Potency decreased from an EC50 of 1.8 nM for Etavirine to 8.1 nm for (**2**). However, at the same time log *P* decreased from >5 for (**1**) to 1.9 for (**2**). This is an example where by structural modification a compound has been moved from out of the "Lipinski rule of 5" space and into it by respective changes of physicochemical properties.

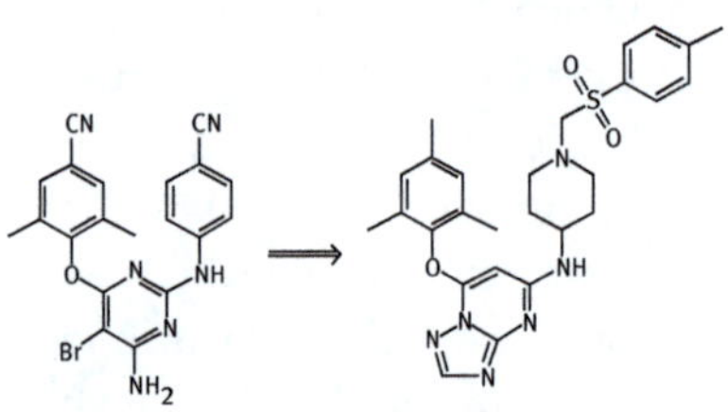

(**1**) (Etavirine)	(**2**)
M=435 g/mol	*M*=521 g/mol
S(pH 7)<<1μg/mL	*S*(pH 7) = 10.31 μg/mL
log *P* >5	log *P* = 1.9
EC50 =1.8 nM	EC50 =8.1 nM

Figure 7.8: Modification of physico-chemical parameters by reduction of planarity and the introduction of basicity to Etavirine.

As a second example where increasing the fraction of sp^3-hybridized atoms was also used as a tool to increase solubility, Press et al. [52] described the optimization of a PDE4 inhibitor for the treatment of inflammatory diseases such as asthma and chronic obstructive pulmonary disease (COPD). Structures of the starting compound and the optimized compound are shown Figure 7.9. The lead compound (**3**) had physicochemical properties, which did not allow further development of this molecule. Solubility was just 2.3 μg/mL at pH 6.8. To increase solubility the planar benzoic acid moiety was replaced by a cyclohexane–carboxylic acid moiety, which increased both the number and the fraction of sp^3-hybridized atoms and decreased the planarity of the molecule. Additionally, the number of aromatic ring systems was further reduced

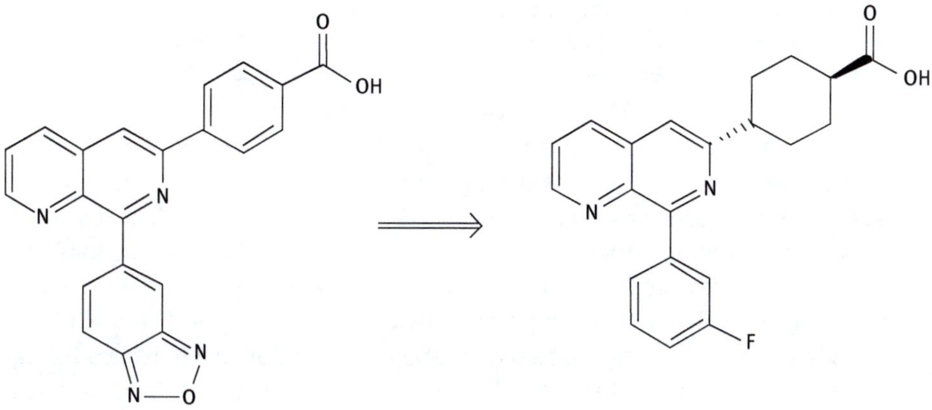

(3)

S(pH 6.8) = 2.3 µg/mL

IC50 = 1.5 nM

(4)

S(pH 6.8) = 920 µg/mL

IC50 =32 nM

Figure 7.9: Optimization of solubility of a PDE4 inhibitor by reduction of planarity.

by replacing the benzoxadiazole system with a fluorophenyl moiety. This structural modification leads to a decrease of π–π interactions and reduces lattice energy. The resulting molecule (**4**) exhibits a solubility of 920 µg/mL, which has increased solubility by about a factor of 400 compared to the starting molecule (**3**). As in the previous example, the structural modifications that increased solubility lead at the same time to a decrease in potency from an IC50 of 1.5 nm for (**3**) to 32 nm for (**4**). With regard to the developability of the molecule, this decrease in potency was more than compensated for by the increase in solubility that led, in total, to a more favourable overall profile of the compound.

In contrast to the previous two examples, the work by Tehler et al. [53] represents an example where lipophilicity was increased, based on structural modifications to increase solubility. Here the goal was to increase solubility and permeability of Ciprofloxacin (**5**), an antibiotic used against bacterial infections, which had been developed by Bayer in the 1980s. According to the BCS scheme, Ciprofloxacin belongs to class IV, suffering from low solubility and low permeability. Ciprofloxacin represents a case where solubility is limited by strong lattice energy. This strong intermolecular interaction is generated by the planar rings in the molecule. At the same time, polar groups such as the piperazine and the carboxylic acid not only contribute to lattice energy, but also contribute to increased hydration energy. Tehler et al. [53] used esterification to introduce flexible aliphatic chains into the molecule to disrupt the crystal lattice. Respective structures are shown in Figure 7.10. With this structural modification, ionic interactions in the crystal lattice were reduced, as in contrast to the original compound no acidic groups are maintained in the resulting, optimized molecules. Derivatization of Ciprofloxacin to its methyl-ester (**6**) leads to a 50-fold increase in solubility. A still more bulky derivate, the *n*-butyl-ester (**7**), leads to a 10-fold increase in solubility. This behaviour reflects the influence of crystal-lattice energy and hydration energy. Obviously for the *n*-butyl-ester, an increase of lipophilicity over-compensates for the reduction of lattice energy, and this results in a decreased solubility compared to the methyl ester. From this example it becomes obvious that compound optimization must be carried out in a balanced way to obtain the best results. Optimizing a BCS class IV compound can be done in two ways: either by increasing solubility or improving permeability. The first approach will lead to a BCS class III compound, whereas the second approach will lead to a BCS class II compound. Only if both approaches are followed in a balanced way, a BCS class I compound will be the result of structural optimization. This example represents a prodrug approach, as esters will be metabolized to Ciprofloxacin in vivo.

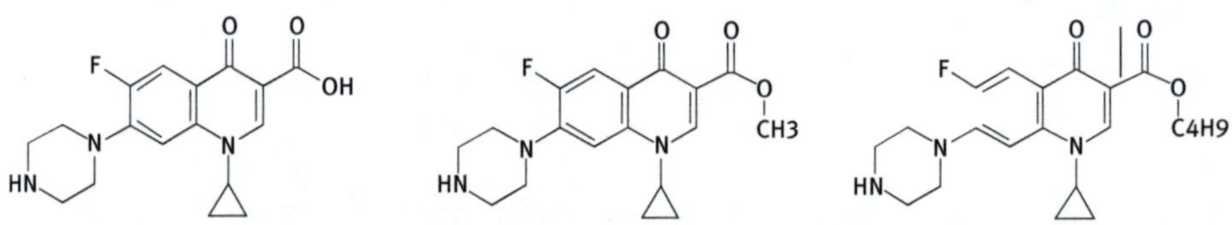

(5) (Ciprofloxazin)	(6)	(7)
Increase in solubility compared to (5)		
1	50	10
BCS classification		
IV	I	II
log*D* 7.4		
−0.9	0.6	2.8
Melting point		
266 °C	255 °C	200 °C

Figure 7.10: Optimization of physicochemical properties of Ciprofloxacin by the introduction of flexible chains by esterification.

Esterification leads to increased flexibility of the molecule and a reduction of planarity. This is reflected in the melting point that decreases from Ciprofloxacin to its methyl-ester and further to its *n*-butyl-ester. At the same time log *D* increases. The solubility of the methyl-ester is highest because the lipophilicity of the *n*-butyl-ester over-compensates for the reduction in the planarity of Ciprofloxacin and adds no further benefit with regard to avoiding Zwitterionic structures in the crystal lattice. The ability to form zwitterions or strong hydrogen bridges between the piperazine moiety and the carboxylic acid group is already eliminated by the methyl-ester.

As already discussed earlier, solubility is governed by solvation energy and lattice energy. The last example has already shown certain effects of the molecular structure on lattice energy and solubility. This effect becomes still more important if molecules contain a large fraction of aromatic ring systems, especially condensed aromatic rings, which lead to larger planar structural parts and strong π–π interactions. However, there are not many examples available in the literature, which describe optimization of solubility based on the crystal structure of research compounds. The number of publications dealing with this topic is relatively low compared to the much larger number of publications describing optimization of solubility based on decreasing lipophilicity. Certainly, one reason for this is that log *P* can be measured easily and can even be calculated by in silico methods. However, still even today melting points are difficult to predict, as they rely on crystal structures [54] and accordingly, to optimize solubility based on lattice energy, an understanding of the crystal structure of a compound is helpful.

However, there are general principles that can be applied to modify research compounds to increase solubility based on lattice energy, which can be applied without the knowledge of crystal structures. The most useful principle is that the disruption of molecular planarity will decrease the intermolecular interactions within the crystal lattice, lower the melting point, and increase solubility. This is also reflected in the high solubility of amorphous phases compared to their crystalline counterparts. Ordered planar structures with strong interactions (π–π interactions in stacks) are typically absent in amorphous phases due to the loss of symmetry.

Generally, the geometry – including the planarity of the molecule – will be influenced by the hybridization of the involved atom. As sp^2-hybridized C-atoms lead to planar structures and sp^3-hybridized C-atoms will have their neighbours in the three-dimensional environment, an increased fraction of sp^3-hybridized C-atoms should lead to a decrease in lattice energy. A respective analysis that confirmed this hypothesis can be found in [39].

How the reduction in the number and fraction of sp^2-hybridized C-atoms increases solubility can be seen from the example shown in Figure 7.11 published by Wang et al. [55]. AMG517 (**8**) is a vanilloid receptor 1 antagonist suffering from low solubility, which is less than 1 µg/mL in 0.01 M HCl. The molecule was moved into clinical development as a co-crystal and represented the first case, where a co-crystal was used for clinical development. However, for this compound solubility could also be optimized based on structural modifications leading to less planar molecules. If the trifluoromethyl-substituted phenyl ring is replaced by a trifluoromethyl-substituted cyclohexene ring, the solubility in the same solvent increases to 13 µg/mL. This can be ascribed not only to breaking the planarity in the molecule, but also to decreasing the lipophilicity with the *c*log *P* decreasing from 4.6 to 3.7. From the melting point of the compounds, which decreases from 219–221 °C to 130–131 °C, the significant contribution of crystal lattice energy to the increased solubility becomes obvious.

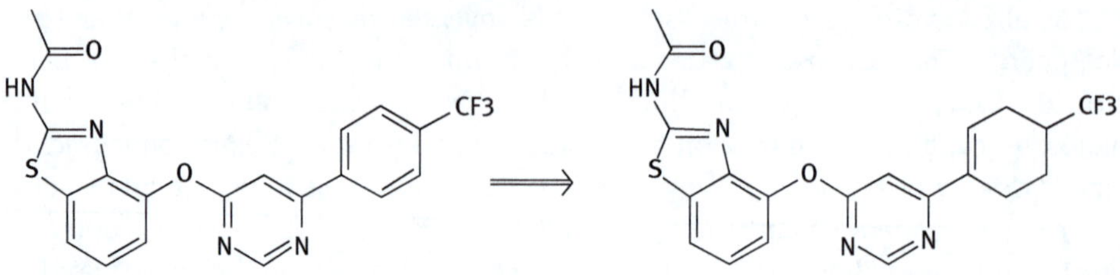

(**8**) AMG517
$S(0.01\ M\ HCl) < 1\ \mu g/mL$
$c \log P = 4.6$
Melting point = 219–221 °C

(**9**)
$S(0.01\ M\ HCl) = 13\ \mu g/mL$
$c \log P = 3.7$
Melting point = 130–131 °C

Figure 7.11: Increase of solubility by reduction of planarity starting from the vanilloid receptor 1 antagonist AMG517 (8). Replacement of the trifluoro-substituted phenyl ring by a cyclohexene ring leads to a significant increase in solubility, which is due to reduced lattice energy and reduced lipophilicity.

There are still additional ways to reduce planarity beyond reducing the fraction of sp^2-hybridized C-atoms. As an early example outside the pharmaceutical world, the solubility of polychlorinated biphenyls (PCBs) was studied in the 1970s. PCBs were used in electrical devices, hydraulics, and as plasticizers, but were later banned because of their toxicological behaviour. PCBs exhibit increased solubility if the dihedral angle between the two phenyl rings is increased. This is the case in ortho-substituted biphenyls: 2,2′-dichlorobiphenyl shows a higher solubility in water compared to 4-chlorobiphenyl and 2,4′-dichlorobiphenyl as is shown from the work of Hoover [56] and Lee [57] in Table 7.2.

Table 7.2: The solubility, dihedral angle, log *P* and melting point of some polychlorinated biphenyls.

Compound	Solubility	Dihedral angle	log *P*	Melting point
2,2′-Dichlorobiphenyl	900 µg/mL	69.17°	4.97	62 °C
2,4′-Dichlorobiphenyl	637 µg/mL	47.71°	5.09	44 °C
4-Chlorobiphenyl	400 µg/mL	42.48°	4.55	77 °C

From this data it can be seen how the dihedral angle influences the melting point. This is highest for 4-chlorobiphenyl, which has the lowest dihedral angle. Even if 4-chlorobiphenyl has the lowest log *P*, its solubility is lower compared to that of the dichlorinated PCBs. This can be ascribed to the more planar structure with increased lattice energy. A similar trend is seen when comparing the density of 2,2′-dichlorobiphenyl (1.386 g/cm^3) and 2,4′-dichlorobiphenyl (1.445 g/cm^3): the first compound has a lower density also reflecting a lower lattice energy and accordingly a higher solubility.

A dual integrin antagonist described by Ishikawa et al. [58] represents an example of how solubility can be enhanced for research compounds using the approach of an increasing dihedral angle to reduce planarity. Compound (**10**) – as depicted in Figure 7.12 – exhibited a solubility of <0.1 mg/mL and was the starting point for the optimization of solubility. Attempts to improve the solubility by introducing hydrophilic groups lead to a decreased activity in cellular assays. However, structural modifications introducing hydrophobic substituents lead to increased potency in receptor binding assays and other cellular assays in parallel with an increase of solubility. As solubility increases from <0.1 to 0.6 mg/mL for (**11**) and finally to 1.3 mg/mL for (**12**), the melting point of the compounds decreases from 252–254 °C to 182–184 °C and finally to 162–164 °C. This indicates a decrease in lattice energy, which is caused by the bulkier substituents that disrupt molecular planarity.

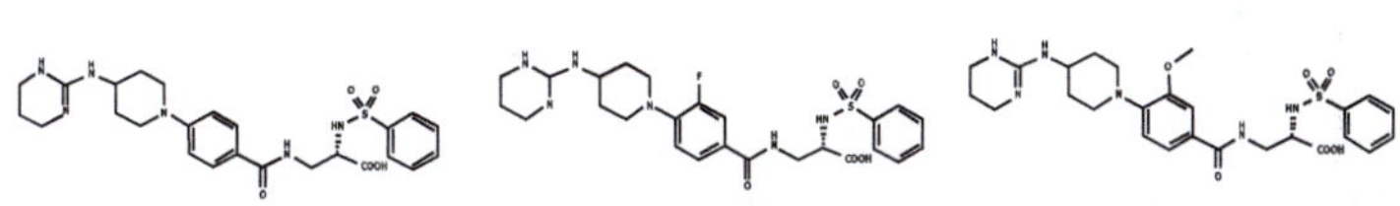

	(**10**)	(**11**)	(**12**)
Solubility (water)	<0.1 mg/mL	0.6 mg/mL	1.3 mg/mL
*C*log *P*	1.5	1.6	1.3
Melting point	252–254 °C	182–184 °C	162–164 °C

Figure 7.12: Optimization of the solubility of integrin $\alpha_v\beta_3/\alpha_{IIb}\beta_3$ dual antagonist by the increasing dihedral angle based on the substitution in the ortho-position of phenyl ring.

Another example for optimization of solubility for research compounds is given by Bachovchin et al. [59]. In this contribution, Lapatinib, which represents an approved epidermal growth factor receptor inhibitor, was used as a starting point for optimization of compound against Trypanosoma brucei that causes human African trypanosomiasis. First activities yielded the compound (**13**), which suffered from poor aqueous solubility. Several strategies such as increasing the fraction of sp^3-hybridized C-atoms, introduction of ionizable groups, ortho-methylation, NH-insertion and formation of pharmaceutical salts were used to increase solubility without jeopardizing potency. Finally compound (**14**) was obtained for which solubility has been increased by a factor of 20 while potency has also been optimized.

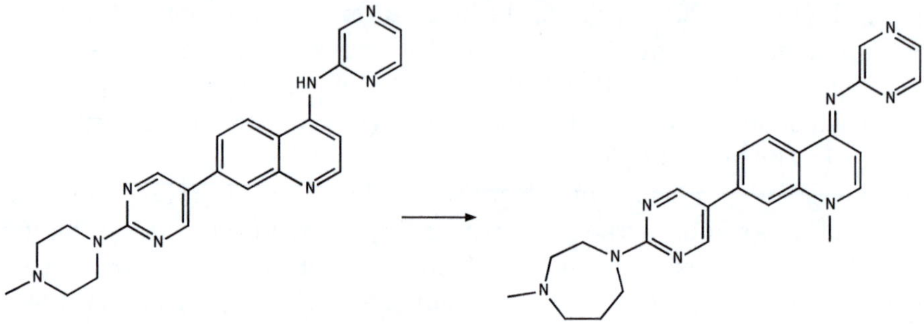

	(13)	(14)
Aqueous solubility	44 μM	880 μM
T. brucei EC50	0.43 μM	0.013 μM

Figure 7.13: Optimization of the solubility of Lapatinib-derived compounds against human African trypanosomiasis using several strategies.

7.5.1 Approaches for optimization of solubility in research programs

Generally, the optimization of solubility in research programs can be done in two ways, and this is also the case for the optimization of other properties.

The first approach represents a stepwise procedure as presented with the previous examples. This means a compound is designed on paper. For this purpose, information on other compounds that have already been designed, synthesized and tested, as well as in silico tools to predict solubility are used together with the experience of the chemist. To move a research project quickly ahead fast cycle times for synthesis and the testing of compounds are key, particularly the solubility measurement. The need for highly standardized, quick assays to measure meaningful solubility data, as described earlier, arises. With regard to the cycle time, the amount of compound required, and effort, kinetic solubility measurements are attractive. However, one must bear in mind that kinetic solubility can be misleading when used in the optimization of solubility and therefore thermodynamic solubility should be used for compound optimization. The process of solubility optimization is schematically depicted in Figure 7.14. This approach will not only be used for the optimization of solubility but also for other compound properties such as potency, selectivity, stability, permeability and others.

The second approach uses compound libraries to assess the optimization of solubility in a systematic way. For this purpose, smaller libraries of compounds are

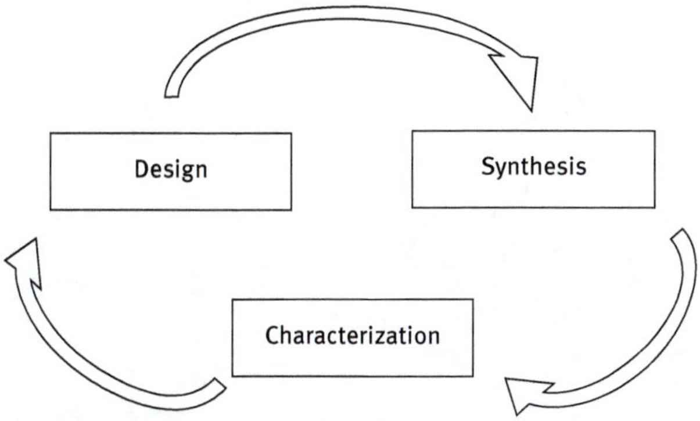

Figure 7.14: Stepwise optimization of research compounds. First a compound is designed by the chemist. In the next step the synthesis of the compound is carried out. The third step in the optimization cycle is the characterization of the compound using several assays including solubility measurement. Based on the results of this characterization, the next compound is designed by the chemist.

designed. Libraries modify a defined scaffold at certain positions by different substituents. In principal, the steps of this approach are like the first approach: the design of compounds using results from previous compounds solubility measurements, in silico tools and personal experience of the chemist, in the synthesis of compounds. However, this approach is a much more parallelized procedure. This means that steps such as the synthesis of compounds can frequently be done more efficiently, as the synthesis of a whole set of compounds can be done using the same precursors and only the last step or last steps of the synthesis are different for different compounds belonging to the same library. This reduces the effort for the synthesis of compounds. Also, the measurement of solubility can be done in parallel, which reduces effort and accelerates the process. Due to these benefits, the library approach leads to a reduction of effort and reduction of time from the design of a compound to having the compound characterized with regard to solubility. The disadvantage of this approach is that more different compounds must be synthesized compared to the previous approach and redundant information might be generated.

Accordingly, the chemist should choose carefully which approach to use for compound optimization based on the stage of the project, synthetic accessibility of the compounds as well as the need for the optimization of other compound properties

Liu et al. [60] presented an example of solubility optimization using compound libraries. Their goal was to optimize thiazolidinone compounds with regard to their cytoselective toxicity towards non-small cell lung cancer (NSCLC) cells and drug-resistant NSCLC cells at the same time showing low toxicity to normal human fibroblasts. In addition to this pharmacological optimization, the solubility should be

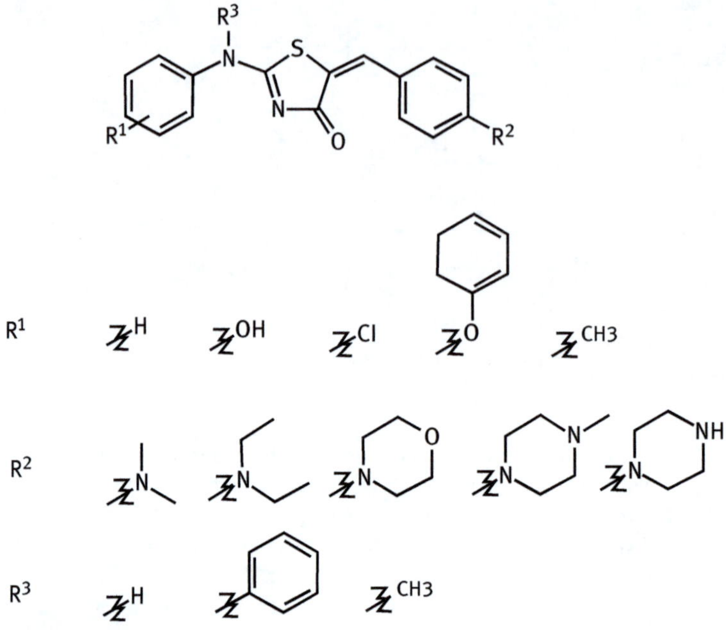

Figure 7.15: Scaffold and substituents upon which the library for compound optimization was based by Liu et al.

increased by derivatization. The scaffold used to design appropriate libraries is shown in Figure 7.15.

This scaffold includes three positions that were considered for derivatization (R^1, R^2, R^3). Therefore, in principal optimization could be carried out based on a three-dimensional library where all three substituents are modified in parallel. As this would lead to many compounds from a huge synthetic effort and consequent compound testing, the authors instead preferred a sequential two-step approach using two two-dimensional libraries. The first library contains 25 compounds that are obtained by modification of R^1 and R^2. At the same time, R^3 is kept constant by always using a hydrogen substituent.

Compounds from this library were characterized for aqueous solubility and for their cytoselective toxicity in NSCLC cells. Results from the solubility measurements showed that solubility was always low, expect for compounds where R^2 was a piperazine: compounds (**16**), (**17**), and (**18**) showed considerably enhanced solubility. This improved solubility was explained by the presence of an H-bond donor in the piperazine substituent. This effect was expected for these compounds as during the design of the library lower calculated log *P* values had been obtained, which was an indicative of higher solubility. On the other hand, better activity in the cellular assay was obtained for compounds with R^2 substituted by a dimethyl-amino or diethyl-amino group. This is a typical example of the diverging needs with regard to the optimization of activity and solubility.

Based on these results, a second library including 15 compounds was designed: where R^2 and R^3 were varied, while R^1 was kept constant using a hydrogen substituent. Results for solubility from this library clearly showed that solubility increased by a factor of more than 10 when a methyl group was used for R^3. Larger phenyl and smaller hydrogen substituents used as R^3 did not lead to a higher solubility.

Finally, compound **(19)** has been obtained (see Figure 7.16) based on the described library-based optimization, which started from compound **(15)**. In this respect optimization was neither done solely focused on potency, nor solely on solubility, but on an approach considering both properties. Therefore, compound **(19)** does not represent the compound with the highest solubility or lowest EC50 within the library. Instead it represents a balanced compromise with fivefold increased solubility and about a 10-fold increase in potency compared to compound **(15)**, which represented the starting point for the optimization.

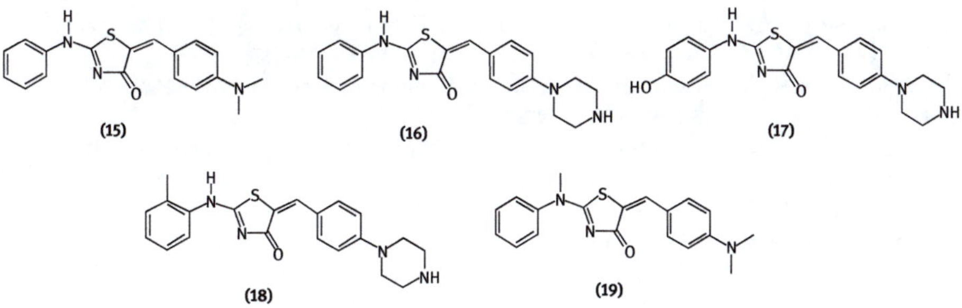

(15): Solubility = 0.1 μg/mL, EC50 = 0.72 μM

(19): Solubility = 0.5 μg/mL, EC50 = 0.08 μM

Figure 7.16: Some of the library compounds used in the optimization of thiazolidinone compounds. Compound (15) represented the starting point. Design of libraries with sequential variation of R^1, R^2, and R^3 lead to compound **(19)** with a balanced profile of improved solubility and potency.

7.6 Conclusion

From this chapter it becomes clear where it is important for medicinal chemist to have a clear understanding of solubility: a major part in the everyday life of a medicinal chemist is designing research compounds that have a sufficient solubility to become bioavailable. As increased potency in many cases leads to more lipophilic and less soluble compounds. Optimizing compounds in research programs to increase solubility is an important goal for medicinal chemists. In this chapter, we have summarized in which situations solubility matters in medicinal chemistry, how medicinal chemists can obtain meaningful solubility data, and what strategies to modify structures of research compounds to increase solubility can be used.

References

[1] Amidon, G. L., Lennernäs, H., Shah, V. P., & Crison, J. R. A theoretical basis for a biopharmaceutic drug classification: the correlation of in vitro drug product dissolution and in vivo bioavailability. Pharm. Res.: Off. J. Am. Assoc. Pharm. Sci. 1995, 12(3), 413–420.

[2] Yalkowsky, S. H., & Valvani, C.S. Physical Chemical Properties of Drugs, Marcel Dekker, 1980.

[3] Lipinski, C. A., Lombardo, F., Dominy, B. W., & Feeney, P. J. Experimental and computational approaches to estimate solubility and permeability in drug discovery and development settings. Adv. Drug Deliv. Rev. 1997, 23, 3–25.

[4] Hamlin, B. W. E., Northam, J. I., & Wagner, J. G. Solubilities of numerous compounds representative of various chemical species. J. Pharnaceutical Sci. 1965, 54(11), 1651–1653.

[5] Rasenak, B. W. M. N. Dissolution rate enhancement by in situ micronization of poorly water-soluble drugs. Pharm Res. 2002, 19(12), 1894–1900.

[6] Hu, J., Johnston, K. P., & Williams, R. O. Nanoparticle engineering processes for enhancing the dissolution rates of poorly water soluble drugs. Drug Dev. Ind. Pharm. 2004, 30(3), 233–245.

[7] Sodeifian, G., & Sajadian, S. A. Solubility measurement and preparation of nanoparticles of an anticancer drug (Letrozole) using rapid expansion of supercritical solutions with solid cosolvent (RESS-SC). J. Supercrit. Fluids 133, October, 2017, 2018, 239–252.

[8] Yalkowsky, S. H., & Valvani, S. C. Solubility and partitioning I: solubility of nonelectrolytes in water. J. Pharm. Sci. 1980, 69(8), 912–922.

[9] Abraham, M. H., & Le, J. The correlation and prediction of the solubility of compounds in water using an amended solvation energy relationship. J. Pharm. Sci. 1999, 88(9), 868–880.

[10] Abraham, M. H., Chadha, H. S., Whiting, G. S., & Mitchell, R. C. Hydrogen bonding. 32. An analysis of water-octanol and water-alkane partitioning and the Δlog p parameter of Seiler. J. Pharm. Sci. Aug 1994, 83(8), 1085–1100.

[11] –Br, A. B. Perlovich, G. L., & Volkova, T. V. Towards an understanding of the molecular mechanism of solvation of drug molecules: a thermodynamic approach by crystal lattice energy, sublimation, and solubility exemplified by paracetamol, acetanilide, and phenacetin. J. Pharm. Sci. 2006, 95(10), 2158.

[12] Bevan, C. D., & Lloyd, R. S. A high-throughput screening method for the determination of aqueous drug solubility using laser nephelometry in microtiter plates. Anal. Chem. 2000, 72(8), 1781–1787.

[13] Sugano, K., Okazaki, A., Sugimoto, S., Tavornvipas, S., Omura, A., & Mano, T. Solubility and dissolution profile assessment in drug discovery. Drug Metab. Pharmacokinet. 2007, 22(4), 225–254.

[14] Guha, R. et al Exploratory analysis of kinetic solubility measurements of a small molecule library. Bioorg. Med. Chem. 2011, 19(13), 4127–4134.

[15] Tan, H., Semin, D., Wacker, M., & Cheetham, J. An automated screening assay for determination of aqueous equilibrium solubility enabling SPR study during drug lead optimization. JALA – J. Assoc. Lab. Autom. 2005, 10(6), 364–373.

[16] Pan, L., Ho, Q., Tsutsui, K., & Takahashi, L. Comparison of chromatographic and spectroscopic methods used to rank compounds for aqueous solubility. J. Pharm. Sci. 2001, 90(4), 521–529.

[17] Rubino, J. T., & Yalkowsky, S. H. Cosolvency and cosolvent polarity. Pharm. Res.: Off. J. Am. Assoc. Pharm. Sci. 1987, 4(3), 220–230.

[18] Kalayani, M. "Drug Discovery MSc project report UCL School of Pharmacy," 2005.

[19] Vemula, V. R., Lagishetty, V., & Lingala, S. Solubility enhancement techniques. Int. J. Pharm. Sci. Rev. Res. 2010, 5(1), 41–51.

[20] Hoelke, B., Gieringer, S., Arlt, M., & Saal, C. Comparison of nephelometric, UV-spectroscopic, and HPLC methods for High-throughput determination of aqueous drug solubility in microtiter plates. Anal. Chem. 2009, 81(8), 3165–3172.

[21] Stuart, M., & Box, K. Chasing equilibrium: measuring the intrinsic solubility of weak acids and bases. Anal. Chem. 2005, 77(4), 983–990.

[22] Yalkowski, S. B. S. H. Aqueous solubility: Methods of estimation for organic compounds, Marcel Dekker, 1992.

[23] Box, K. J., Völgyi, G., Baka, E., Stuart, M., Takács-Novák, K., & Comer, J. E. A. Equilibrium versus kinetic measurements of aqueous solubility, and the ability of compounds to supersaturate in solution – A validation study. J. Pharm. Sci. 2006, 95(6), 1298–1307.

[24] Box, K. J., Comer, J. E., Gravestock, T. & Stuart, M. New Ideas about the Solubility of Drugs, Chemistry & Biodiversity, 2009, 6, 1767–1788.

[25] a Avdeef Physicochemical profiling (solubility, permeability and charge state). Curr. Top. Med. Chem. 2001, 1(4), 277–351.

[26] Bhattachar, S. N., Wesley, J. A., & Seadeek, C. Evaluation of the chemiluminescent nitrogen detector for solubility determinations to support drug discovery. J. Pharm. Biomed. Anal. 2006, 41(1), 152–157.

[27] Wenlock, M. C., Austin, R. P., Potter, T., & Barton, P. A highly automated assay for determining the aqueous equilibrium solubility of drug discovery compounds. J. Lab. Autom. 2011, 16(4), 276–284.

[28] Alsenz, J., & Kansy, M. High throughput solubility measurement in drug discovery and development. Adv. Drug Deliv. Rev. 2007, 59(7), 546–567.

[29] Baka, E., Comer, J. E. A., & Takács-Novák, K. Study of equilibrium solubility measurement by saturation shake-flask method using hydrochlorothiazide as model compound. J. Pharm. Biomed. Anal. 2008, 46(2), 335–341.

[30] Banerjee, S., Yalkowsky, S. H., & Valvani, S. C. Water solubility and octanol/water partition coefficients of organics. limitations of the solubility-partition coefficient correlation. Environ. Sci. Technol. 1980, 14(10), 1227–1229.

[31] Ran, Y., & Yalkowsky, S. H. Prediction of drug solubility by the general solubility equation (GSE). J. Chem. Inf. Comput. Sci. 2001, 41(2), 354–357.

[32] Ran, Y., Jain, N., & Yalkowsky, S. H. Prediction of aqueous solubility of organic compounds by the general solubility equation (GSE). J. Chem. Inf. Comput. Sci. 2001, 41(5), 1208–1217.

[33] Dannenfelser, R. M., & Yalkowsky, S. H. Predicting the total entropy of melting: Application to pharmaceuticals and environmentally relevant compounds. J. Pharm. Sci. 1999, 88(7), 722–724.

[34] Chickos, J. S., Nichols, G., & Ruelle, P. The estimation of melting points and fusion enthalpies using experimental solubilities, estimated total phase change entropies, and mobile order and disorder theory. J. Chem. Inf. Comput. Sci. 2002, 42(2), 368–374.

[35] Wassvik, C. M., Holmén, A. G., Bergström, C. A. S., Zamora, I., & Artursson, P. Contribution of solid-state properties to the aqueous solubility of drugs. Eur. J. Pharm. Sci. 2006, vol. 29(no. 3–4), SPEC. ISS., 294–305.

[36] Schultes, S. et al "A medicinal chemistry perspective on melting point: matched molecular pair analysis of the effects of simple descriptors on the melting point of drug-like compounds,". Medchemcomm 2012, 3(5), 584–591.

[37] Hussain, C., & Rea, J. Computationally effiecient algorithm to identify matched molecular pairs (MMPs) in large data sets. J. Chem. Inf. Model. 2010, 4, 339–348.

[38] Zhang, L., Zhu, H., Mathiowetz, A., & Gao, H. Deep understanding of structure-solubility relationship for a diverse set of organic compounds using matched molecular pairs. Bioorganic Med. Chem. 2011, 19(19), 5763–5770.

[39] Lovering, F., Bikker, J., & Humblet, C. Escape from flatland: Increasing saturation as an approach to improving clinical success. J. Med. Chem. 2009, 52(21), 6752–6756.
[40] Ritchie, T. J., & Macdonald, S. J. F. The impact of aromatic ring count on compound developability – are too many aromatic rings a liability in drug design?. Drug Discov. Today 2009, vol. 14(no. 21–22), 1011–1020.
[41] Ritchie, T. J., MacDonald, S. J. F., Young, R. J., & Pickett, S. D. The impact of aromatic ring count on compound developability: Further insights by examining carbo- and hetero-aromatic and -aliphatic ring types. Drug Discov. Today 2011, vol. 16(no. 3–4), 164–171.
[42] Hill, A. P., & Young, R. J. Getting physical in drug discovery: A contemporary perspective on solubility and hydrophobicity. Drug Discov. Today 2010, vol. 15(no. 15–16), 648–655.
[43] Wuitschik, G. et al. Oxetanes in drug discovery: structural and synthetic insights. J. Med. Chem. 2010, 53(8), 3227–3246.
[44] Delaney, J. S. Predicting aqueous solubility from structure. Drug Discov. Today 2005, 10(4), 289–295.
[45] Dearden, J. C. In silico prediction of aqueous solubility. Expert Opin. Drug Discov. 2006, 1, October, 31–52.
[46] Wang, J., & Hou, T. Recent advances on aqueous solubility prediction. Comb. Chem. High Throughput Screen. 2011, 14(5), 328–338.
[47] Wang, J., Krudy, G., Hou, T., Zhang, W., Holland, G., & Xu, X. Development of reliable aqueous solubility models and their application in druglike analysis. J. Chem. Inf. Model. 2007, 47(4), 1395–1404.
[48] Huuskonen, J., Livingstone, D. J., & Manallack, D. T. "Prediction of drug solubility from molecular structure using a drug-like training set,". SAR QSAR Environ. Res. 2008, vol. 19(no. 3–4), 191–212.
[49] Llinàs, A., Glen, R. C., & Goodman, J. M. Supporting information solubility challenge : can you predict solubilities of thirty-two molecules using a database of one hundred reliable measurements ?. J. Chem. Inf. Model. 2008, 48(7), 1289–1303.
[50] Hewitt, J. C., Cronin, M., Enoch, M.T.D., Madden, S. J., Roberts, J. C., & Dearden, D. W. In silico prediction of aqueous solubility. Expert Opin. Drug Discov. 2006, 1, October, 31–52.
[51] Huang, B. et al Fused heterocycles bearing bridgehead nitrogen as potent HIV-1 NNRTIs. Part 4: Design, synthesis and biological evaluation of novel imidazo[1,2-a]pyrazines. Eur. J. Med. Chem., 2015, 93, 330–337.
[52] Press, N. J. et al Solubility-driven optimization of phosphodiesterase-4 inhibitors leading to a clinical candidate. J. Med. Chem. 2012, 55(17), 7472–7479.
[53] Tehler, U., Fagerberg, J. H., Svensson, R., Larhed, M., Artursson, P., & Bergström, C. A. S. Optimizing solubility and permeability of a biopharmaceutics classification system (BCS) class 4 antibiotic drug using lipophilic fragments disturbing the crystal lattice. J. Med. Chem. 2013, 56(6), 2690–2694.
[54] Gavezotti, A. Are crystal structures predictable?. Acc. Chem. Res. 1994, 27, 309–314.
[55] Wang, H. L., Katon, J., Balan, C., Bannon, A. W., Bernard, C., Doherty, E. M., Dominguez, C., Gavva, N. R., Gore, V., Ma, V., Nishimura, N., Surapaneni, S., Tang, P., Tamir, R., Thiel, O., Treanor, J.J., & Norman, M. H. Novel vanilloid receptor-1 antagonists: 3. The identification of a second-generation clinical candidate with improved physicochemical and pharmacokinetic properties. J. Med. Chem. 2007, 50(15), 3528–3539.
[56] Hoover, T. B. Water solubility of PCB isomers. PCB Newsl. 1971, 3, 4–5.
[57] Lee, M. C., Chian, E. S. K., & Griffin, R. A. Solubility of polychlorinated biphenyls and capacitor fluid in water. Water Res. 1979, 13(12), 1249–1258.
[58] Ishikawa, M., Hiraiwa, Y., Kubota, D., Tsushima, M., Watanabe, T., Murakami, S., Ouchi, S., & Ajito, K. Tricyclic pharmacophore-based molecules as novel integrin αvβ3antagonists. Part III:

Synthesis of potent antagonists with αvβ3/αIIbβ3dual activity and improved water solubility. Bioorganic Med. Chem. 2006, 14(7), 2131–2150.
[59] Bachovchin, K. A., Sharma, A., Bag, S., Klug, D.M., Schneider, K.M., Singh, B., Jalani, H.B., Buskes, M.J., Mehta, N., Tanghe, S., Momper, J.D., Sciotti, R.J., Rodruguez, A., Mensa-Wilmot, K., Pollastri, M.P., & Ferrins, L. "Improvement of aqueous solubility of lapatenib-derived analogues: Identification of a quinolinimine lead for human African trypanosomiasis drug development". J. Med. Chem. 2019, 62, 665–687.
[60] Liu, B., Li, Y., Wu, F., Wang, L., Zhu, W., Zhang, H., Zhou, Q., & Yan, H. Improving both aqueous solubility and anti-cancer activity by assessing progressive lead optimization libraries. Bioorg. Med. Chem. Lett. 2015, 25(9), 1971–1974.

René Holm

8 The role of solubility in optimizing drug products – a pharmaceutical development perspective

8.1 Introduction

Solubility is one of the most important physical–chemical properties that define a formulation – irrespective of the intended administration route. The formulation route and its associated physiology define both the formulation variables and excipients that can be used. For oral administration, solubility and permeability are the two key parameters for drug absorption. For parenteral dosage forms, that is, formulations for injection, it is important from a safety perspective to obtain a sufficient solubility, whereas for other dosage forms as eye drops solubility is important for efficacy. Thus, for either biopharmaceutical or technical reasons, solubility is important for all administration routes such as transdermal, sublingual, topical and for all dosage forms.

This chapter focuses on how solubility data are used in drug product formulation, with selected examples important for common dosage forms. The chapter will not provide an exhaustive discussion of all drug delivery systems or cover all administration routes but aims at educating the reader about the chemical considerations in the use of solubility data in drug product development. In general, the formulation strategy is to minimize the use of excipients, while ensuring that the formulation is robust from an efficacy, safety, chemical, physical and manufacturing point of view. This chapter is organized into three main sections, focusing on how solubility (1) is used to define the formulation strategy for oral dosage forms, (2) influences the selection for bio-enabling formulations, and (3) considerations on liquid formulation from a solubility perspective.

The scope of the chapter is to provide solubility-based perspectives that can be applied to all delivery systems and all modalities.

8.2 Aqueous-based liquid formulations

Liquid formulations can be used for many applications, including oral solutions, injectables, eye drops, nasal spray and liquid for inhalation. This field is relatively

René Holm, Drug Product Development, Janssen Research and Development, Johnson & Johnson, Beerse, Belgium; Department of Science and Environment, Roskilde University, Roskilde, Denmark

https://doi.org/10.1515/9783110559835-008

broad. However, the elements of solubilization and how solubility determines the formulation work are the common denominators for these liquid formulations. In other words, it is one of the key properties of the compound when making an aqueous dosage form.

Pharmaceutical excipients have a prominent role in liquid formulations as they are needed to formulate water-insoluble compounds. Their use varies considerably between the different aqueous dosage forms. The restrictions in choosing an excipient for oral administration versus intravenous administration will differ, although conceptually the formulation work is similar and aims for complete solubilization in the formulation, for example, choice of cosolvent or preservatives used. Strickley [1] suggested a flow chart for selecting the solubilization strategy for different administration routes, which can function as general guidance for the choice of a solubilization strategy. Different pharma-companies have different preferences and the use of the flow diagram may vary as a function of formulation, toxicological experience, regulatory strategy, indication, patient population and so on.

The approach proposed by Strickley [1] does not take the physicochemical properties of the compound to be solubilized into consideration, but focuses solely on the effect and safety of the solubilization strategy. For compounds with an ionizable group, that is, acids or bases, pH adjustment is normally investigated first and if sufficient, a relatively simple formulation can be achieved. If the compound has a molecular fit for inclusion complexes, for example, cyclodextrins, these are frequently investigated as a second approach, depending upon concentrations needed and patient populations. If solubility is still not sufficient and the log P is not too high, then cosolvent systems combined with pH adjustment can be used. Finally, for compounds with a high lipophilicity, more complex systems, such as micellar solutions/dispersions, emulsions, liposomes, or other more suitable methods might be needed. The various strategies to solubilize a drug compound in an aqueous formulation will be discussed below, and lipid-based systems are discussed in Section 8.3.2.2.

8.2.1 Adjustment of pH

pH adjustment is probably the most commonly used method to alter the solubility of ionizable compounds. The two main reasons behind this strategy are (i) it is safe, and (ii) it is a very effective approach from a solubilization perspective. Changing pH will dramatically change the solubility of ionizable compounds, and the relationship between solubility, pH, and the dissociation constant is the most well understood of the solubilization strategies.

When a compound has a pK_a value within the pH range in which a formulation can be dosed, for example, pH 3-8, the first thing to investigate is its solubility dependency as a function of pH, that is, determine the pH-solubility profile. Most drug compounds are weak electrolytes and can in principle be perceived as a buffering agent that is in equilibrium with its corresponding base or acid. For an acid in an aqueous solution,

$$HA \rightleftharpoons A^- + H^+ \tag{8.1}$$

where HA is a weak acid and A^- the corresponding weak base. The equilibrium constant from this equation is the acid dissociation constant K_a. It relates to the degree of dissociation of the acid and is calculated based on the concentrations of the reacting species:

$$K_a = \frac{[H^+][A^-]}{[HA]} \tag{8.2}$$

This value is related to pH of the solution and the Henderson-Hasselbalch equation (eq. 8.3) describes how pH depends on the concentration of the buffer components and pK_a:

$$pH = pK_a + \log\frac{[A^-]}{[HA]} \tag{8.3}$$

From a drug solubilization perspective, the charged form (here A^-) is often soluble in water, whereas the neutral species (here HA) are usually less soluble. As the relationship is logarithmic, the correlation between the solubility and the pH indicates that the solubility increases by a factor of 10 when changing pH by one unit, that being decreasing the pH for bases and increasing for acids. In principle, eq. (8.3) indicates that the solubility of a weak acid or base can be increased indefinitely. However, there are limitations based upon the solubility product of the salt formed with the acid or base that is used to adjust pH. The acid or base selected to adjust the pH can therefore be a critical formulation parameter and would often need to be explored experimentally.

In order to maintain a defined pH throughout both the product process of the formulation and over the shelf life of the drug product, a buffer is normally added to the formulation. Buffers improve both solubility and stability along with reducing irritation. When defining the properties of a liquid pharmaceutical formulation, it is important to keep robustness in mind, which includes defining a pH range that extends beyond the desired solubility. A close monitoring of pH is possible in small-scale experiments, but with batch sizes of hundreds of litres or more, achieving an accuracy of even 0.2 pH units may be challenging. Determining the solubilities at the extremes of the formulation is thus also an important part of defining a liquid formulation. The goal is hence to select the optimal pH range over which the API shows the desired solubility and stability.

pK_a depends on temperature, pressure, and ionic strength and as described by the Henderson–Hasselbalch equation (8.3) these parameters will cause a change in the pH of the solution. In practice, pressure has a limited influence on pK_a over the ranges that pharmaceutical processing occurs unless, for example, high-pressure homogenization is applied during formulation [2]. Ionic strength also has an influence on some buffer pH [3], but in pharmaceutical liquids especially parenterals, the ionic strength is normally relatively constant and adjusted to be an isotonic solution. Thus, in principle, a well-designed formulation design can omit this parameter.

Most formulations are exposed to variations in temperature, for example, during transportation, storage, or a pharmaceutical processing such as autoclaving. If the formulated API has a solubility that is highly pH dependent within the pH range defined in the formulation, then one needs to consider (and potentially evaluate) how this may influence the performance of the formulation. The influence on pK_a as a function of temperature depends on the chemical structure of the buffer. Carboxylic acid buffers are least affected by temperature [4–6]. However, carboxylic acid-based buffers may not always be compatible with the formulation and hence phosphate buffer, amino acid-based buffers, or biological buffers such as Tris may be considered. The highest variation in pK_a of buffers is reported for Tris, where a decrease of around ≈2 pH units is observed in the temperature range 25-125 °C [4]. When formulating a buffered liquid formulation, it is therefore important to consider the solubility not only as a function of pH, but also of temperature. Considering variations of both temperature and pH at the same time is necessary to ensure a robust formulation design.

In cases where an antimicrobial preservative is needed, for example, multiple use formulations for oral administration, the selection of pH may also be influenced by the preservative that needs to be added. Most preservatives for oral applications are more effective in the acidic range. Hence, if a more neutral pH is defined, the concentration of the preservative must be adjusted to ensure sufficient antimicrobial effect.

8.2.2 Cosolvents

If pH adjustment is not sufficient to achieve the desired level of solubility, then the use of cosolvents can be explored. A cosolvent is a water-soluble organic solvent that is added to an aqueous system to increase the solubility of an API by altering solution parameters such as polarity, surface tension and dielectric constant. For oral administration, pure cosolvents can be administered; for example, in a soft gelatine capsule or by using a syringe in early phases of development. It should though be kept in mind that the use of pure cosolvents such as PEG400 and propylene glycol is not optimal from a taste perspective. For parenteral administration, cosolvents are often blended into aqueous systems. Typically used cosolvents include propylene glycol and ethanol, while even DMSO can be found in commercial injectable formulations. However, there is a restriction regarding which cosolvents can be used for injectable formulations versus oral formulations, due to the high osmotic pressure that can be associated with cosolvents.

A solubility enhancement of several orders of magnitude may be observed by using pure organic cosolvents, which may be relevant for an oral solid formulation which can be encapsulated in a soft gelatine capsule. There is a nonlinear relationship between solubility and cosolvent fraction in an aqueous solution. Thus, the addition of small quantities of water can reduce the solubility benefit significantly. This may lead to biopharmaceutical considerations for oral products solubilized in cosolvents, as there is an inherent risk that the compound may precipitate into a crystalline form in the intestine when the capsule dissolves and the cosolvent becomes diluted into the aqueous phase. For some compounds, this can be avoided by addition of a crystallization/precipitation inhibitor to the formulation [7] (see Chapter 2). For parenteral formulations, this option does not exist and the use of cosolvents to enhance the solubility becomes a compromise between solubility and osmotic pressure and consequently the potential for patient discomfort associated with the administration of the formulation.

A combination of different cosolvents in the same formulation can be beneficial and it should be noted that a higher proportion of cosolvents will not lead to a higher solubility in all cases, as demonstrated by the nonpolar compound tioconazole [8] (see Table 8.1).

Table 8.1: Measured solubilities for tioconazole as a function of cosolvent and cosolvent concentrations (% v/v).

% Ethanol	% Propylene glycol	% PEG400	log (measured solubility)
20	40	0	2.019
40	20	0	1.645
15	0	25	3.031
25	0	15	2.921
0	20	20	3.210
0	30	30	2.325
15	15	15	2.795
30	10	10	2.285
10	30	10	2.673

Data from Gould et al. [8].

The data from Gould and co-workers [8] demonstrate the advantages in combining different cosolvents to adjust the formulation to the specific compound. Solubility investigations based on design of experiments can therefore sometimes be a very useful tool to optimize a cosolvent-based formulation. Beyond the data published by Gould et al. [8], which demonstrates that combining different cosolvents in a system can be beneficial, it is worth noticing that benzyl alcohol, also a cosolvent, is used as a preservative in several commercial products for intravenous administration, for example, with diazepam and lorazepam. Benzyl alcohol is an effective preservative having limited interactions with drug compounds. However, it is interesting to notice also the excipients ability to function as a cosolvent why it can be used with dual purposes in a formulation, as reported by Yalkowsky [9]. If the compound that needs to be solubilized can be ionized, it may be beneficial to combine formulation selection with pH adjustments, which can be relevant for cosolvent, surfactant and inclusion complex systems (see Section 8.2.4) [10].

As noted for pH-adjusted formulations, a robustness investigation needs to be conducted for cosolvent-based formulations, given the relationship between solubility and cosolvent concentration. This implies, it is important to define the formulation in a way that small variations in the cosolvent concentration will not lead to product performance failure. A very careful measurement of the solubility as a function of cosolvent concentration is therefore an important element in defining this class of formulation.

8.2.3 Micellar solubilization using surfactants

Surfactants are amphiphiles that have both polar and nonpolar regions. They form multicomponent systems and will accumulate at surfaces with different polarities. Above their critical micelle concentration (CMC), they form micelles, that is, a disperse phase that can solubilize an API. Good drug candidates for surfactant-based liquid formulations are often lipophilic, non-polar and non-ionizable compounds with very poor water solubility.

Non-ionic surfactants are often used as an excipient in pharmaceutical formulations because of their relatively low toxicity and their solubilizing power. Examples of surfactants used in commercial formulations include polysorbate 20, polysorbate 80 and cremophor EL. Given that the chemical structure of surfactants can vary, different surfactants will have different solubilization properties for different compounds. This implies that determination of solubility to identify the most suitable surfactant for a given compound is crucial. Above the CMC, the relationship between surfactant concentration and compound solubilization is linear. Hence, identification of a robust formulation range with respect to this excipient is more straightforward than cosolvent systems. As mentioned earlier, potential synergistic effects using surfactants and pH adjustment should be considered as well.

For formulations containing surfactants at a concentration above the CMC, it should be noted that a preservative used in the formulation may have an affinity for the micellar phase, and thus, the relative fraction of the preservative in the aqueous phase needs to be considered. Hence, investigating the solubility of the preservative in the selected vehicle may also be a part of the investigations needed to get an experimental insight into the relative affinity of the preservative for the micellar phase and thereby provide guidance in selecting the desired preservative concentration. As stated in Section 8.2.1, this factor should also be considered when the preservation may be ionized at the pH used.

8.2.4 Cyclodextrins

Cyclodextrins form reversible non-covalent inclusion complexes with many drug compounds and can be a very effective approach for solubilization of poorly aqueous soluble compounds in cases where the physical shape and form of the drug molecule fits the cyclodextrin cavity. Cyclodextrins are cyclic oligosaccharides, and pharmaceutically relevant cyclodextrins contain six to eight dextrose units connected in a circular manner by 1,4-bonds (see Figure 8.1). Due to the orientation of the hydroxyl groups, a cone-shaped structure is formed with a hydrophilic exterior and a lipophilic core. For oral applications, both natural and modified cyclodextrins can be used, although for parenteral applications, the most frequently used cyclodextrins are the two modified, hydroxypropyl-β-cyclodextrin and sulphobutylether-6-cyclodextrin, as these have a very high aqueous solubility (>400 mg/mL) and a favourable toxicological profile. The hydroxyl groups of the natural cyclodextrins are chemically modified, so up to three molecules can be added per dextrose group. In practice, the substitution is less controlled and the cyclodextrins used are a mixture with variable substitutions, due to which the mean degree of substitution is often used. For both hydroxypropyl-β-cyclodextrin and sulphobutyl-β-cyclodextrin, the substitution occurs

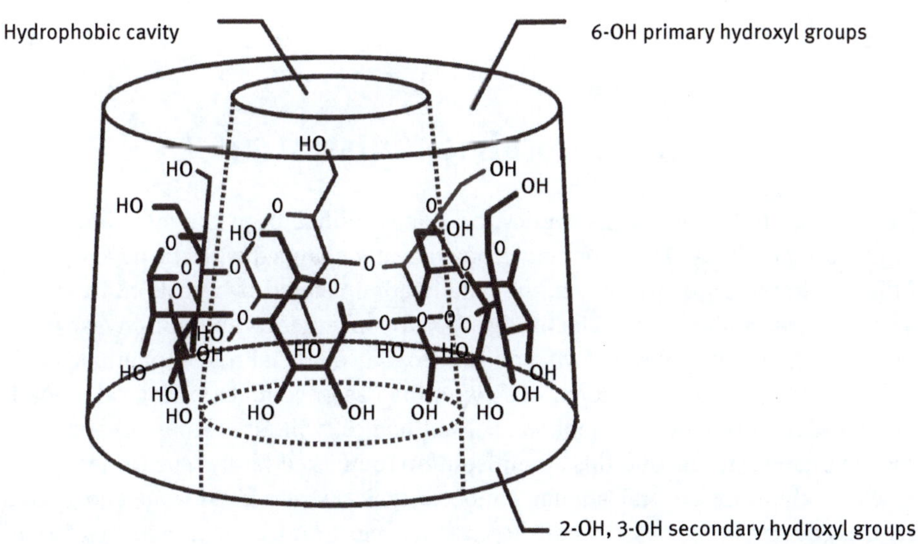

Figure 8.1: Chemical structure of β-cyclodextrin, in which the 2, 3, and 6-hydroxyls of a glucopyranose unit may have a substituent attached.

mostly at the 2- and 3-hydroxyl groups and only to a limited extent on the 6-hydroxyl group [11, 12].

The complex between cyclodextrins and a drug compound is often a 1:1 interaction, though exceptions have been published [13]. The different cyclodextrins have different stability constants with different APIs, and different stability constants for the charged and uncharged species have also been observed [14]. Constructing the phase solubility profiles for different cyclodextrins with a drug compound can thereby support the formulator in selecting the best solubilizing cyclodextrin, but also in selecting the most stable region of solubilization in cases where nonlinear behaviour is observed. In principle, the correlation between the cyclodextrin concentration and drug solubility should be determined over the relevant pH range as the solubility can vary as a function of both parameters. Determining the stability constant for both the charged and uncharged species of the compound may provide the possibility to support the experimental values with theoretical observations.

The stability constant and therefore the solubility of a drug compound in cyclodextrins can vary as a function of the degree of cyclodextrin substitution and the position of the substitution [15, 16]. As a part of the formulation work, the influence of this excipient on the solubility therefore needs to be evaluated to ensure formulation robustness.

With respect to preservation of cyclodextrin containing systems, the same issue as for surfactant-based systems exists which is – the solubilizing excipient can inactivate the preservative by complexation. From a theoretical perspective, models can be developed to guide the concentration of preservatives needed [17]; however, experimental proof is still needed.

8.3 Formulations for oral administration

Since the beginning of pharmaceutics, oral delivery has been the most important route of drug delivery. This dominance has recently changed with the development of the biopharmaceuticals, which are usually administered by injection. However, due to cost, established manufacturing plants, convenience and non-invasive administration, oral formulations will continue to play a crucial role in pharmaceutics in the future. Oral administration, even with the use of complex biological modalities as antisense, sRNAs and monoclonal antibodies, will also dominate approved drugs in the near future and this administration route is certainly here to stay.

While choosing an oral administration is typically straightforward and occurs very early as part of the target product profile, it is more difficult to decide on a formulation principle or a process technology for modern drug candidates that often exhibit challenging molecular properties for oral delivery [18, 19]. Companies differ in their decision making regarding not only the formulation strategy, but also on manufacturing. Hence, scientific arguments are not decisive strategic aspects, company traditions and past investments also play a role. For new drug candidates, the results of pharmaceutical profiling provide an early basis for initial decision making for the design of first-in-man formulations. Once a drug candidate has been dosed in phase I clinical trials, the availability of human pharmacokinetic data helps define the formulation strategy in later clinical phases and in the market.

8.3.1 Defining a formulation strategy

The Biopharmaceutics Classification System (BCS) [20] emphasizes the API properties of aqueous solubility and permeability. The solubility limit takes into account the highest dosage strength and a reference water volume (250 mL) at a pH range spanning the entire gastrointestinal tract, that is, pH 1-6.8 (see Chapter 11). This classification scheme was originally intended to correlate in vitro drug dissolution with in vivo drug bioavailability and minimize the need for bioequivalence studies where the outcome could be predicted by the in vitro surrogate measurements for highly soluble compounds, that is, for compounds classified as class I. For class III compounds (i.e. high solubility and low permeability) a biowaiver is allowed in most of the world provided that the formulation does not contribute to the permeability, whereas in vivo studies are required to show equivalence between formulations containing classes II and IV (i.e. both low solubility and permeability) compounds. Despite this, the BCS can be used to define the oral formulation strategy in early development and is widely implemented in the pharmaceutical industry as it conceptually describes the science well.

In principle BCS is a regulatory tool, so drugs of classes II, III, and IV cannot generically be considered as problematic with respect to oral absorption and bioavailability. BCS classes II and IV are naturally scientifically more challenging from a formulation perspective in order to achieve a consistent bioavailability. However, in general a more refined view than the BCS system needs to be defined, based on the understanding of the physiology and physico-chemical conditions at the site of absorption, for example, the "high solubility" definition of the FDA guidance can also serve for acidic drugs [21]. These compounds benefit from a pH-induced solubility increase at the relevant intestinal site of absorption and some acids can therefore behave biopharmaceutically rather similar to BCS I or III compounds when defining the formulation strategy, that is, with no limitations in absorption due to solubility issues. While the BCS is a regulatory tool, it can also be used when taking the decision on a formulation strategy. If the solubility of the API is so good that the entire dose can be solubilized in the pH range defined in the BCS, then a conventional solid dosage form such as immediate release tablets, powder-filled capsules or similar dosage forms would normally be selected.

Defining the solubility for a compound according to the BCS system may be considered as conservative as it uses water as the solvent, which have led to obvious challenges and considerations on what would be a biorelevant decision point. The initial work of Galia et al. [22] and later Vertzoni et al. [23] stressed the importance of biorelevant drug solubility, as also discussed in Chapters 5 and 6 on GI solubility. For this reason, drug solubility is typically measured using simulated gastric fluids, fasted state simulated intestinal fluid (FaSSIF), or fed state simulated intestinal fluids using bile salt and lecithin as biorelevant solubilizers [24]. Hereby a different solubility product is obtained, which is more relevant for the decision to be taken, that is, if conventional dosage forms can be used for the given compound or if enabling formulations are needed.

As an extension of the work with biorelevant media, Butler and Dressman [25] greatly contributed to the field of selecting the formulation strategy based on BCS. The authors defined the term "Developability Classification System" (DCS) (see Figure 8.2), for their approach, which considered biorelevant solubility (FaSSIF) as well as a dissolution rate that was expressed as a target drug particle size. Moreover, a solubility-limited absorbable dose was used in the DCS. This concept modified the idea of a maximal absorbable dose [26–28]. The DCS further differentiates class II compounds according to either a dissolution limitation (IIa) or a solubility limitation (IIb) for drug absorption, in contrast to the BCS. Although DCS has been introduced as a developability assessment tool, it can serve as a guide to define the formulation strategy. For example, BCS II compounds with some dissolution limitation may be developed with rather simple formulations based on

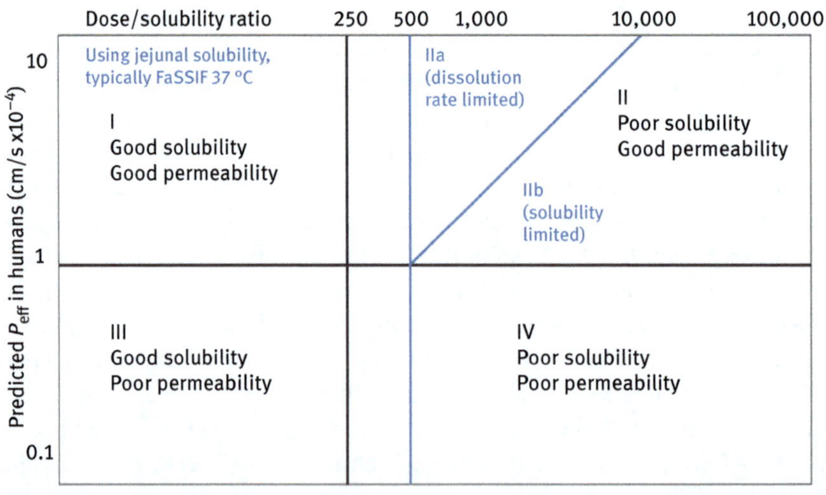

Figure 8.2: The biopharmaceutics classification system (BCS) is shown in black and the modifications from the BCS to the DCS are shown in blue. Modified from Butler and Dressman [25].

particle size reduction, wet granulation, and/or addition of surfactants. However, other BCS II compounds in the subcategory DCS IIb may need enabling formulations, such as lipid-based formulations (LBFs) or amorphous solid dispersions (ASDs). It seems that when selecting the early formulation strategy based on biopharmaceutical classes, the DCS provides a modern starting point that is more detailed than BCS, but still has simplicity as a key strength.

In addition to solubility, dissolution rate data and permeability data, in vivo nonclinical data are frequently used to define if a conventional formulation is sufficient or if an enabling formulation is needed. This in vivo work is part of the investigations conducted to identify the drug absorption hurdles. Comparison of a drug solution with a suspension formulation using different drug species has been suggested by Brewster [29] (see Figure 8.3). The decision tree depicts the possible outcomes of the pharmacokinetic experiments [29]. Similar areas under the curve (AUC) for both formulations could either mean that the compound shows extensive absorption as for a BCS I (-like) compound or there is barely a formulation difference observed due to a first pass or a strongly limiting drug permeability. Figure 8.3 also highlights the alternative scenarios of AUC differences due to formulation in order to differentiate the case of BCS II/IV (-like) compounds from other drugs that exhibit a chemical stability issue. It is recommended to compare the formulations in more than one species (e.g. rat vs dog) to better assess the biological effects on oral drug exposure. Depending on the given scenario, Figure 8.2 suggests different formulation approaches that can be also considered as part of the four-step plan described in Figure 8.1. This plan is flexible enough that it can be combined with various concepts. For example, the investigation of drug absorption hurdles (step 2) can make use of advanced biopharmaceutical modelling.

8.3.2 Enabling formulations

In cases where solubility seems suboptimal and in vivo data confirms issues achieving the desired bioavailability from a suspension (as a surrogate for a conventional formulation), the formulator may need to evaluate bio-enabling formulations. As suggested by Brewster [29], this brings the formulator into a more complex decision tree, which the DCS also eluded to in the separation of dissolution and solubility-limited drug absorption. Many different technologies can be used for a compound with solubility-limited absorption, including LBFs and ASDs. There are limitations for the different technologies that can define if they can be used or not. However, due to the complexity of these formulations, they will normally only be selected as a last resort. The typical approach to formulate a drug product with limited solubility includes formation of pharmaceutical salts, particle size reduction, LBF, and ASDs. In the following sections the importance of solubility when defining these formulation systems are discussed.

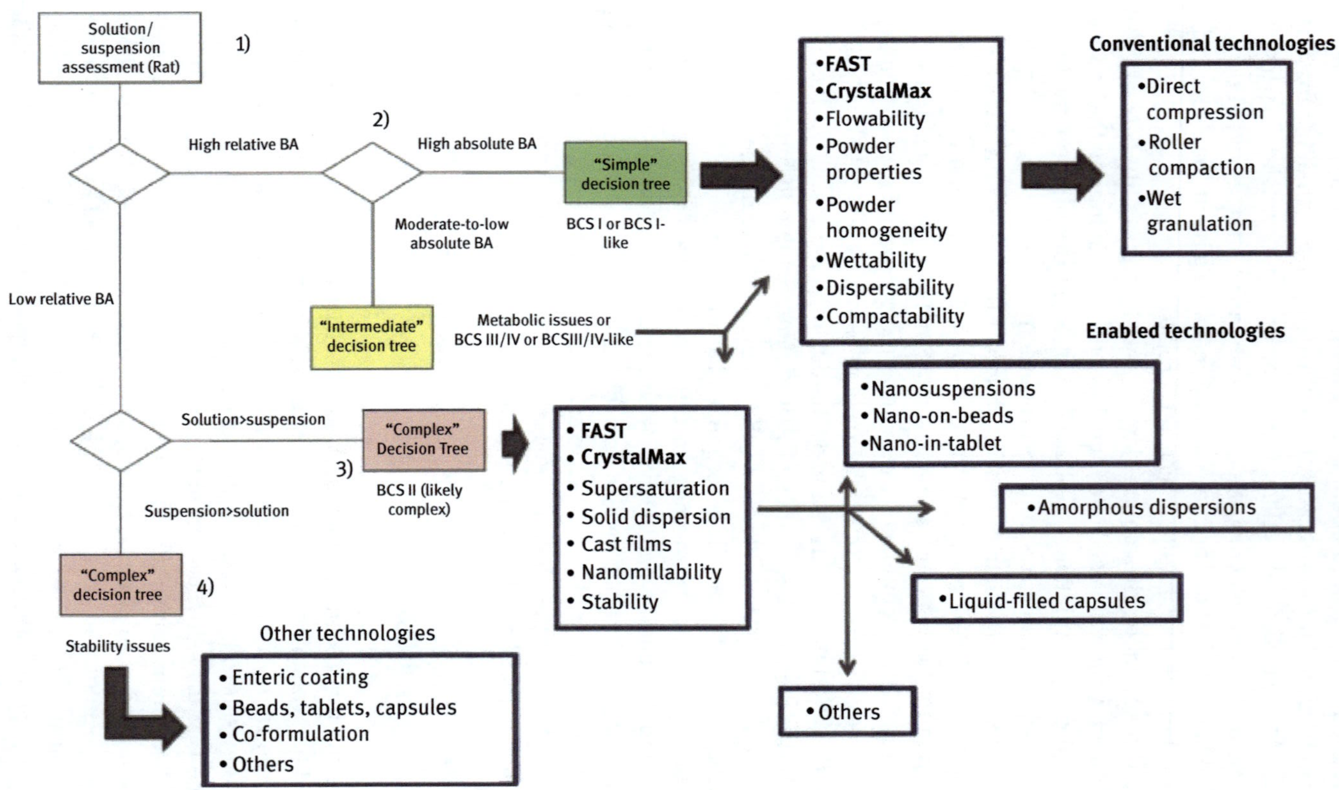

Figure 8.3: Outline of a strategy to compare a drug solution and suspension to study hurdles of drug exposure regarding selection of the formulation technology (adapted from [29]). Further details are explained in the text.

8.3.2.1 Particle size reduction

The development of micro- and nano-scale drug particles is one of the more promising strategies to improve the oral bioavailability of drugs with insufficient solubility, that is, compounds for which dissolution is the rate-limiting step. Academia and the pharmaceutical industry have extensively investigated the potential of particle size reduced systems, including nano-scaling for therapeutic agents belonging to BCS class II (or BCS class IV in some cases) with improved bioavailability in mind [30–32] – or Class 2a according to the DCS. The commercial availability of several oral pharmaceutical products containing nano-sized drug is sufficient to indicate the utility and translational potential of this formulation approach. Some key products include Rapamune® and Emend®, but in addition to these oral applications, suspensions can also be used for topical and injectable applications.

Obtaining particles in the micro-range can be achieved through dry milling, unless the intention is to make a formulation for parenteral use. For a suspension obtained by dry milling, physical properties such as melting point are important. However, if the suspension is produced by media milling, in either an aqueous or non-aqueous media, then solubility becomes an important parameter in formulation selection due to its potential influence on the physical stability of the suspension. In the pharmaceutical arena, the terms suspension and nano-suspension generally refer to a colloidal dispersion of nano-sized API particles in a liquid medium that contains a suitable stabilizer or a mixture of stabilizers. The liquid media used for generation of nano-suspensions can be aqueous or non-aqueous (polyethylene glycol or oils) depending on its end application [33]. These suspensions contain a relatively small amount of the stabilizer(s) that adsorbs onto the API surface to prevent particle agglomeration during manufacturing and storage.

Ideally, the API should be insoluble in the continuous phase when formulating a suspension. If this is the case and the compound is physically and chemically stable in the continuous phase, then the formulation should be relatively uncomplicated. However, many compounds have some solubility in the continuous phase, which may lead to physical instability of the suspension. This physical instability can be a consequence of storage temperature variations leading to super-saturation and thereby potential crystal growth, that is, Ostwald ripening (see Chapter 2). This phenomenon can be counteracted by cold storage, or through the addition of polymers and/or surfactants inhibiting the crystal growth [34]. The selection of these stabilizers is compound dependent. Nevertheless, it is important for the physical stability that these excipients associate with the particles, while they solubilize the API as little as possible. Solubility or lack of the same therefore becomes a formulation optimization parameter for suspensions.

8.3.2.2 Lipid-based formulations

Lipid-based systems comprise drug delivery technologies that use a solid and/or a liquid lipid to pre-solubilize the drug compound before administration to improve the bioavailability of the API. Lipid-based systems have been traditionally classified into four main categories: type I through type IV also known as lipid formulation classification system. Readers are referred to other sources for a more detailed understanding of this classification system [35–37]. In terms of improving the bioavailability of a poorly soluble compound, LBFs are generally more suited than particle size reduction for very hard to formulate compounds, that is, compounds described as Class 2b according to the DCS system.

From a mechanistic perspective, lipid-based systems may work on two parameters; they alter the solubility of the API by solubilizing it in the lipid phase and they alter its permeability by enhanced absorption due to lipid digestion and bile-salt-mediated lipid solubilization/micellization in the gastrointestinal tract. Lipid-based drug delivery systems normally present the compound pre-solubilized, that is, in a state ready for absorption in the intestine. These formulations may consist of several different excipients and generally they can be categorized into triglycerides, mixed glycerides and polar oils, cosolvents, and water-insoluble and water-soluble surfactants [38]. Triglycerides can be further classified as long-chain triglycerides, medium-chain triglycerides and short-chain triglycerides. Mixed glycerides obtained by partial hydrolysis of vegetable oils generally have greater solvent capacity and promote emulsification. Cosolvents, including ethanol, glycerol and propylene glycol, can also be added to improve the solvent capacity of these formulations.

When defining an LBF, several considerations need to be included. Among these, API solubility and particle size upon dispersion in the gastrointestinal content, chemical stability of the system, capsule compatibility and naturally, the solubility of the compound to be incorporated into the formulation [39]. All these factors can be modulated by adjusting the composition, hence a statistical design of experiment optimization approach can be considered. The solubility of the compound in mixtures of lipid-based excipients can largely be calculated by knowing the solubility in the individual components. However, synergistic effects can be observed, so the true solubility needs to be determined experimentally in the desired formulation space. In order to define the formulation space, several solubility studies need to be conducted to define the relevant excipients. As the formulation approach is based on presenting the API in a pre-solubilized stage, solubility determinations are a critical part of defining an LBF.

LBFs can contain excipients such as triglycerides and various surfactants that are digested in the intestine by co-lipase-dependent pancreatic lipase. The formulation dosed is therefore bio-converted in the intestine after ingestion [40]. In vitro lipolysis models have been developed to mimic this effect [41–47], where the concentration in the aqueous phase have been defined as a particularly critical parameter, given that this is the available phase for absorption. Solubility in the digested aqueous phase depends upon the formulation composition, adding an additional layer of solubility determinations to the formulation work for a LBF, to provide a prediction on the bioperformance of the formulation.

8.3.2.3 Solid dispersions

Solid dispersions are generally amorphous systems that are considered to be disordered with random molecular configurations and packing of the components. The formulations may possess local and short-range order crystallinities, residual crystallinity and different density regions that add to formulation complexity. From a process perspective, various production methods exist, including spray drying which has been used for several commercial products. In order to get a spray-dried process defined, a common solvent for the polymer and the drug compound needs to be defined. As compounds are often rather lipophilic, these solvent systems are most often based on organic solvents, such as ethanol and dichloromethane.

The formulation system is designed to be thermodynamically unstable and the success of the formulation includes the ability of the composition not only to prevent the drug compound from crystallizing during storage but also upon dissolution in the intestine, as rapid crystallization upon dissolution of the dosage form may hamper its bio-performance. Defining a stable and robust formulation in a thermodynamically unstable space is not a trivial exercise. However, it can be considered as a potential strategy for compounds with a very low aqueous solubility, because this method is designed to utilize the higher solubility that an amorphous system relative to the crystalline forms in order to improve the bioavailability of the compound. Investigation of the ability of a formulation to maintain a supersaturated solution for an extended period of time sufficient to allow absorption of the drug compound is therefore a critical parameter, that is, kinetic solubility determinations. These may require formulation strategies such as the use of precipitation inhibitors to maintain supersaturation, the spring and parachute effect [48, 49] (see Chapter 2).

8.4 Use of dissolution data to optimize formulation design

It is of immense value to be able to predict the in vivo performance of a formulation or a batch based on in vitro dissolution data. An in vitro–in vivo correlation (IVIVC) allows one to distinguish between good and bad formulations and to identify changes in bioavailability in case of changes in formulation or manufacturing. Defining the dissolution method for the different dosage forms in a biorelevant and predictive manner is a specialized scientific discipline, where continuous learning from past products and projects can be highly beneficial. The methods should be defined with insights into the formulation type and the relevant physiology – and these should be linked to the variation options allowed by the dissolution method.

The selection of an appropriate dissolution medium is critical to the development of a suitable discriminating dissolution method. The earliest media tested are often simple aqueous solutions defined to mimic gastric (0.1 M HCl, pH 1.2), jejunal (acetate buffer, pH 4.5) or small intestine (phosphate buffer, pH 6.8) fluids and it is a regulatory expectation that the dissolution in these media is investigated for new drug products. Such simple aqueous systems may not offer sufficient solubilizing capacity for poorly soluble compounds, which is why surfactants like sodium lauryl sulfate, may need to be added to the medium, or biorelevant media can be used. For robust dissolution methods it is generally preferred that the selected medium meets sink conditions, that is, the solubility of the drug compound in the medium should be at least 3 times greater than the concentration obtained when the dosage form is fully dissolved. Hence, solubility is an important driver for the selection of a dissolution medium. From a formulation perspective, a predictive dissolution method is extremely valuable for the support of formulation development. Different dosage forms have different requirements for the dissolution method, so just a few examples will be mentioned here, based upon some of the dosage forms described earlier.

For systems such as solid dispersions and LBFs with a high content of cosolvents, dissolution testing is a very useful tool to provide an insight into the potential in vivo performance of a formulation. An important parameter to investigate for these two formulation types is the ability of the formulation to maintain a supersaturated solution for a extended period of time sufficient to allow absorption of the API; hence, the conditions should be defined as non-sink to evaluate this specific parameter. The use of biorelevant media and performing dissolution study for 3-4 h can be considered to simulate intestinal conditions as well as the potential residence time in the small intestine where the formulations should ensure that the drug compound is in a supersaturated state. Another often used biorelevant dissolution method is combination with a pH shift approach, that is, first conductance of the dissolution in a media simulating the conditions in the stomach followed by a change in pH and media to simulate the conditions in the intestine. Using this approach captures both gastric and intestinal pH conditions and the potential precipitation that may occur in the intestine during these physiological changes. The final dissolution method for these formulation systems may deviate from this and still provide sink conditions. The above is just an important biopharmaceutical parameter to investigate in the first part of formulation development, that is, different dissolution methods may be used as the development programme progresses.

For immediate release tablets and suspensions for oral intake, the important elements to investigate are formulation parameters that may influence bio-performance. This could be the amount of added surfactant or the influence of drug substance particle size. In addition to dissolution, an insight into the intrinsic solubility of the API, supportive in silico predictions and non-clinical investigations can help define the optimal particle size. Of course, the availability of human data with variations in what are considered to be the critical formulation parameters (with investigation of extremes) is the strongest fundament to define the dissolution method; – no matter what the dosage form.

8.5 Concluding remarks

Solubility is a critical parameter for all pharmaceutical formulations, both solid and liquid. Determination of the solubility is therefore a key and critical activity of drug formulation. The relevant media in which to determine solubility is naturally dependent upon the question asked, that is, when selecting the solid dosage form strategy, bio-relevant solubility is the most important. However, when defining a liquid formulation, solubility in the vehicle is the first determinant and a secondary determinant may be in lipolysis models for LBFs and others. Thus, conducting high-quality, accurate solubility determinations is of enormous importance for developing a robust formulation design.

References

[1] Strickley, RG. Solubilizing excipients in oral and injectable formulations. Pharm. Res. 2004, 21, 201–230.

[2] Samuelsen, L., Holm, R., Lathuile, A., & Schönbeck, C. Buffer solutions in drug formulation and processing: How pKa values depend on temperature, pressure and ionic strength. Int. J. Pharm. 2019, 560, 357–364.

[3] Ferra, MIA., Graca, JR., & Marques, AMM. Ionization of acetic acid in aqueous potassium chloride solutions. J. Chem. Eng. Data 2011, 56, 3673–3678.

[4] Goldberg, RN., Kishore, N., & Lennen, RM. Thermodynamic quantities for the ionization reactions of buffers. J. Phys. Chem. Ref. Data 2002, 31, 231–370.

[5] Karow, AR., Bahrenburg, S., & Garidel, P. Buffer capacity of biologics-from buffer salts to buffering by antibodies. Biotechnol. Prog. 2013, 29, 480–492.

[6] Mauger, JW. Physicochemical properties of buffers used in simulated biological fluids with potential application for in vitro dissolution testing: A mini-review. Dissolut. Technol. 2017, 24, 38–51.

[7] Gao, P., Rush, BD., Pfund, WP., Huang, T., Bauer, JM., Morozowich, W., Kuo, MS., & Hageman, MJ. Development of a supersaturable SEDDS (S-SEDDS) formulation of paclitaxel with improved oral bioavailability. J. Pharm. Sci. 2003, 92, 2386–2398.

[8] Gould, PL., Goodman, M., & Hanson, PA. Investigation of the solubility relationships of polar, semi-polar and non-polar drugs in mixed co-solvent systems. Int. J. Pharm. 1984, 19, 149–159.

[9] Yalkowsky, SH. Solubility and solubilization in aqueous media, USA Oxford University Press, New York, 1999.

[10] He, Y., Tabibi, SE., & Yalkowsky, SH. Solubilization of two structurally related anticancer drugs: XK-469 and PPA. J. Pharm. Sci. 2006, 95, 97–107.
[11] Holm, R., Madsen, JC., Shi, W., Larsen, KL., Städe, LW., & Westh, P. Thermodynamics of complexation of tauro- and glyco-conjugated bile salts with two modified -cyclodextrins. J. Incl. Phenom. Macrocycl. Chem. 2011, 69, 201–211.
[12] Holm, R., Østergaard, J., Schönbeck, C., Jensen, H., Shi, W., Peters, GH., & Westh, P. Determination of stability constants of tauro- and glyco-conjugated bile salts with the negatively charged sulfobutylether--cyclodextrin: Comparison of affinity capillary electrophoresis and isothermal titration calorimetry and thermodynamic analysis of the interaction. J. Incl. Phenom. Macrocycl. Chem. 2014, 78, 185–194.
[13] Schönbeck, C., Madsen, TL., Peters, GH., Holm, R., & Loftsson, T. Soluble 1:1 complexes and insoluble 3:2 complexes – Understanding the phase-solubility diagram of hydrocortisone and γ-cyclodextrin. Int. J. Pharm. 2017, 531, 504–511.
[14] Zia, V., Rajewski, RA., & Stella, VJ. Effect of cyclodextrin charge on complexation of neutral and charged substrates: Comparison of $(SBE)_{7M}$--CD to HP--CD. J. Pharm. Sci. 2001, 18, 667–673.
[15] Schönbeck, C., Westh, P., Madsen, J-C., Larsen, KL., Städe, L., & Holm, R. Hydroxypropyl substituted β-cyclodextrins: Influence of degree of substitution on the thermodynamics of complexation with tauro- and glyco-conjugated bile salts. Langmuir 2010, 26, 17949–17957.
[16] Schönbeck, C., Westh, P., Madsen, J., Larsen, KL., Städe, L., & Holm, R. Methylated β-cyclodextrins: Influence of degree and pattern of substitution on the thermodynamics of complexation with tauro- and glyco-conjugated bile salts. Langmuir 2011, 27, 5832–5841.
[17] Holm, R., Olesen, NE., Alexandersen, SD., Dahlgaard, BN., Westh, P., & Mu, H. Thermodynamic investigation of the interaction between cyclodextrins and preservatives – application and verification in a mathematical model to determine the needed preservative surplus in aqueous cyclodextrin formulations. Eur. J. Pharm. Sci. 2016, 87, 22–29.
[18] Lipinski, CA., Lombardo, F., Dominy, BW., & Feeney, PJ. Experimental and computational approaches to estimate solubility and permeability in drug discovery and development settings. Adv. Drug Deliver Rev. 1997, 23, 3–25.
[19] Veber, DF., Johnson, SR., Cheng, HY., Smith, BR., Ward, KW., & Kopple, KD. Molecular properties that influence the oral bioavailability of drug candidates. J. Med. Chem. 2002, 45, 2615–2623.
[20] Amidon, G., Lennernäs, H., Shah, V., & Crison, J. A theoretical basis for a biopharmaceutical drug classification: The correlation of in vitro drug product dissolution and in vivo bioavailability. Pharm. Res. 1995, 12, 413–420.
[21] Yazdanian, M., Briggs, K., Jankovsky, C., & Hawi, A. 2004. The "high solubility" definition of the current FDA guidance on biopharmaceutical classification system may be too strict for acidic drugs. Pharm. Res. 2004, 21, 293–299.
[22] Galia, E., Nicolaides, E., Hörter, D., Löbenberg, R., Reppas, C., & Dressman, JB. Evaluation of various dissolution media for predicting in vivo performance of class I and II dugs. Pharm. Res. 1998, 15, 698–705.
[23] Vertzoni, M., Fotaki, N., Kostewicz, E., Stippler, E., Leuner, C., Nicolaides, E., Dressman, JB., & Reppas, C. Dissolution media simulating the intralumenal composition of the small intestine: Physiological issues and practical aspects. J. Pharm. Pharmacol. 2004, 56, 453–462.
[24] Kloefer, B., van Hoogevest, P., Moloney, R., Kuentz, M., Mathew, LSL., & Dressman, JB. Study of a standardized taurocholate-lecithin powder for preparing the biorelevant media FaSSIF and FeSSIF. In: SIF powder references: How it's been used to make biorelevant media. Dissolut. Technol. 2010, 17, 6–13.

[25] Butler, JM., & Dressman, JB. The developability classification system: Application of biopharmaceutics concepts to formulation development. J. Pharm. Sci. 2010, 99, 4940–4954.
[26] Johnson, KC., & Swindell, AC. Guidance in the setting of drug particle size specifications to minimize variability in absorption. Pharm. Res. 1996, 13, 1795–1798.
[27] Curatolo, W. Physical chemical properties of oral drug candidates in the discovery and exploratory development settings. Pharm. Sci. Technol. Today 1998, 1, 387–393.
[28] Sun, D., Yu, LX., Hussain, MA., Wall, DA., Smith, RL., & Amidon, GL. 2004. In vitro testing of drug absorption for drug developability assessment. Curr. Opin. Drug Discov. Devel. 2004, 7, 75–85.
[29] Brewster, M. Modified from a presentation at the world conference on drug absorption transport and delivery, Uppsala, 2013, 24–27 June.
[30] Malamatari, M., Taylor, KM., Malamataris, S., Douroumis, D., & Kachrimanis, K. Pharmaceutical nanocrystals: Production by wet milling and applications. Drug Disc Today 2018, 23, 534–547.
[31] Chen, ML., John, M., Lee, SL., & Tyner, KM. Development considerations for nanocrystal drug products. AAPS J. 2017, 19, 642–651.
[32] Gao, L., Liu, G., Ma, J., Wang, X., Zhou, L., Li, X., & Wang, F. Application of drug nanocrystal technologies on oral drug delivery of poorly soluble drugs. Pharm. Res. 2013, 30, 307–324.
[33] Chin, WWL., Parmentier, J., Widzinski, M., Tan, EH., & Gokhale, R. A brief literature and patent review of nanosuspensions to a final drug product. J. Pharm. Sci. 2014, 103, 2980–2999.
[34] Moreton, RC. Commonly used excipients in pharmaceutical suspensions, Kulshreshtha, AK, Singh ON, Wall GM., eds, Pharmaceutical suspensions, Springer, New York, USA, 2010, 67–102.
[35] Feeney, OM., Crum, MF., McEvoy, CL., Trevaskis, NL., Williams, HD., Pouton, CW., Charman, WN., Bergström, CAS., & Porter, CJH. 2016. 50 years of oral lipid-based formulations: Provenance, progress and future perspectives. Adv. Drug. Deliver Rev. 2016, 101, 167–194.
[36] Pouton, CW., & Porter, CJ. Formulation of lipid-based delivery systems for oral administration: Materials, methods and strategies. Adv. Drug Deliver. Rev. 2008, 60, 625–637.
[37] Savla, R., Browne, J., Plassat, V., Wasan, KM., & Wasan, EK. Review and analysis of FDA approved drugs using lipid-based formulations. Drug Dev. Ind. Pharm. 2017, 43, 1743–1758.
[38] Kalepu, S., Manthina, M., & Padavala, V. Oral lipid-based drug delivery systems – An overview. Acta Pharm. 2013, 3, 361–372.
[39] Holm, R., Jensen, IHM., & Sonnergaard, J. Optimization of self-microemulsifying drug delivery systems (SMEDDS) using a D-optimal design and the desirability function. Drug Dev. Ind. Pharm. 2006, 32, 1025–1032.
[40] Porter, CJH., Trevaskis, NL., & Charman, WN. Lipids and lipid-based formulations: Optimizing the oral delivery of lipophilic drugs. Nat. Rev. Drug. Disc. 2007, 6, 231–248.
[41] Williams, HD., Sassene, P., Kleberg, K., Bakala-N'Goma, JC., Calderone, M., Jannin, V., Igonin, A., Partheil, A., Marchaud, D., Jule, E., Vertommen, J., Maio, M., Blundell, R., Benameur, H., Carrière, F., Müllertz, A., Porter, CJH., & Pouton, CW. Toward the establishment of standardized in vitro tests for lipid-based formulations, part 1: Method parameterization and comparison of in vitro digestion profiles across a range of representative formulations. J. Pharm. Sci. 2012, 101, 3360–3380.
[42] Williams, HD., Anby, MU., Sassene, P., Kleberg, K., Bakala-N'Goma, JC., Calderone, M., Jannin, V., Igonin, A., Partheil, A., Marchaud, D., Jule, E., Vertommen, J., Maio, M., Blundell, R., Benameur, H., Carrière, F., Müllertz, A., Pouton, CW., & Porter, CJH. Toward the establishment of standardized in vitro tests for lipid-based formulations. 2. The effect of bile salt concentration and drug loading on the performance of type I, II, IIIA, IIIB, and IV formulations during in vitro digestion. Mol. Pharm. 2012, 9, 3286–3300.

[43] Williams, HD., Sassene, P., Kleberg, K., Calderone, M., Igonin, A., Jule, E., Vertommen, J., Blundell, R., Benameur, H., Müllertz, A., Pouton, CW., & Porter, CJH. Toward the establishment of standardized in vitro tests for lipid-based formulations, part 3: Understanding supersaturation versus precipitation potential during the in vitro digestion of type I, II, IIIA, IIIB and IV lipid-based formulations. Pharm. Res. 2013, 30, 3059–3076.
[44] Williams, HD., Sassene, P., Kleberg, K., Calderone, M., Igonin, A., Jule, E., Vertommen, J., Blundell, R., Benameur, H., Müllertz, A., Porter, CJH., & Pouton, CW. Toward the establishment of standardized in vitro tests for lipid-based formulations, part 4: Proposing a new lipid formulation performance classification system. J. Pharm. Sci. 2014, 103, 2441–2455.
[45] Bakala-N'Goma, JC., Williams, HD., Sassene, P., Kleberg, K., Calderone, M., Jannin, V., Igonin, A., Partheil, A., Marchaud, D., Jule, E., Vertommen, J., Maio, M., Blundell, R., Benameur, H., Müllertz, A., Pouton, CW., Porter, CJH., & Carrière, F. Toward the establishment of standardized in vitro tests for lipid-based formulations. 5. Lipolysis of representative formulations by gastric lipase. Pharm. Res. 2015, 32, 1279–1287.
[46] Sassene, P., Kleberg, K., Williams, HD., Bakala-N'Goma, JC., Carrière, F., Calderone, M., Jannin, V., Igonin, A., Partheil, A., Marchaud, D., Jule, E., Vertommen, J., Maio, M., Blundell, R., Benameur, H., Porter, CJH., Pouton, CW., & Müllertz, A. Toward the establishment of standardized in vitro tests for lipid-based formulations, part 6: Effects of varying pancreatin and calcium levels. AAPS J. 2014, 16, 1344–1357.
[47] Thomas, N., Holm, R., Müllertz, A., & Rades, T. Characterizing lipid lipolysis and its implication in lipid-based formulation development. AAPS J 2012, 14, 860–871.
[48] Brouwers, J., Brewster, ME., & Augustijns, P. Supersaturating drug delivery systems: The answer to solubility-limited oral bioavailability?. J. Pharm. Sci. 2009, 98, 2549–2572.
[49] Guzman, HR., Tawa, M., Zhang, Z., Ratanabanangkoon, P., Shaw, P., Gardner, CR., Chen, H., Moreau, JP., Almarsson, O., & Remenar, JF. Combined use of crystalline salt forms and precipitation inhibitors to improve oral absorption of celecoxib from solid oral formulations. J. Pharm. Sci. 2007, 96, 2686–2702.

Christoph Saal

9 The relevance of solid-state forms for solubility

9.1 Solid-state forms for pharmaceutical research, development, and commercial manufacturing

9.1.1 Parameters relevant for solid-state form selection

Selection of solid-state forms during pharmaceutical research, development, and finally for market-products is a key step which nowadays is typically carried out during the late research phase i.e before nomination of a candidate for clinical development or at an early state of clinical development. Selection of solid-state forms for drugs today, is a topic which requires much more attention than 20 years ago; and there are good reasons for this, which are discussed within this chapter. The main aspects to be considered for solid-state form selection of a drug are:
- Which solid-state forms are available for selection?
- What should be achieved by selecting a certain solid-state form?
- At which phase does solid-state selection takes place i.e. during research, development, or even marketing?

Generally, the selection of a certain solid-state form for an API during research and development must fulfill a multitude of requirements. Sometimes – if not frequently – these requirements are even conflicting. This will become obvious from the following list of properties which are influenced by solid-state forms.

9.1.1.1 Solubility

As already discussed in the introduction and other chapters in this book, solubility has become a key challenge for modern research compounds. We must distinguish between "thermodynamic solubility", "kinetic solubility", and related properties such as the "dissolution rate". Solubility is governed by the hydration energy or solvation energy, bearing in mind also organic solvents as well as by the lattice energy considering also amorphous solid-state forms intermolecular interactions – of an API. These intermolecular interactions in the solid-state are the reason, why solid-state selection influences solubility.

Christoph Saal, Merck KGaA, Frankfurter Strasse 250, 64293 Darmstadt, Germany

https://doi.org/10.1515/9783110559835-009

9.1.1.2 Dissolution rate

Like solubility, the solid-state form will also have consequences for the dissolution rate. In contrast to solubility dissolution rate represents not a thermodynamic but a kinetic parameter. Therefore, the key parameter is activation energy: which is the activation energy to liberate an API molecule from the surface of a crystal to dissolve it in water or another solvent. This depends of course also on intermolecular interactions within the crystal lattice but also on other parameters such as wettability of the surface of a crystal.

Both solubility and dissolution rate influences bioavailability for orally administered drugs, as discussed in respective chapters of this book. Therefore, selection of solid-state forms is also closely linked to bioavailability. Pharmacokinetics represents one of the disciplines which must be considered for solid-state form selection. Of course, other disciplines such as manufacturing and formulation development also have certain requirements on solubility and dissolution rate. In both cases, uncontrolled supersaturation, might lead to uncontrolled precipitation has to be strictly avoided. For this reason, it becomes obvious that even in the case of a liquid formulation, for example, an intravenous infusion, an understanding of the solid-state behaviour of an API is necessary. Otherwise one might, for example, use a metastable solid-state form in a formulation, which has a higher solubility compared to the thermodynamically stable solid-state form. The latter might crash out in an uncontrolled way from a solution-based formulation which is supersaturated with regard to the thermodynamic stable solid-state form.

9.1.1.3 Melting point

The melting point reflects the lattice energy of a solid: The stronger the interactions within the crystal lattice of a substance, the more difficult it will be to break down the lattice. As the destruction of the crystal lattice represents the first step of the dissolution process – followed by hydration or more generally solvatation of the substance – a high melting point will generally result in low solubility. This is discussed in further detail in other parts of this book, based on the general solubility equation as introduced in chapters 1 and 7 of this book. As different solid state-forms will have different melting points, solubility of a substance will be linked to its solid-state form. This holds true for solid-state forms with the same chemical composition. Different polymorphs of a substance will have different solubility. As mentioned earlier on, the thermodynamically stable polymorph will have the lowest solubility. However, as the thermodynamically stable polymorph is not always the polymorph showing the highest melting point, this does not mean that the polymorph with the highest melting point exhibits the lowest solubility. For example, for enantiotropically related polymorphs where the low-temperature form is thermodynamically stable at room temperature, its melting point will be lower compared to the high temperature form, see Figure 9.1.

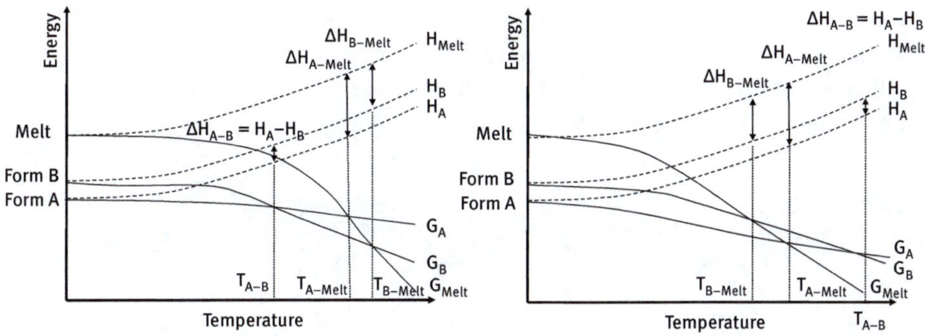

Figure 9.1: Enthalpy diagrams for enantiotropically (left) and monotropically related polymorphs (right). At a certain temperature the thermodynamically stable polymorph exhibits the lower free enthalpy. $T_{A\text{-}B}$ represents the transition temperature between forms A and B. $T_{A\text{-}Melt}$ and $T_{B\text{-}Melt}$ are the melting temperatures of form A and B. $\Delta H_{A\text{-}B}$ represents the transition enthalpy between form A and B and $\Delta H_{A\text{-}Melt}$ and $\Delta H_{B\text{-}Melt}$ represent the melting enthalpies.

Beyond true polymorphs, solid-state forms such as pseudo-polymorphs (hydrates, solvates), co-crystals and pharmaceutical salts show a different chemical composition. They also exhibit a melting point which might be higher or lower compared to the original substance and its polymorph. Melting points of pharmaceutical salts are frequently higher compared to the free base or free acid of the substance [1]. This is caused by the fact that in pharmaceutical salts ionic-interactions also (Coulomb-interactions) contribute to the lattice energy. This is not the case for free bases and free acids. On the other hand, counterions of pharmaceutical salts upon hydration also contribute to an enhanced hydration energy. Consequently, in many cases pharmaceutical salts show a higher melting point but at the same time an enhanced solubility compared to the free base or free acid of an API. The same holds true for their dissolution behaviour. Clearly this represents one of the reasons for the more widespread use of a variety of different counterions to produce pharmaceutical salts. This trend is seen since the past decades [2] and goes hand in hand with the advent of APIs with low solubility.

An approach to improve solubility and dissolution rate based on lowering the lattice energy and the melting point is the usage of "weakly coordinating" counterions for pharmaceutical salts. These counterions carry a charge which is in a big and frequently symmetric ion. Accordingly, despite being ionic compounds, Coulomb interactions in the crystal lattice get weak as the distance (r) in eq. (9.1) between positive and negative charges (q_1 and q_2) is increased compared to conventional pharmaceutical salts such as, for example, chlorides, sulphates, phosphates, or others.

$$F = \frac{1}{4\pi\varepsilon_0}\frac{q_1 q_2}{r^2} \tag{9.1}$$

Descriptions of such salts can be found in [3–7]. In extreme cases, even liquid pharmaceutical salts with a melting point below room temperature can be obtained. This leads to greatly enhanced solubility and dissolution rate.

Figure 9.2 shows the consequence of increased radius of "weakly coordinating" counterions together with an example.

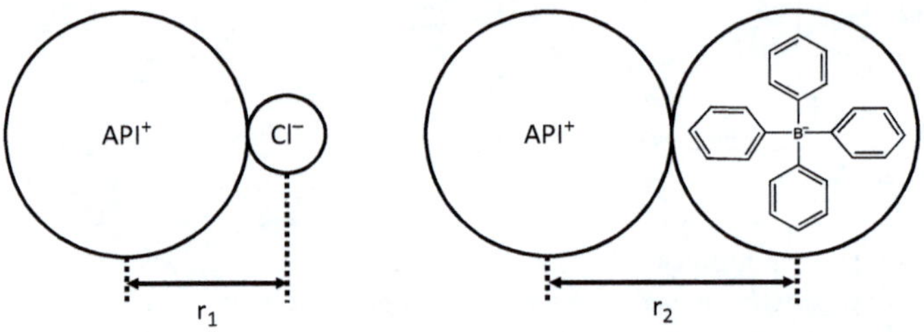

Figure 9.2: Principle of "weakly coordinating" counterions and example for a weakly coordinating counterion. The chloride ion is considerably smaller compared to the tetraphenylborate ion ($r_1 < r_2$). Accordingly, the Coulomb-force and lattice energy for the tetraphenylborate salt are weaker compared to the chloride salt.

Similar aspects as discussed for pharmaceutical salts apply for hydrates, solvates, and co-crystals. Instead to an ionic contribution to the crystal lattice energy and an ion-dipole interaction to the solvation energy, in such cases, we have additional non-ionic contributions to the lattice energy and solvation energy.

Beyond its effect on solubility, the melting point of a solid-state form used for development of an API is also important for manufacturing of drug-substance and drug product. Relevant aspects are:

- The melting point of the solid-state form should be sufficiently high to allow drying the API after crystallization, filtration or centrifugation to get rid of residual solvents and match the ICH limits [8].
- The melting point of the solid-state form should match formulation requirements. For standard formulation approaches such as blending, granulation, and tableting, the melting point should be sufficiently high (e.g. above 110 °C). Enabling formulation techniques will have special requirements. As two examples, nano-milling of a certain solid-state form of an API will require a high melting point as the process induces much thermal stress [9]. On the other hand, hot-melt extrudation [10, 11] will require a low melting point, as high melting solid-state forms might represent challenges for this formulation technique.

9.1.1.4 Hygroscopicity

Generally, pharmaceutical development of an API is facilitated by a low hygroscopicity. Hygroscopicity can be measured and classified according to Pharmaceopoeias [12]. A more detailed and useful picture is obtained from dynamic-vapor-sorption measurements. During these measurements the substance is placed onto a sensitive balance and equilibrated in an atmosphere of well-defined relative humidity and temperature. Using this approach, the water uptake of the substance is measured over a broad range of relative humidity, for example, 0% to 98% relative humidity. Again, water uptake will depend on the solid-state form. Water uptake can be due to physisorption at the surface of the substance. Water uptake might also be due to chemisorption which might result in the formation of stoichiometric or non-stoichiometric hydrates, for example, channel-hydrates. A good overview on the assessment of hygroscopicity is provided in [13] and beyond the scope of this book. However, hygroscopicity and aqueous solubility rely on the same effect: the interaction of a substance with water. Accordingly, in many cases, high solubility is accompanied by increased hygroscopicity. Improving solubility and dissolution rate on the one hand and keeping hygroscopicity at a manageable level on the other hand represents taking care of two opposed properties. Thus this can be a challenge and requires careful selection of useful solid-state forms.

9.1.1.5 Particle-shape (habit)

Different solid-state forms will show different particle shape also called habit. This could be, for example, cubes, platelets, rods or needles. The crystal habit has an important influence on the processability of an API. For example, the speed of a filtration step can depend on the crystal habit: here platelets can be a problem if they tend to clog the filter. This can result in slow filtration steps or in filtration steps which have a high variability with regard to speed. Particle habit will also directly influence flowability of the API. Generally, more isotropic habits – ideally spheres or cubes – will result in optimal flow properties. On the contrary, more anisotropic particles will have less optimal or even problematic flowability. An extreme case might be a situation where particles might be hair-like. Flowability is a key property during formulation of an API, including steps like blending of the API with excipients or tableting. Finally, particle habit also influences the bulk density of an API. More isotropic habits such as cubes will result in higher bulk density compared to platelets, rods or even needles. Bulk density also matters for formulation aspects. For example, filling an API into capsules requires a certain bulk density to realize a certain dose of the API in a certain size of a capsule.

Particle habit results from the speed of growth of a crystal along certain crystallographic axis. If crystal growth along all crystal axis occurs at the same speed, this will lead to isotropic particles, such as cubes. If crystal growth along one axis is much faster compared to growth along other directions, this will result in needle shaped crystals. Accordingly, the crystal habit is a result of the ratio of the speeds of crystal growth for all surfaces of a crystal as defined by the Miller indices. As mentioned in the beginning of this paragraph, different solid-state forms will have different crystal habits. However, even a well-defined solid-state form can occur in different crystal habits. The reason for this is that the speed of crystal growth along a certain axis is not only a function of the solid-state form but also of the crystallization process. This means that crystal habit can be influenced by temperature during the crystallization process, the choice of the solvent or solvent mixture, stirring or even additives used to modify crystal habit. During recent years, more attention has been paid to this topic, emerging in the field of “particle design” [14].

Solubility will not be influenced by particle habit. However, dissolution rate can be influenced by particle habit. The dissolution process mainly represents a reverse crystallization process. Accordingly, different faces of a crystal will also dissolve with a different speed and consequently dissolution rate will be influenced by particle habit.

9.1.1.6 Particle-size distribution

As crystal habit, particle-size will also influence processability. Parameters such as flowability, bulk-density, and processability during filtration also depend on the particle size distribution of an API.

Particle size also directly influences dissolution rate, as smaller particles possess a larger specific surface from which molecules of the API can dissolve into solution. The relation between particle size and dissolution is described by the Noyes-Whitney eq. (9.2) [15].

$$\frac{dc}{dt} = \frac{D^{*}S}{h}(c_S - c_t) \tag{9.2}$$

c = concentration of solute,
dc/dt = dissolution rate,
D = diffusion coefficient,
S = surface of particle which is in contact with solvent,
h = thickness of diffusion layer,
c_S = thermodynamic solubility,
c_t = concentration of solute in solution.

Accordingly, particle size reduction is a common formulation approach for BCS class II APIs to realize higher bioavailability.

Beyond influencing the dissolution rate of an API, particle size can also have an influence on solubility. Solubility as a function of particle size can be described by the Kelvin equation as given by eq. (9.3) [16]

$$RTln\frac{c}{c_\infty} = \frac{2\gamma M}{\rho r} \tag{9.3}$$

R = unviversal gas constant,
T = absolute temperature,
c = solubility of spherical particle with radius r,
c_∞ = solubility of infinite large particle,
γ = surface tension,
M = molar mass,
ρ = density,
r = radius of particle

According to the Kelvin equation, solubility will increase with smaller particle size. The mechanism behind the phenomenon is the curvature of a crystal surface which increases with decreasing particle size as shown in Figure 9.3. The number of molecules in the neighbourhood of a molecule within the same crystal will decrease with decreasing particle size. Accordingly, intermolecular forces which retract a molecule and slow down dissolution, decreases with decreasing particle size.

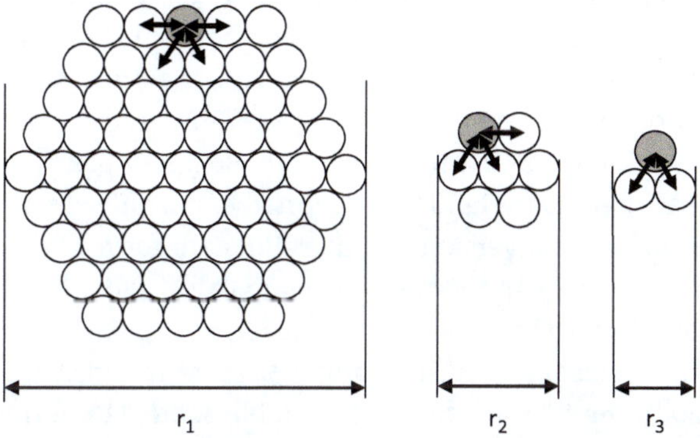

Figure 9.3: Intermolecular forces in 2-dimensional crystals of three different particle sizes: $r_1 > r_2 > r_3$. With decreasing particle size the average number of next neighbour molecules and intermolecular interactions with them decreases. r_1 = 4 interactions with neighbour molecules, r_2 = 3 interactions with neighbour molecules, r_3 = 2 interactions with neighbour molecules. Accordingly, with decreasing particle size it gets easier for molecules to leave the crystal and dissolve.

However, a pronounced effect of particle size on solubility is only observed for very small particles, typically nanoparticles.

Crystal size can be defined using processing steps such as crystallization, micronization or nano-milling . The first represents a bottom-up process, whereas micronization and nano-milling are top-down approaches.

Particle size cannot be described by a single parameter, as API particles are never all the same size. Instead, an API batch exhibits a particle size distribution which might be narrow or broad and might be mono-modal, bi-modal, or even multi-modal.

Additionally, expect for cubes and spheres, which are very uncommon shapes for API crystals, the particle size also reflects the habit of the particles. This must be taken into account when measuring particle size using methods such as laser-diffraction which are based on the assumption of isotropic particles.

It has also to be born in mind that both, the Noyes-Whitney equation and the Kelvin equation rely on the same assumption of isotropic particles. As this assumption is almost never fulfilled for APIs, both relations between particle size and dissolution rate as well as between particle size and solubility represent only approximations of real scenarios.

9.1.1.7 Stability

Stability of APIs as well for drug products is key as this is the basis for a sufficient shelf-life. Guidance how to carry out stability studies in provided in the ICH Q1A-F [17]. The term “stability” is frequently just referred to chemical stability which means the way in which chemical impurities are formed during storage by degradation, oxidation, hydrolysis, or other processes. However, beyond chemical stability, physical stability also matters: solid-state forms as present in an API or a drug product must usually not convert into other solid-state forms over shelf-life under relevant conditions. Relevant conditions include a certain range of temperature and relevant humidity in a certain packaging material. If the thermodynamically stable solid-state form is chosen, the risk for a conversion of this solid-state form to another solid-state form upon storage is low, as the thermodynamically stable form already represents the form which is lowest in energy. However, if a metastable solid-state form is selected as API for usage in a drug product, there is a risk for conversion of this solid-state form to the thermodynamically stable solid-state form or to another metastable solid-state form which is lower in energy, as this will lower the free enthalpy of the system. This conversion can happen spontaneously and might be difficult to control. This is one of the main reasons, why in most cases the thermodynamically stable solid-state form of an API is selected for pharmaceutical development. With regard to polymorphs, conversion of a metastable polymorph including an amorphous form to a lower energy polymorph might occur. Other scenarios are the conversion of an anhydrate to a hydrate at elevated humidity levels. There is also a general increased tendency for formation of hydrates at constant humidity, if the temperature is lowered. Vice versa, hydrates might convert to anhydrates upon reduced humidity or increased temperature [18]. Situations in which dynamic conversion between anhydrates and hydrates occur at conditions close to room temperature and ambient humidity might be especially difficult to handle. Finally, with regard to polymorphs, there might be also conversions from hydrates/solvates to anhydrates/ansolvates due to the loss of solvent. This risk is triggered by the fact that usually the partial pressure of the respective solvent in the surrounding atmosphere is zero. However, usage of solvates as APIs represents an exceptional case. Only a few solvates are used in APIs, for example, Trametinib which represents a DMSO solvate [19] and Darunavir representing an ethanol solvate [20]. Similar considerations as for the stability of solvates are relevant for co-crystals.

Beyond the fact that stability is considered critical with regard to polymorphs – at least since the famous "Ritonavir case" which will be discussed later in this chapter – stability is also relevant for pharmaceutical salts. Pharmaceutical salts might hydrolyze upon exposure to high humidity and yield either the free base or the free acid. The risk for hydrolysis of pharmaceutical salts is especially pronounced for salts of weak bases and weak acids, where the difference in pKa of the free base or free acid of the API and the salt former is low. Prasugrel represents a case where such a hydrolysis of a pharmaceutical salt – a hydrochloride – occurred and this was discovered only in late-stage clinical trials [21]. The reason to select the hydrochloride salt for clinical development was improved solubility and dissolution of this pharmaceutical salt compared to the free base. However, it turned out that the salt converted to the free base upon storage. Accordingly, tablets contained a mixture of hydrochloride salt and free base. This example – like the "Ritonavir case" illustrates that care must be taken about stability of pharmaceutical salts and that stability does not only matter for polymorphs but also for pharmaceutical salts.

The above example shows that solid-state selection as mentioned in the beginning of this chapter deals with conflicting properties. In both examples – Ritonavir and Prasugrel metastable solid-state forms have been used (unintentionally). The benefit was an improved solubility. However, this induced an imminent risk of low or non-existing stability of the drug products.

Humidity and temperature do not present the only two parameters which matter for stability. ICH guidance also deals with light stress. However, the author is not aware of any example where light induces a conversion of a solid-state form. Pressure is a parameter which is not mentioned in ICH Q1, but is relevant for stability of solid-state forms. The pressure induced conversion from graphite to diamond is well known. Similar pressure induced solid-state conversions can occur also for APIs. This is usually not relevant for stability upon storage during shelf life, but this matters for processing steps which include a high mechanical stress, including tableting, roller-compaction or even steps like sieving.

Finally, conversions of a solid-state form within drug substance or drug product are not strictly forbidden. Conversions of polymorphs can be non-critical if they do not affect processability and bioavailability. To fulfill the latter, the respective polymorphs must exhibit similar solubility and dissolution rate which is not affecting bioavailability or pharmacokinetics, efficacy, or safety in a critical way. A description of how such equivalence must be assessed is beyond the scope of this book and can be found in ICH Q6A [22].

9.1.1.8 Molar mass of the API

The molar mass of an API is the next property to be discussed in this chapter which is relevant for solid-state form selection. The molar mass of an API is determined by the solid-state form in so far, as a true polymorph (anhydrate, ansolvate) shows the lowest possible molar mass. All other solid-state forms, such as pharmaceutical salts, hydrates, solvates or co-crystals, will exhibit higher molar masses. The increase of molar mass in these solid-state forms is caused by a component which is pharmacologically inactive but might improve other properties of the API such as dissolution rate or solubility.

Molar mass of the API can become important in two scenarios:

- The required dose of the API is high, for example, an oral tablet which requires 500 mg or more of the active moiety. In such cases the additional mass which is introduced by the salt-former, water, solvate-former or co-crystal former is added to the mass of the active moiety. This increases the dose of the API to be administered. This becomes especially critical, if the molar mass of the API is low. For example, if an active moiety has a molar mass of 250 g/mol and a dose of this moiety of 500 mg is required, salt formation using a counterion with a molar mass of 100 g/mol (which is in the range of a sulfate salt) will increase the dose to 700 mg. Of course, this only holds true if the pharmaceutical salt exhibits the same bioavailability as the free base and no improvement of the pharmacokinetics is observed.
- The required does of the API is low, for example, an oral tablet which requires the active moiety in the µg range. In this case an increase of the molar mass of the API due to salt-formers, water, solvate-former, or co-crystal former can be beneficial. For such low-dose formulations frequently process steps such as blending can be critical as it is difficult to reach a homogeneous distribution in the blend and finally to fulfill content-uniformity requirements. Here an increase of the molar mass can be beneficial to develop a simpler and more robust blending process.

9.1.1.9 Toxicity and pharmacological properties

The properties which have been discussed so far with regard to selection of solid-state forms mainly refer to the physical and chemical behaviour of an API. Beyond this, pharmacological and toxicological properties of salt-formers and co-crystal formers as well as solvate formers also matter for solid-state selection. In the following section we use the example of a pharmaceutical salt to discuss the relevant aspects. The same holds true also for other solid-state forms such as solvates, co-crystals and hydrates.

Generally, one must distinguish two types of toxicological and pharmacological effects of the salt former:

- Direct influence of the salt-former on toxicology and pharmacology

In this case the counterion used for salt formation itself possesses certain toxicological or pharmacological properties. Pfannkuch et al. [23] defined three classes of salt-formers regarding their toxicological and pharmacological effect:

- 1st Class Salt-Formers: These are salt-former which can be used without restriction because they form physiologically ubiquitous ions, or because they occur as intermediate metabolites in biochemical pathways. For example, chlorides and sodium salts belong to this class.
- 2nd Class Salt-Formers: These are salt-formers that are not occurring naturally, but, so far, their use has shown low toxicity and good tolerability.
- 3rd Class Salt-Formers: These are salt-formers which are used in particular situations in order to achieve special effects.

Additionally, one must distinguish between general toxicological properties of salt-formers and properties which are relevant for a certain indication. If patients within an indication show a disease relevant for a specific organ which is at the same time the target of the toxicological profile of the counter ion, this can be critical. As an example, the use of oxalates as salt-formers can be problematic in indications where renal function in impaired or where patients suffer from renal disease [24, 25].

For the direct effect of salt formers on toxicology and pharmacology the pharmacokinetic of the counterions matters. This will be influenced by the dose, formulation, and administration route, for example oral, intravenous, pulmonary, and topical route. An overview of maximum doses of salt-formers used in approved drugs for several administration routes is provided in [26]. This data cannot replace an in-depth pharmacokinetic, pharmacodynamic and toxicological assessment, but it can be helpful for the design of salt-screens during the research phase.

- Indirect influence of the salt-former on toxicology and pharmacology

The indirect influence of the salt-former is based on its effect on the pharmacokinetics of the API, for example, an improved dissolution rate of the pharmaceutical salt compared to the free base and therefore an earlier t_{max}, higher c_{max}, and higher AUC (Area Under the Curve) and bioavailability. These pharmacokinetic parameters will influence pharmacodynamic and toxicological properties of the API, for example, is the c_{max} becomes higher compared to the level, where toxicological effects of the API become relevant.

9.1.1.10 Processability

Processability represents another aspect which is relevant for selecting a solid-state form for a drug. Processability is relevant API manufacturing as well as for drug-product manufacturing. Whereas in API manufacturing usually only the final steps starting with crystallization of the API are relevant but for drug product manufacturing typically all steps can be affected by solid-state properties of the API. Steps which must be considered for API manufacturing are crystallization, filtration, or centrifugation to separate the API from the mother liquor, drying, and milling or micronization as well as filling of the API into bins. For drug product manufacturing the variety of process steps is much broader including steps such as blending, dry or wet granulation, roller compaction, compression, film-coating but also special process steps used for enabling formulation such as hot-melt extrusion, spray drying, co-precipitation, or nano-milling.

Whilst there are many process steps which theoretically can be affected by solid-state properties, in most API and drug product manufacturing processes, only some of them – but not all – are critical with regard to solid-state properties.

Main parameters which will have an influence on processability have been already mentioned in previous parts of this chapter and other chapter of this book: aspects of crystallization will be discussed in chapter 10. Filtration, flowability, and bulk density are mainly influenced by particle habit of the solid-state form as mentioned in this chapter earlier on. Generally, flowability and represents the most important parameter, as it is not only the flowability of the API itself which matters, but also the flowability of many pharmaceutical intermediates if influenced by the flowability of the API, especially for high-dose drugs.

Finally, especially for bio-enabling formulations which are intended to improve the bioavailability of low-soluble (e.g. BCS class II or IV) drugs, thermal behaviour matters, and there is not a one-fits-all solution. Different enabling formulations will require different solid-state properties. Just as an example, hot-melt-extrusion will profit from a lower melting point, whereas for nano-milling such a melting point can be critical as it will lead to softening, partial melting, and aggregation of the API.

9.1.2 Overview on solid-state forms

There are many different solid-state forms, which can be used for pharmaceutical development and marketing of drugs. The plethora of solid-state forms includes pharmaceutical salts, polymorphs, co-crystal, amorphous phases and others. To structure these solid-state forms a system as shown in Figure 9.4 is useful.

Classification of solid-state forms is not always straight-forward, as there are several principles according to which classification can be carried out, for example, whether solid-state forms show a long-range order or not, whether they are ionic or not, which chemical composition they exhibit and others. Accordingly, the classification as shown in Figure 9.4 represents just one way how to structure solid-state forms, but not the only one.

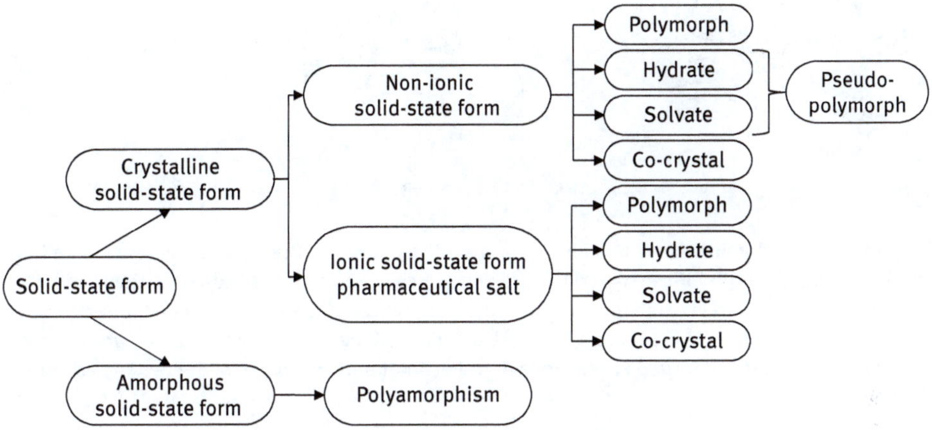

Figure 9.4: Classification of solid-state forms.

Before we discuss relevant aspects with regard to solubility of these solid-state forms, first a brief description of the respective solid-state forms will be provided.

As first principle, solid-state forms can be classified as either being crystalline or amorphous. Crystalline means that within the solid-state form, there is a long-range order within the solid material: Distances in a certain direction between constituents of the solid-state form – such as molecules or ions – are constant. Generally, this holds true in all three dimensions. However, cases exist, where a long-range order exists just in one or two dimensions, but not in three dimensions. Such phases are called liquid-crystalline phases or mesophases, for example, smectic phases, nematic-phases, or cholesteric phases. However, for APIs this represents a very special case, which is only encountered rarely. As intermolecular interactions along the disordered directions within the liquid-crystalline phase are weaker compared to an ordered state, solubility of liquid-crystalline phases is generally higher compared to crystalline phases. Solid-state forms which do not show any long-range order in any direction are called amorphous phases. They represent merely a chaotic, disordered structure of their constituents within the solid-state form. As disorder occurs in a three-dimensional way – so even more pronounced compared to liquid crystalline phases – intermolecular interactions in an amorphous phase are considerably weaker compared to a crystalline phase. Together with the fact that hydration or solvation energy depends only on the species which undergoes hydration or solvation and thus does not depend on the solid-state form, amorphous phases show a significantly enhanced solubility compared to crystalline phases. Even if amorphous phases are disordered, there can be different ways to realize disorder. This results in different amorphous phases of an API and is called polyamorphism. The effect is schematically outlined in Figure 9.5.

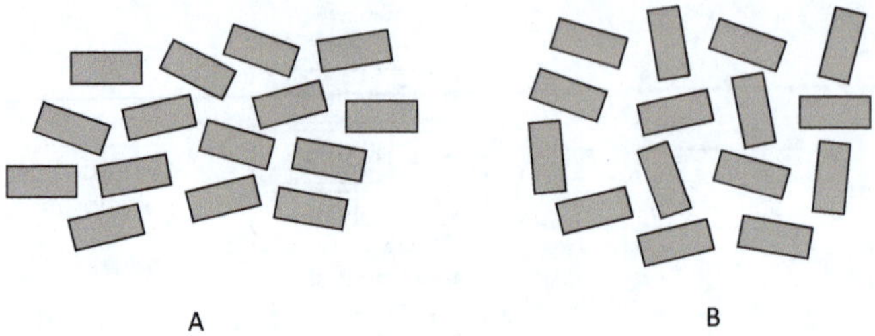

Figure 9.5: Two different amorphous phases of the same molecule. The molecule is represented by the grey rectangle. Both forms – A and B – are disordered, so amorphous. However, they exhibit different structures and accordingly different physicochemical properties. In form A interactions between the short sides of the rectangles and between the long sides of the rectangle are dominant. In contrast, in form B interactions between short and long sides of the rectangles are dominant.

The distinction whether a solid-state form is either ionic or non-ionic can be used as a second classification criterion. Ionic APIs which are composed of the active moiety and a counterion which is used for salt formation are called pharmaceutical salts. They can be obtained either by acid-base reactions where a basic API undergoes salt formation with an acid, or by reactions where an acidic API undergoes salt-formation with a base which is used as salt-former. To obtain stable pharmaceutical salts, which are not labile with regard to hydrolysis, a sufficient difference in pKa between API and salt former should be given, for example, this difference should be more than 2 pKa units. A comprehensive overview of pharmaceutical salts can be found in [1]. The distinction between non-ionic and ionic solid-state forms has pronounced effect on the physico-chemical properties of solid-state forms since for ionic compounds intermolecular interactions are largely governed by Coulomb-interactions, which are strong and do not exist in non-ionic compounds. Beyond the presence of Coulomb-interactions, the presence of charged species within the crystal lattice also gives rise to ion-dipole interactions. These are also considerably strong forces and allow an efficient interaction between the ions and water as solvent. This ion-dipole interaction between the API and water as solvent allows an efficient interaction between a pharmaceutical salt and water resulting, for example, in improved wettability and dissolution rate of pharmaceutical salts compared to a non-ionic API as the corresponding free base or free acid. For pharmaceutical salts, the presence of charged species increases both, the lattice energy as well as the hydration energy. Accordingly, these two effects compensate each other to a certain extent and either the one or the other will dominate.

The polymorphic behaviour of a solid-state form can be used for classification of solid-state forms. Polymorphic forms of a compound have the same chemical composition but differ with regard to the spatial arrangement of the constituents of the crystal lattice. Most prominent examples come from inorganic chemistry such as graphite, diamond and several fullerenes as polymorphic forms of graphite or rutile, anatase and brookit as polymorphic modifications of titanium dioxide. However, polymorphism is also very widespread amongst APIs, with the vast majority of APIs showing several different polymorphic forms. A schematic two-dimensional picture, how polymorphic forms differ with regard to the order of the molecules within the crystal lattice is provided in Figure 9.6.

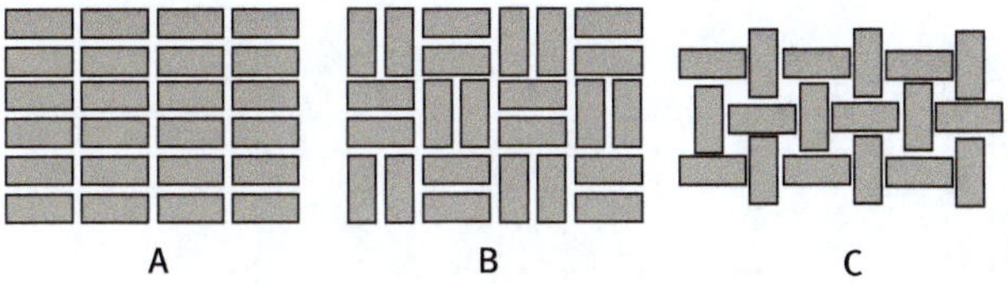

Figure 9.6: Two-dimensional scheme for three crystalline polymorphs (A, B, and C) of a compound. Molecules of the compound are represented by grey rectangles. Even if the molecule is the same, its arrangement in the lattice is different in the three forms.

From Figure 9.6 it can be seen that the chemical composition of the three different polymorphs is exactly the same. It also becomes obvious that intermolecular interactions in the three polymorphs are different: In polymorph A, there are just interactions between the short faces and short faces of the molecules and between the long faces and long faces of the molecules. Contrary, in polymorph B there are interactions between long faces and long faces as well as between short faces and long faces. Still in contrast in polymorph C, there are only interactions between short-faces of the molecules and long faces. Accordingly, intermolecular interaction in all three polymorphs will be different and lead to different physico-chemical properties of theses polymorphs including melting point, solubility and dissolution rate. It is also readily seen that polymorphs A and B will have the same density whereas polymorph C shows a lower density, as it has more voids in the crystal lattice. These three exemplified polymorphs are not the only ones which are possible. Even for this simple, two-dimensional example the reader will easily find more polymorphs. Accordingly, in the real, three-dimensional world of APIs many more different polymorphs can exist for APIs.

Beyond such "true polymorphs", the term "pseudo-polymorphism" describes cases where the solid-state form does not show the same chemical composition, but where additional molecules such as water and solvents contribute to the crystal lattice. Respective solid-state forms are named hydrates or solvates. Solvates and hydrates are frequently obtained from crystallization processes from organic solvents or solvent-water mixtures or – in the case of hydrates also from storage of the API under ambient conditions. In these cases, water or solvents is used to build a stronger crystal lattice with a reduced number of voids in it. The term “co-crystal” which has gained more and more attraction during the recent decade, in principle describes systems which are very similar to solvates. The only difference between a solvate and a co-crystal is the physical state of the pure co-former: For a solvate the co-former – a solvent – is liquid under ambient conditions. For a co-crystal the co-former is a solid material under ambient conditions. Accordingly, the distinction between solvate and co-crystals is somewhat arbitrary and the physcio-chemical principles governing their behaviour are very similar.

Figure 9.7 depicts two-dimensional schematic examples for pseudo-polymorphs.

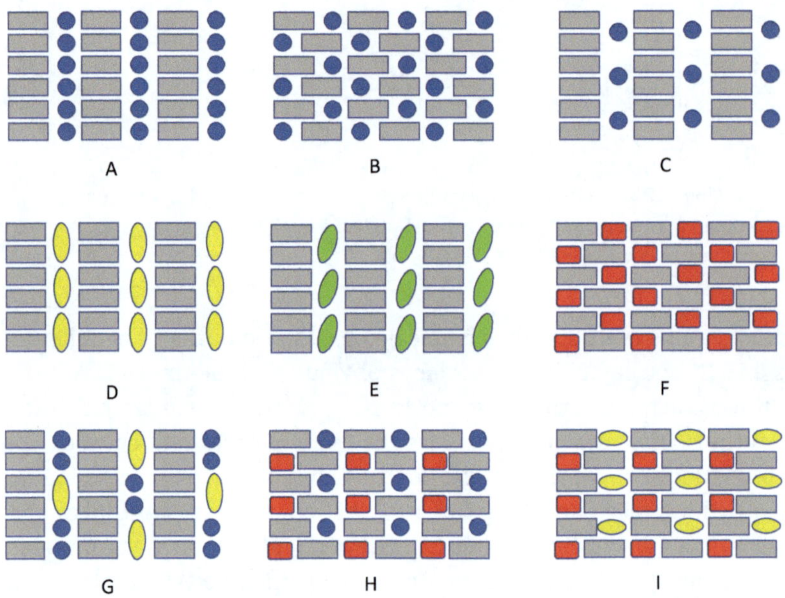

Figure 9.7: Two-dimensional scheme of different hydrates, solvates, and co-crystals of a compound. Molecules of the compound are represented by grey rectangles. Blue circles represent water molecules. Yellow and green ellipses represent solvent molecules. Red squares represent co-crystal formers. A = monohydrate, B = second form of monohydrate, C = hemi-hydrate, D = hemi-solvate, E = hemi-solvate with different solvate former, F = co-crystal with 1:1 stoichiometry, G = mixed solvate – hydrate, H = co-crystal hydrate, I = co-crystal solvate.

From this scheme it is easily recognized that a manifold of possibilities exists, how pseudo-polymorphs can occur. Starting with the simple case of a hydrate with a fixed stoichiometry, for example, a monohydrate, one can realize several different crystal lattices with this stoichiometry. If stoichiometry is varied, this can lead to other hydrates such as dihydrates, hemihydrates, trihydrates, and others. All of them can again exhibit different geometrical orders of the active moiety and water within the crystal lattice leading to a plethora of hydrates.

If water is replaced by a solvent to contribute to the crystal lattice, this allows the realization of still more pseudo-polymorphs: Many different solvents can be built into the crystal lattice. For each different solvent there might be different stoichiometry, for example, monosolvate, disolvate, hemisolvate, and others. Each of these again might exhibit different geometrical orders of active moiety and solvent within the crystal lattice leading to another plethora of solvates.

Finally, cases can occur where not only one solvent contributes to the crystal lattice but this is the case for two or more solvents. This leads to mixed solvates or mixed solvate-hydrates. Of course, for these crystal forms the same aspects with regard to stoichiometry and geometrical order within the crystal lattice apply as mentioned earlier.

Still another special case is the existence of non-stoichiometric pseudo-polymorphs. These are cases, where positions in the crystal lattice can be filled partially with water or solvents leading to a variable stoichiometry.

According to this discussion from a theoretical standpoint a manifold of different polymorphs and pseudo-polymorphs can exist for a certain API in its neutral form. However, usually, nature realizes only a few of these which facilitate solid-state form selection, pharmaceutical development, and manufacturing.

To conclude the overview on solid-state forms, we still must mention that all the discussion with regard to polymorphism, pseudo-polymorphism, and co-crystals which has been discussed for non-ionic compounds can be applied to pharmaceutical salts, too. There might be polymorphs of pharmaceutical salts, hydrates of pharmaceutical salts, solvates of pharmaceutical salts, mixed hydrate-solvates of pharmaceutical salts, co-crystals of pharmaceutical salts, and others. Nature realizes only few of the theoretically existing possibilities.

9.2 Why does solubility matter for solid-state forms?

In this section we will discuss the effect on solid-state forms on solubility in the same order as usually solid-state form selection is carried out. This means the first step represents the pharmaceutical salt selection which is followed by polymorph selection or even selection of a co-crystal.

Pharmaceutical salts can be a very attractive way to improve bioavailability of low soluble APIs, which are either basic or acidic. The improvement of bioavailability based on pharmaceutical salts can be either due to higher thermodynamic solubility or due to improved dissolution rate or due to both effects. A general overview on the topic is provided in [27].

The basis to understand solubility enhancement of APIs using pharmaceutical salts is the pH dependent solubility of protolytically active species. This is schematically shown in Figure 9.8 for a basic API, an acidic API and an amphoteric API.

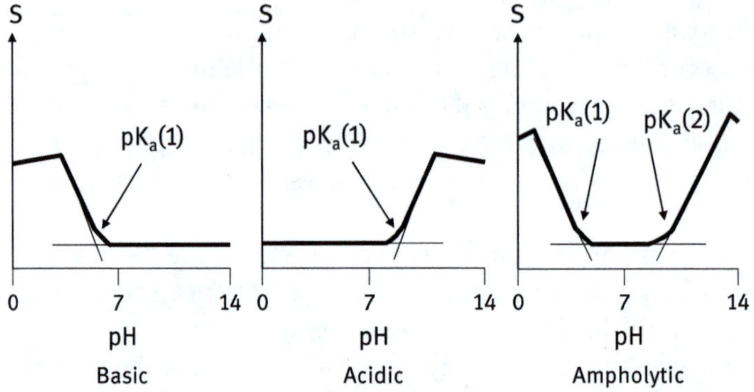

Figure 9.8: pH dependent solubility for a basic (left), acidic (middle) and ampholytic (right) API.

In all cases solubility is lowest in the pH region where the API is not ionized. Accordingly, bases exhibit a low solubility at high pH, and solubility is increasing as pH is decreased. For acidic APIs exactly, the opposite behaviour is observed. APIs which represent ampholytes are neutral – which means either neutral without any atom carrying a charge or zwitterionic – and show a solubility minimum in a medium pH range. In all cases the increase in solubility is caused by an increasing concentration of charged species which are more efficiently hydrated and consequently lead to better solubility. As ionization starts to get important as the pH approaches the pKa of the API, this is the pH where solubility starts to significantly increase. The ratio between unionized and ionized species as a function of pH is described by the Henderson-Hasselbalch eq. (9.4).

$$\mathrm{pH} = \mathrm{pKa} - \log\frac{[\mathrm{HA}]}{[\mathrm{A}^-]} \tag{9.4}$$

pKa = Acid-Base constant of the API
[HA] = concentration of undissociated acid
$[A^-]$ = concentration of conjugated base

From this the pH dependent solubility can be derived as described by eq. (9.5).

$$S(\mathrm{pH}) = S_0{}^*\left(1 + 10^{\mathrm{pKa}-\mathrm{pH}}\right) \tag{9.5}$$

$S(\mathrm{pH})$ = pH dependent solubility
S_0 = Thermodynamic solubility of the free base

Equation (9.5) describes pH dependent solubility as shown in Figure 9.8. However, below a certain pH - called pH_{max} – another phenomenon becomes relevant. For a basic API, below pH_{max} and for an acidic API above pH_{max} solubility is governed by the solubility product of the respective pharmaceutical salt. As an example, the solubility product for a hydrochloride salt of an API is given by eq. (9.6).

$$K = [API^+]*[Cl^-] \tag{9.6}$$

K = solubility product
$[API^+]$ = concentration of protonated API
$[Cl^-]$ = concentration of chloride

To illustrate what happens at pH_{max}, we show a theoretical titration experiment starting with a suspension of a basic API with pKa 5 in aqueous solution at pH 14. Be aware that the titration is carried out with a suspension and not as generally done in analytical chemistry using a solution! Using a suspension for this experiment will allow the establishment of an equilibrium between dissolved API and undissolved API in a certain solid-state form. The concentration of the API in solution represents the thermodynamic solubility at every pH during the titration. To allow this equilibration, the titration must be carried out extremely slowly. The first addition of a small portion of an acid – for example hydrochloric acid – will lead to a small decrease of pH. The free base represents the solid-state form in equilibrium with the solution and accordingly thermodynamic solubility refers to this solid-state form. The next addition of a small portion of acid has the same effect: pH is slightly lowered, there is almost no change in thermodynamic solubility as almost all molecules of the API are not protonated according to the Henderson-Hasselbalch equation. The situation changes, as the pH is further lowered by addition of hydrochloric acid and is approaching the pKa of the API. In this pH region more and more molecules of the API in solution become protonated by the addition of hydrochloric acid. As these molecules carry a charge, they are more easily hydrated, and solubility starts to increase considerably. At pH = pKa, 50% of the API molecules in solution are ionized and 50 % of the API molecules in solution are not ionized. Still the solid-state form which is in equilibrium with the solution is the free base and accordingly solubility at this pH still refers to the free base. Now the addition of each drop of hydrochloric acid protonates more API molecules and dissolves them from the solid residue. At pH_{max} addition of hydrochloric acid does not lower the pH anymore. Instead, each portion of hydrochloric acid converts the solid residue which is in equilibrium with the solution from the free base to the hydrochloride salt. Accordingly, the concentration of protons in solution does not increase upon addition of hydrochloric acid and pH keeps constant. At pH_{max} the suspension represents a heterogeneous buffer. After the solid residue is completely converted from the free base to the hydrochloride salt, addition of hydrochloric acid again leads to a decrease of pH, as there is no solid base available anymore which can act as proton-acceptor. At this pH the solubility is not governed anymore by eq. (9.5) but by the solubility product of the hydrochloride salt as described by eq. (9.6). The addition of more hydrochloric acid increases the chloride concentration and consequently due to the solubility product solubility decreases towards lower pH. As can be seen from this experiment, pH_{max} represents the pH where an API exhibit highest solubility. Additionally, pH_{max} for a basic API represents also the pH where an important change in solid-state form occurs: above pH_{max}, the free base is in equilibrium with the solution. Below pH_{max}, the pharmaceutical salt is in equilibrium with the solution. This further implies thermodynamic solubility: below pH_{max}, the pharmaceutical salt is thermodynamically stable in the suspension. Above pH_{max}, the free base is thermodynamically stable. This theoretical experiment has been carried out for a basic API. The same can be done with an acidic API, leading to a mirrored pH dependent solubility. Finally, the titration experiment might be carried out using other acids instead of hydrochloric acid. In this case pH dependent solubility for a basic API will be the same above pH_{max}. Below pH_{max}, different pharmaceutical salt will have different solubility products and consequently different solubility. Also pH_{max} will be different for different pharmaceutical salts. The higher the solubility of a pharmaceutical salt, the lower pH_{max} will be. This is schematically depicted in Figure 9.9 for three pharmaceutical salts.

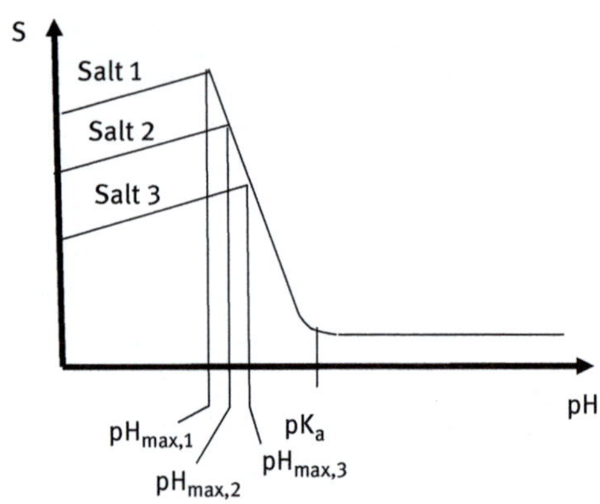

Figure 9.9: pH dependent solubility for three pharmaceutical salts of a basic API.

This can have important implications for pharmaceutical salts, for example, in formulations. If the formulation has a microenvironment which is higher in pH compared to pH_{max}, this means that the pharmaceutical salt is not thermodynamically stable. Instead, the free base represents the thermodynamically stable solid-state form. Consequently, a conversion of the salt-form from the pharmaceutical salt towards the free base might occur. Prasugrel represents a weakly basic API which has been used in oral formulations as a hydrochloride salt. During clinical development it was observed that such a conversion of the hydrochloride to the free base occurred in the formulation and this might affect drug product performance and bioavailability. A detailed discussion of this case can be found in [21].

As mentioned above, pharmaceutical salts can also improve bioavailability by enhancing dissolution rate. After the introduction of the Biopharmaceutical Classification Scheme (BCS) [28], it was recognized that for low-soluble and low-permeable drugs (BCS class II), the parameter which limits bioavailability can be either thermodynamic solubility or dissolution rate. This led to the introduction of the Development Classification System (DCS) [29]. Within the DCS the BCS class II is subdivided into two classes, DCS class IIa and class IIb: For DCS class IIa APIs bioavailability is limited by thermodynamic solubility, whereas for DCS class IIb APIs dissolution rate represents the limiting parameter.

Due to their more polar structure with charged species constituting the crystal lattice and consequently also the surfaces of crystals, pharmaceutical salts generally dissolve faster compared to un-ionized APIs, as interaction between water molecules and such ionized species is stronger compared to interactions between water and neutral species from the crystal lattice.

A typical example comparing dissolution rate for a neutral API and its pharmaceutical salts is shown in Figure 9.10 [30]. Albendazole represents a weak base with a pKa of 2.8 [31].

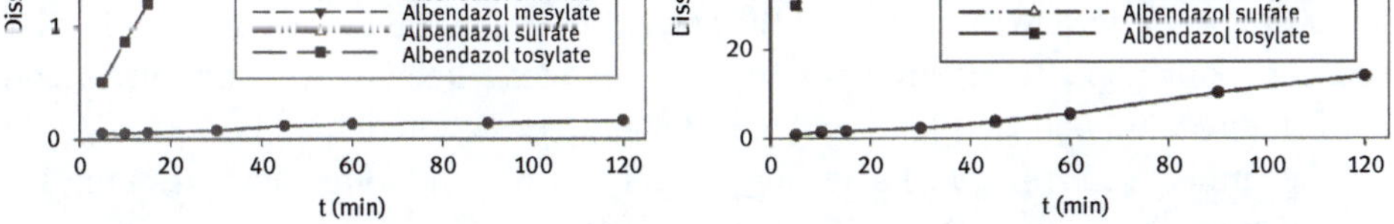

Figure 9.10: Dissolution profiles of albendazole and four pharmaceutical salt of albendazole (chloride, mesylate, sulphate, and tosylate) in 0.002 M acetate-buffer pH 4.5 (left) and SGFsp pH 1.2 (right).

Dissolution of the free base of albendazole in simulated gastric fluid without pepsin (SGFsp) pH 1.2 and 0.002 M acetate buffer pH 4.5 was considerably lower compared to dissolution of all the pharmaceutical salts – the hydrochloride, sulfate, tosylate, and mesylate – which were investigated. In both dissolution media after 2 h there is still a slow and steady increase of albendazole concentration, indicating that the system is still far away from equilibrium solubility.

In contrast, dissolution of the pharmaceutical salts is much faster and realizes much higher concentrations of albendazole in solution. The order of dissolution rate is different in SGFsp and 0.002 M acetate buffer for the tosylate and the chloride salt. In acetate buffer the chloride salt shows a superior dissolution profile compared to the tosyalte. In SGFsp the order has changed, and the chloride salt dissolves slower and leads to a lower final concentration. This effect is due to the presence of a high concentration of chloride in SGFsp which contributes via the solubility-product of the albendazol chloride to a common-ion effect which decreases performance of the chloride salt. Another example – haloperidol – with in depth discussion can be found in [32].

In both dissolution media pharmaceutical salts from sulfonic acids exhibit very good dissolution profiles. The albendazole mesylate represents an outstanding example. For this complete dissolution or even supersaturation is reached within the first 5 min of the dissolution experiment. Very good dissolution profiles of sulfonic-acid salts are frequently observed for pharmaceutical salts using sulfonic acids, such as methane sulfonic acid (mesylates), toluene sulfonic acid (tosylates), benzene sulfonic acid (besylates), naphtene sulfonic acid (napsylates) and ethanedisulfonic acid (edisylates). For this reason, the utilization of sulfonic acid salts to increase bioavailability of BCS class II APIs has increased during recent years [2]. The phenomenon has been described in detail in [33]. However, a detailed physical explanation of it is still missing.

Finally, we discuss the influence of polymorphs on solubility. As explained above different polymorphs show different crystal lattices, consequently different lattice energies and finally different solubilities. The same holds true for pseudo-polymorphs such as hydrates and solvates. Here in addition to the different crystal lattices formed from the API, energy contributions to the crystal lattice coming from water or solvent also play a role. Still in addition, hydration or solvation energy and entropy it not only due to the API but also due to water and the solvent from the pseudo-polymorph. Similar considerations apply to dissolution rate and even bioavailability. An early discussion of the influence of polymorphism on bioavailability can be found in [12].

Generally, with regard to solubility of polymorphs the following questions are of relevance:

- How is solubility related to thermodynamic stability of a polymorph?
- How big is the difference in solubility of different polymorphs and pseudo-polymorphs?

With regard to the first question, the answer – and dilemma – is that always the thermodynamically stable solid-state form will exhibit the lowest solubility. With regard to the second question the answer can be given by calculating the ratio of the solubility of different polymorphs. As discussed in chapter 10, the ratio of solubilities of different polymorphs in different solvents will be constant as the solvation energy only depends on the molecule and not on the polymorph. Accordingly, to answer this question data from determination of solubility of different polymorphs of the same compound in a variety of solvents can be used. In most cases solubility of different polymorphs shows a solubility ratio in the range 1–4. A good overview can be found in [34]. A few examples for solubility of different true polymorphs are given in Table 9.1.

Table 9.1: Solubility ratio of polymorphs.

API (Forms)	Solubility Ratio
Acetazolamide	1.1
Chloramphenicol palmitate	4.2
Cyclopenthiazide (II/III)	3.6
Cyclopenthiazide(I/III)	2.0
Diflunisal (I/IV)	1.4
Diflunisal (II/IV)	1.3
Diflunisal (III/IV)	1.3
Fluconazole (AII/AI)	1.1
Glibenclamide (II/I)	1.6
Glibenclamide (IV/II)	1.6
Indomethacin (α/γ) (at 45 °C)	1.1
Losartan	2.3
Mebendazole (B/A)	3.5
Mebendazole (C/A)	2.0
Methylprednisolone	1.7
Piroxicam (I/III)	1.3
Piroxicam (II/III)	1.4
Propranolo	1.4
Ranitidine (I/IV)	1.4
Ranitidine (II/IV)	1.4
Ranitidine (III/IV)	1.3
Ritonavir (at 50 °C)	4.0
Sulphathiazole (III/I)	1.7

For pseudo-polymorphs we focus the discussion on hydrates, as these have the highest relevance from a pharmaceutical perspective. There are only very few solvates which have been approved as drugs on the market such as the BRAF-inhibitor Trametinib [19] which represents a DMSO solvate or the protease-inhibitor Darunavir representing an ethanol solvate [20]. In contrast, hydrates play a much more important role as APIs. A good discussion of this topic can be found in [13]. If we calculate the solubility ratio between hydrates and anhydrates, this ratio is usually higher compared to the solubility ratio for true polymorphs. In many cases we find solubility ratios in the range 1.5 – 10. Some examples can be found in Table 9.2.

In most of cases, the aqueous solubility of hydrates is lower compared to the respective anhydrate. This basically means that anhydrates when suspended and

Table 9.2: Anhyrate/hydrate solubility ratio.

API (Forms)	Solubility Ratio
Ampicillin (A/Trihydrate)	1.6
Caffeine	1.7
Carbamazepine (A/Dihydrate)	1.6
Erythromycin (A/Dihydrate)	2.2
Fluconazole (A/H I)	1.2
Glutethemide	1.6
Niclosamide	22.9
Succinyl sulphathiazole (A I/HI)	3.2
Succinyl sulphathiazole (A II/HI)	5.7
Succinyl sulphathiazole (A III/H1)	6.2
Succinyl sulphathiazole (A IV/H1)	9.3
Succinyl sulphathiazole (A V/H 1)	12.7
Theophylline	1.9

slurried in water will convert to the thermodynamically stable hydrate as this possesses a lower solubility. However, there are about ten examples in the literature, where an anhydrate pseudo-polymorph exhibits a lower solubility in aqueous systems compared to the hydrate. This ends up in the strange situation that the hydrate is suspended in water and it converts to the anhydrate form which represents a dehydration procedure using water. Such examples where the solubility of the anhydrate form in water is lower compared to the hydrate form can be found in [35–42]. A similar behaviour for hydrates and anhydrates is usually observed for the dissolution rate: typically, anhydrates will dissolve faster in aqueous systems compared to hydrates. There are only very few examples reported in the literature where the opposite behaviour was observed. One of them is described in [35]. The development compound LY334370 represented a hydrochloride salt, which formed an anhydrate and a hydrate form. The intrinsic dissolution rate of the hydrate form was approximately six times higher compared to the anhydrous form.

These aspects show that solubility of polymorphs and pseudo-polymorphs systematically must be considered when assessing the bioavailability for APIs. For oral drugs, the BCS and DCS classification scheme are helpful tools, where especially APIs falling in BCS classes II and IV need a close consideration of polymorphism in light of bioavailability. For such orally administered APIs situations might also become relevant, where there is a conversion of a morphic form within the gastrointestinal tract. For example, there we might have a conversion from an anhydrate form to a hydrate form. Even if the anhydrate form is stable under ambient conditions as described in ICH Q1 [43] and has been selected as solid-state form for this reason, the hydrate form might become the thermodynamically stable form in the aqueous environment of the gastrointestinal tract. Accordingly, such a conversion can take place resulting in a lower solubility in vivo and a decreased bioavailability.

With regard to the influence of polymorphism to solubility, Ritonavir represented the first example where the importance became obvious [44]. The difficult behaviour which was suddenly observed there, triggered a huge amount of work in this specific case and the implementation of systematic assessments and solid-state form selection in the research and development process in the pharmaceutical industry. Ritonavir was marketed since 1996 as Norvir oral liquid and Norvir semi-solid capsules for treatment of Acquired Immunodeficiency Syndrome (AIDS). Both formulations were based on solutions of ritonavir in water-ethanol mixtures, as bioavailability from the solid state was very low. Administration of the pre-dissolved API to patients increased bioavailability. 240 lots of Norvir capsules were produced without any problems. Thereafter, certain lots of capsules failed the dissolution test. The root cause for failing the dissolution test was quickly identified as the occurrence of a new polymorph of ritonavir, named form II. Form II represented a conformational polymorph which means that the conformation of the ritonavir molecule in the crystal lattice was different compared to the conformation of the API molecule in form I. Ritonavir form II turned out to be thermodynamically more stable compared to form I. The water-ethanol solution which was used in the liquid-filled capsules was not saturated with regard to form I, but 400% supersaturated with regard to form II. Solubility data for both forms is given in Table 9.3.

Table 9.3: Solubility profile of ritonavir polymorphs in various water-ethanol solvent systems at 5°C. Solubility is given in mg/mL.

Ethanol/water	99/1	95/5	90/10	85/15	80/20	75/25	
Form I		90	188	234	294	236	170
Form II		19	41	60	61	45	30
Solubility ratio	4.7	4.6	3.9	4.8	5.2	5.7	

This supersaturation caused the risk of precipitation of form II out of the solution in the liquid-filled capsules and severe manufacturing problems, as the form II – the new polymorph – began to appear throughout drug-substance and drug-product manufacturing areas.

This sudden appearance of a new polymorph, especially if this new polymorph represents the thermodynamic stable polymorph, can represent major challenges during clinical development and even during commercial manufacturing. In a worst-case scenario, people might be unable to manufacture the metastable form which was used before. Stunning examples for such behaviour – disappearing polymorphs – are described in [45]. Another example from clinical development of a drug – telmisartan – is described in [46]. Still another interesting example can be found in [47]. Here clinical development was started with a drug substance and drug product which turned out to be a mixture of two polymorphs. This had considerable implications with regard to bioavailability of the API based on the solubility and dissolution rate of both polymorphs.

An example of a system where an anhydrate form exhibits lower aqueous solubility compared to a hydrate form is the integrin-inhibitor cilengitide, which was in clinical development as intravenous infusion against glioblastoma [48]. The solubility of the tetrahydrate form of cilengitide in physiological NaCl solution was about 20 mg/mL, whereas the solubility of the anhydrate in the same solvent was only about 7 mg/mL. The reason for this uncommon behaviour is based on the crystal structure as shown in Figure 9.11.

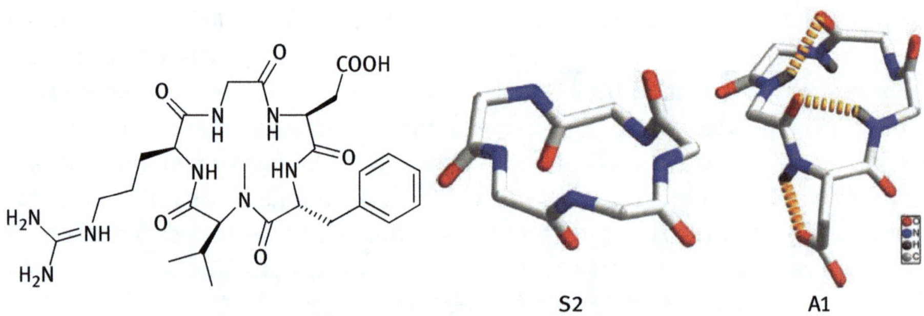

Figure 9.11: Cilengitide – cyclo-(Arg-Gly-Asp-DPhe-N-MeVal) and conformation of the cyclic peptide backbone in the tetrahydrate form S2 and the anhydrate form A1.

Within the tetrahydrate form (S2) there are no intramolecular hydrogen-bridges within the cyclic pentapeptide which stabilize the crystal form. However, the anhydrate form (A1) builds a crystal lattice which exhibits three strong intramolecular hydrogen bridges. These intramolecular hydrogen bridges result in stronger crystal-lattice energy and accordingly in a lower solubility of the anhydrate form compared to the tetrahydrate form. Accordingly, cilengitide represents one of those few strange examples, where an API is "dried" upon exposure to liquid water.

The cilengitide tetrahydrate turned out to show an even more complicated but useful behaviour, as within the crystal lattice of the tetrahydrate some of the positions of water molecules could even be replaced by small alcohol molecules – such as methanol and ethanol – or even by voids resulting in a non-stoichiometric mixed water-alcohol solvate pseudo-polymorph. This enabled a safe way to manufacture the metastable pseudo-polymorph and making use of its increased solubility for formulation purposes.

The last part of this chapter is dedicated to amorphous solid-state forms of APIs. As amorphous solid-state forms do not possess any long-range order, they exhibit a decreased lattice energy and accordingly an enhanced solubility. On one hand side, this represents a huge advantage with regard to improved solubility and bioavailability for low-soluble drugs. On the other hand, special care must be taken to avoid recrystallization during manufacturing and in vivo settings. Based on this, amorphous forms of APIs are mainly used in enabling formulations such as hot-melt extrudates, spray dried dispersions or co-precipitates [49]. In such formulations, the amorphous form is stabilized using appropriate excipients such as polymers which prevent recrystallization. This can either be done by realizing a solid solution of the API in the polymer. In such a solution, the API is molecularly dispersed in the polymer avoiding direct contact between API molecules and consequently avoiding crystallization. Instead, there are only interactions between API molecules and the stabilizing polymer which has a low mobility.

Even if solubility of amorphous solid-state forms is greatly enhanced compared to crystalline solid-state forms and especially the thermodynamically stable crystal-form, pure amorphous APIs representing small molecules are generally not used for clinical development or commercial supply due to the inherent risk of recrystallization which is difficult to tackle. Nevertheless, a good overview of how solubility of amorphous solid-state forms compares to crystalline solid-state forms can be found in [50–52].

9.3 Conclusion

Within this chapter the relevance of solid-state forms on solubility has been discussed on a theoretical basis and using several examples. One must bear in mind that solubility is a result of lattice energy and solvation energy. As lattice energy depends on the respective solid-state form – including pharmaceutical salts, polymorphs, pseudo-polymorphs, co-crystals, and amorphous phases – when measuring solubility and using solubility results for developing a formulation, assessing bioavailability, or even drug-substance and drug product manufacturing, solid-state forms have always to be considered.

References

[1] Stahl, P.H., & Wermuth, C.G. "Handbook of pharmaceutical salts", Wiley-VCH, 2011.
[2] Paulekuhn, S., Dressman, J., & Saal:, C. „Trends in active pharmaceutical ingredient salt selection based on analysis of the orange book database". J. Med. Chem. 2007, 50(26), 6665–6672.
[3] Balk, A., Widmer, T., Wiest, J., Rybak, J.-C., Matthes, P., Müller-Buschbaum, K., Sakalis, A., Lühmann, T., Berghausen, J., Holzgrabe, U., Galli, B., & Meinel, L. „Ionic liquid versus prodrug strategy to address formulation challenges". Pharm. Res. 2015a, 32, 2154–2167.
[4] Balk, A., Holzgrabe, U., & Meinel, L. „Pro et contra' ionic liquid drugs – Challenges and opportunities for pharmaceutical translation". Eur. J. Pharm. Biopharm 2015, 94, 291–304.
[5] Shamshina, J.L., Barber, P.S., & Rogers, R.D. "Ionic liquids in drug delivery". Exp. Op. Drug Deliv. 2013, 10(10), 1367–1381.

[6] Stoimenovski, J., MacFarlane, D.R., Bica, K., & Rogers, R.D. "Crystalline versus ionic liquid salt forms of active pharmaceutical ingredients: A position paper". Pharm. Res. 2010, 27(4), 521–526.
[7] Cojocaru, O., Andreea, S., Amal, G., Gurau, G., & Rogers, R.D. "Pharmaceutically active supported ionic liquid" in "Supported ionic liquids", edited by R. Fehrmann, A. Riisager, M. Haumann, 2014, 387–405.
[8] ICH International Council for Harmonisation, Quality Guidelines Q3C, http://www.ich.org
[9] Peltonen, L., & Hirvonen, J. „Pharmaceutical nanocrystals by nanomilling: Critical process parameters, particle fracturing and stabilization methods". J. Pharm. Pharm. 2010, 62, 1569–1579.
[10] Crowley, M. M., Zhang, F., Repka, M.A., Thumma, S., Upadhye, S.B., Battu, S.K., McGinity, J.W., & Martin, C. "Pharmaceutical applications of hot-melt extrusion: Part I". Drug Dev. Ind. Pharm. 2007, 33(9), 909–926.
[11] Repka, M.A., Battu, S.K., Upadhye, S.B., Thumma, S., Crowley, M.M., Zhang, F., Martin, C., & McGinity, J.W. "Pharmaceutical applications of hot-melt extrusion: Part II". Drug Dev. Ind. Pharm. 2007, 33(10), 1043–1057.
[12] European Pharmacopoeia, 9th Edition, Chapter 5.11 "Characters Section in Monographs". 2017.
[13] Griesser, U.J. "The importance of solvates", R. Hilfiker, editor, "Polymorphism", Wiley-VCH, 2006.
[14] Kim, S., Lotz, B., Lindrud, M., Girard, K., Moore, T., Nagarajan, K., Alvarez, M., Lee, T., Nikfar, F., Davidovich, M., Srivastava, S., & Kiang, S. "Control of the particle properties of a drug substance by crystallization engineering and the effect on drug product formulation". Org. Process Res. Dev. 2005, 9(6), 894–901.
[15] Noyes, A.A., & Whitney, W.R. The rate of solution of solid substances in their own solutions. J. Am. Chem. Soc. 1897, 19, 930–934.
[16] Skinner, L.M., & Sambles, J.R. "The Kelvin equation—A review", J. Aerosol Science 1972, 3(3), 199–210.
[17] ICH International Council for Harmonisationm Quality Guidelines Q1A-F, http://www.ich.org
[18] R. Hilfiker editor, "Polymorphism in the pharmaceutical industry", Wiley VCH, 2006.
[19] Center for Drug Evaluation and Research, Application Number 204114Orig1s000, Chemistry Review, http://www.accessdata.fda.gov/drugsatfda_docs/nda/2013/204114Orig1s000ChemR.pdf
[20] Center for Drug Evaluation and Research, Application Number 202895Orig1s000, Chemistry Review, http://www.accessdata.fda.gov/drugsatfda_docs/nda/2011/202895Orig1s000ChemR.pdf
[21] European Medicines Agency Evaluation of Medicines for Human Use, "Assessment Report for Efient, International Nonproprietary Name: Prasugrel", Doc.Ref.: EMEA/117561/2009.
[22] ICH International Council for Harmonisation, Quality Guidelines Q6A, http://www.ich.org
[23] Pfannkuch, F., Rettig, H., & Stahl, P.H. "Biolocial effects of the API salt form", P.H. Stahl, C.G. Wermuth, "Handbook of pharmaceutical salts", Wiley-VCH, 2011, 125–145.
[24] Miller, C., Kennington, L., Cooney, R., Kohjimoto, Y., Cao, L.C., Honeyman, T., Pullman, J., Jonassen, J., & Scheid, C. "Oxalate toxicity in renal epithelial cells: Characteristics of apoptosis and necrosis". Tox. Appl. Pharm. 2000, 162(2), 132–141.
[25] Cao, L.-C., Honeyman, T.W., Cooney, R., Kennington, L., Scheid, C.R., & Jonassen, J.A. Julie. "Mitochondrial dysfunction is a primary event in renal cell oxalate toxicity". Kidney International 2004, 66(5), 1890–1900.
[26] Saal, C., & Becker, A. "Pharmaceutical salts: A review on doses of salt formers form the orange book". Eur. J. Pharm. Sci. 2013, 49(4), 614–623.

[27] Serajuddin, A.T.M. "Salt formation to improve drug solubility". Adv. Drug Deliv. Rev. 2007, 59, 603–616.
[28] Amidon, G.L., Lennernaes, H., Shah, V. P., & Crison, J.R. "A theoretical basis for a biopharmaceutic drug classification: The correlation of in vitro drug product dissolution and in vivo bioavailability". Pharm. Res. 1995, 12(3), 413–20.
[29] Butler, J., & Dressman, J.B. "The developability classification system: Application of biopharmaceutics concepts to formulation development". J. Pharm. Sci. 2010, 99(12), 4940–4954.
[30] Paulekuhn, S., Dressmann, J.B., & Saal, C. „Salt Screening and Characterization for poorly soluble, weak basic Compounds: Case Study Albendazole". Die Pharmazie 2013, 68, 555–564.
[31] Jung, H., Medina, L., García, L., Fuentes, I., & Moreno-Esparza, R. Absorption studies of albendazole and some physicochemical properties of the drug and its metabolite albendazole sulphoxide. J. Pharm. Pharm. 1998, 50, 43–48.
[32] Li, S., Doyle, P., Metz, S., Royce, A.E., & Serajuddin, A.T.M. "Effect of chloride ion on dissolution of different salt forms of haloperidol, a model basic drug". J. Pharm. Sci. 2005, 94, 2224–2231.
[33] Elder, D. P., Delaney, E., Teasdale, A., Eyley, S., Reif, V. D., Jacq, K., Facchine, K. L., Schulte-Oestrich, R., Sandra, P., & David, F. "The utility of sulfonate salts in drug development". J. Pharm. Sci. 2009, 99(7), 2948–2961.
[34] Pudipeddi, M., & Serajuddin, A.T.M. "Trends in solubility of polymorphs". J. Pharm. Sci. 2005, 94(5), 929–939.
[35] Reutzel-Edens, S.M., Kleemann, R.L., Lewellen, P.L., Borghese, A.L., & Antoine, L.J. "Crystal forms of LY334370 HCl: Isolation, solid-state characterization, and physicochemical properties". J. Pharm. Sci. 2003, 92(6), 1196–1205.
[36] Katdare, A.V., & Bavitz, J.F. "Hydrate related dissolution characteristics of norfloxacin". Drug Dev. Ind. Pharm. 1984, 10(5), 789–807.
[37] Hu, T.C., Wang, S.L., Chen, T.F., & Lin, S.Y. "Hydration-induced proton transfer in the solid state of norfloxacin". J. Pharm. Sci. 2002, 91(5), 1351–1357.
[38] Petrova, R.I., Peresypkin, A., Mortko, C.J., Mckeown, A.E., Lee, J., & Williams, J.M. Rapid conversion of API hydrates to anhydrous forms in aqeous media. J. Pharm. Sci. 2009, 98(11), 4111–4118.
[39] Allen, P.V., Rahn, P.D., Sarapu, A.C., & Vanderwielen, A.J. "Physical characterization of erythromycin: Anhydrate, monohydrate, and dihydrate crystalline solids". J. Pharm. Sci. 1978, 67(8), 1087–1093.
[40] Gardner, G.L. "Kinetics of the dehydration of calcium oxalate trihydrate crystals in aqueous solution". J. Colloid. Interface Sci. 1976, 54(2), 298–310.
[41] Wu, L.S., Gerard, C., Koval, C., Rowe, S., & Hussain, M.A. "Crystallinity, hygroscopicity and dissolution of moricizine hydrochloride hemihydrate". J. Pharm. Biomed. Anal. 1994, 12(8), 1043–1046.
[42] Shibata, M., Kokubo, H., Morimoto, K., Morisaka, K., Ishida, T., & Inoue, M. X-ray structural studies and physicochemical properties of cimetidine. J. Pharm. Sci. 1983, 72(12), 1436–1442.
[43] ICH International Council for Harmonisation, Quality Guidelines Q1A(R2), http://www.ich.org
[44] Bauer, J., Spanton, S., Henry, R., Quick, J., Dziki, W., Porter, W., & Morris, J. „Ritonavir: An extraordinary example of conformational polymorphism". Pharm. Res. 2001, 18(6), 859–866.
[45] Dunitz, J.D., & Bernstein, J. „Disappearing polymorphs". Acc. Chem. Res. 1995, 28(4), 193–200.

[46] Dinnebier, R. E., Sieger, P., Nar, H., Shankland, K., & David, W. I. F. „Structural characterization of three crystalline modifications of telmisartan by single crystal and high-resolution X-ray powder diffraction". J. Pharm. Sci. 2000, 89(11), 1465–1479.
[47] Cimarosti, Z., Castagnoli, C., Rossetti, M., Scarati, M., Day, C., Johnson, B., & Westerduin, P. "Development of drug substances as mixture of polymorphs: Studies to control form 3 in casopitant mesylate". Org. Proc. Res. Dev. 2010, 14(6), 1337–1346.
[48] Saal, C., Lange, M., Kuehn, C., Untenecker, H., Jonczyk, A., Petersen, S., Scholz, G., Buback, V., Dotzauer, M., Bauer, H., Foerster, J., Schumacher, J., Metz, A., Schmidt, M., & Seemann, K. Cilengitide - Exceptional pseudopolymorphism of a cyclic pentapeptide. Eur. J. Pharm. Sci. 2015, 71, 1–11.
[49] Liu, R. "Water-insoluble drug formulation", CRC Press, 2008.
[50] Murdande, S.B., Pikal, M.J., Shanker, R.M., & Bogner, R.H. "Solubility advantage of amorphous pharmaceuticals: I. A thermodynamic analysis.". J. Pharm. Sci. 2010, 99(3), 1254–1264.
[51] Murdande, S.B., Pikal, M.J., Shanker, R.M., & Bogner, R.H. "Solubility advantage of amorphous pharmaceuticals: II. Application of quantitative thermodynamic relationships for prediction of solubility enhancement in structurally diverse insoluble pharmaceuticals.". Pharm. Res. 2010a, 27(12), 2704–2714.
[52] Murdande, S.B., Pikal, M.J., Shanker, R.M., & Bogner, R.H. "Solubility advantage of amorphous pharmaceuticals: III. Is maximum solubility advantage experimentally attainable and sustainable?". J. Pharm. Sci. 2011, 100(10), 4349–4356.

Wolfgang Beckmann

10 Solubility and phase behaviour from a drug substance manufacturing perspective

10.1 Introduction – solubility and phase diagrams in process design and optimization

A crystallization process for an organic drug substance, an intermediate, or any other species requires the substance to be completely dissolved in a given solvent. Following this, the solubility of the moiety is decreased in such a way that, the solution gets supersaturated, and subsequently the substance crystallizes and can be isolated. The solubility of the moiety in the solvent of crystallization determines both the conditions for the complete dissolution and the maximum attainable yield via the concentration at the start and at the end of the crystallization process. Knowledge on solubility is thus of high value in crystallization process development and optimization. The solubility itself does not only depend on the state variables such as temperature or solvent composition, but also on the solid-state form of the crystals obtained by the crystallization, being it either a polymorph or a solvate as discussed in Chapter 9 of this book. The solid-state form might change over the course of the crystallization as temperature or solvent composition is altered. Thus, all data on solubility needs to be accompanied with information on the solid-state form for which it has been determined. These data are collated into what is also called a phase diagram of the substance. This phase diagram should note all possible phases and relate it.

It is the solid-state form of the solid residue that determines the solubility. Thus, it is evident that solubility measurements should be based on the results from a polymorph screen. Further, results on the phases observed during solubility measurement should be supplemented to the screen.

The experimental determination of solubilities and the respective phase diagram of a substance are straightforward. Several simple rules apply to such phase diagrams. These rules will be discussed in detail as they can be exploited in the determination of solubility and used for a check of the data obtained.

Solubility phenomena of moieties, their phase behaviour, and the relation to crystallization phenomena have for a long time attracted attention in the community. A wealth of experimental data and theoretical considerations exists. Measurements of solubility and phase behaviour are regarded as prerequisite and starting point of every sincere work in crystallization. The reader is referred to some reference works

Wolfgang Beckmann, Bayer AG, Wuppertal, Germany

https://doi.org/10.1515/9783110559835-010

[1–4]. Solubility phenomena and the phase behaviour of organic and of inorganic systems have many parallels.

The determination of sustainable solubility data and phase behaviour should find more room in the pharmaceutical industry from the beginning. Process design and optimization should be based on measured and corroborated solubility data.

10.2 Solubility – basics

10.2.1 Definition of solubility

Throughout this chapter, solubility is defined as the concentration of a certain moiety in a solution that is in thermodynamic equilibrium with a certain crystalline solid-state form of that moiety (see Figure 10.1).

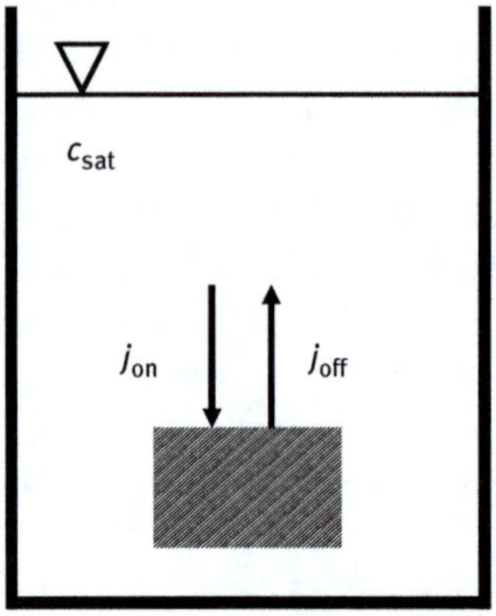

Figure 10.1: Equilibrium of a moiety in solution with its crystalline solid. The net flow of molecules onto the crystal and leaving the crystal is zero. This equilibrium depends on several variables of the system, besides temperature namely on the solid-state form of the undissolved moiety and the purity of the system.

In a thermodynamic equilibrium, the solid residue, which might be either crystalline or amorphous, does not show a net dissolution or growth. This is a dynamic equilibrium, i.e. in case of a crystalline solid residue, the number of molecules leaving the crystal lattice equals the number of molecules attaching to the lattice (see Figure 10.1). Any system will require some time to reach this equilibrium, being it either

- a neat solvent brought into contact with a solid,
- a supersaturated solution from which the substance crystallizes, prepared by cooling a saturated solution,

or

- a solution that is prepared by dissolving a metastable polymorph.

Figure 10.2 shows the concentration in the supernatant of a stable and metastable solid-state form over time. The solubility of the stable form corresponds to the true

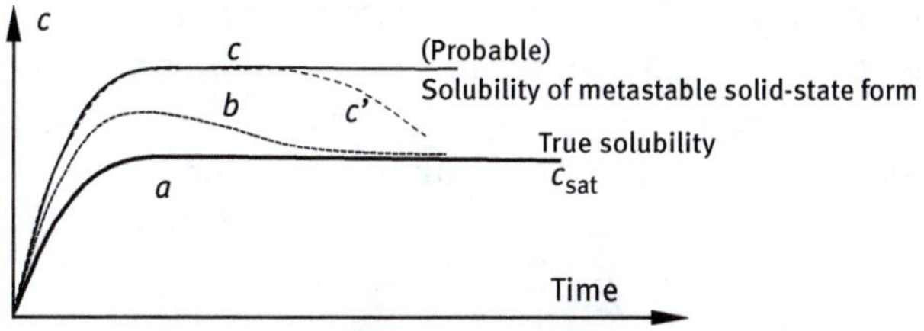

Figure 10.2: Concentration in a solution prepared by dissolving a thermodynamically stable solid-state form, "a", or by dissolving a metastable solid-state form, "b" and "c". In all cases, the solution will require a certain time to reach equilibrium. For solutions prepared using a metastable solid-state form, "b" and "c", the nucleation of the stable form will most probably occur after some time. If this induction time is sufficiently long, the solubility of metastable form can be assessed (case c).

equilibrium. However, the solubility of the metastable solid-state form can also be assessed, if the nucleation of a more stable form is inhibited for a sufficiently long time.

Thermodynamically speaking, the chemical potential μ of the moiety in both the solid phase and in solution are equal:

$$\mu_1 = \mu_2 \tag{10.1}$$

The subscript 1 refers to the undissolved solid and 2 to the solution. The chemical potential is related to the activity a of the moiety $a_i = \gamma_i x_i$ with x and γ as the molar fraction and activity coefficient:

$$\mu = RT \ln x_i \gamma_i \tag{10.2}$$

R is the universal gas constant and T the absolute temperature. The change in free enthalpy G during the dissolution process is

$$\Delta_d G = \Delta_d H - T\Delta_d S \tag{10.3}$$

G represents the free Gibbs enthalpy, H enthalpy, and S entropy. The subscript d is used for dissolution. Combining eqs. (10.1) and (10.2) leads to

$$-\Delta_d G = RT \ln \frac{x_1 \gamma_1}{x_2 \gamma_2} \tag{10.4}$$

For a pure solid, the activity of the moiety is unity so that the van Laar equation for solubility as a function of temperature is obtained:

$$\ln x_1 \gamma_1 = -\frac{\Delta_d G}{RT} \tag{10.5}$$

$\Delta_d G$ depends on temperature via the difference in heat capacity of the (crystalline) solid and the melt, Δc_p. Values for the difference in heat capacity between a crystalline solid and its melt are available in the literature for several organic compounds [5]. However, the dependence of $\Delta_d G$ on temperature can be neglected in eq. (10.5).

In case that solvent–solute interactions are comparable to the solvent–solvent and solute–solute interactions, the free enthalpy of dissolution, $\Delta_d G$, can be approximated by the free enthalpy of melting of the solute, $\Delta_f G$. As $\Delta_f G = 0$ at the temperature of fusion, T_f, eq. (10.6) is obtained for the free enthalpy of fusion:

$$\Delta_f G = \Delta_f H - \frac{\Delta_f H}{T_f} \tag{10.6}$$

Approximating dissolution with fusion, eq. (10.7) is finally obtained:

$$\Delta_d G = \Delta_f H - \frac{\Delta_f H}{T_f} \tag{10.7}$$

Equation (10.7) relates solubility, that is, the free enthalpy of dissolution, left-hand side, with fusion, right-hand side. $\Delta_f H$ and T_f are accessible through thermal analysis. This relation is useful in both, understanding the solubility and in relating it to thermodynamic data.

All dependencies of solubility on thermodynamic parameters can be derived from eq. (10.4) and (10.5). These dependencies are namely

- temperature;
- pressure, not discussed here as the influence of pressure on the free enthalpy is negligible;
- purity of the solid form as given by the activity of the target compound in the solid residue, that is, $x_2 \gamma_2$; and
- presence of third components in the solution that will change the activity or activity coefficient of the target compound in solution.

10.2.2 Parameters influencing solubility

The parameters influencing solubility discussed in the section, should guide the actual measurements of the solubility and process design. They will also help in getting a closer understanding of solubility and assure that important dependencies are recognized. Equally, sanity checks that can be derived from these dependencies will be noted.

10.2.2.1 Influence of temperature on solubility

The most important and most prominent dependence of solubility is that on temperature. Figure 10.3 shows the typical temperature dependence of solubility of an organic moiety in an organic solvent. Almost all systems show an increasing solubility with

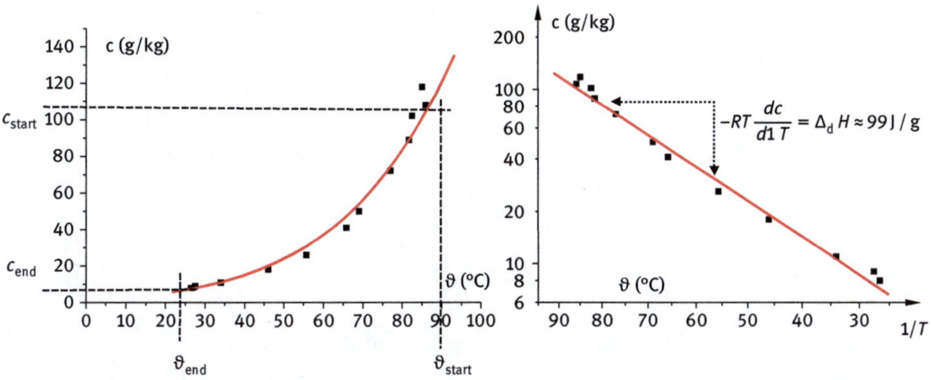

Figure 10.3: Dependence of solubility on temperature. The linear plot of the solubility versus temperature, left, should always be supplemented by a plot of the logarithm of solubility versus $1/T$, that is, the inverse of the absolute temperature. The *x*-axis in the right plot has been scaled as $1/T$, but the temperature is expressed in degree centigrade, which makes reading of the plot easier. The initial and end conditions of a cooling crystallization are shown.

temperature. Inverse dependencies (a decrease in solubility with temperate, retrograde solubility behaviour) are rare, both for organic and inorganic systems.

Figure 10.3 also depicts the starting and endpoints of a cooling crystallization. The maximum yield attainable η_{max} for a certain starting concentration c_o and a saturation concentration (c_{sat}) at the end temperature for the cooling process, $c_{\omega o}|_{\vartheta\omega}$, is given as follows:

$$\eta_{max} = \frac{c_{start} - c_{end}}{c_{sat}} \tag{10.8}$$

Thus, a certain steepness of the solubility curve with temperature is required for a cooling crystallization to afford an acceptable yield. The dependence of solubility on temperature as shown in Figure 10.3 is directly related to the enthalpy of dissolution, $\Delta_d H$, and can be obtained from eqs. (10.3) and (10.5) as

$$\frac{d\ln x_i \gamma_i}{d1/T} = -\frac{\Delta_d H}{RT} \tag{10.9}$$

Thus, neglecting the dependence of the activity coefficient on temperature, a plot of the log of solubility versus the inverse of the absolute temperature should yield a straight line with a slope corresponding to the heat of dissolution, as shown on the right-hand side of Figure 10.3. In most cases, the line will be linear or close to linear only over a limited range of temperature. As the slope corresponds to the heat of dissolution which in turn is often close to the heat of fusion of the substance, the slope can be directly compared with the thermo-analytical data of the enthalpy of fusion. This makes the heat of dissolution a valuable check for plausibility. In addition, the van't Hoff plot of solubility makes a perfect check of the data and for outliers.

If the solid-state form of the solute changes, this will manifest itself in a sudden change in slope of the straight line or a slightly bended line. Such solid-state transitions might otherwise go unnoticed (see Figures 10.11–10.13). As will be discussed below, the change in slope is directly correlated to the changes in heats of dissolution of the different solid-state forms. In case of polymorphs, this is the difference in heat of fusion of both polymorphs, which again can serve as a check for plausibility.

10.2.2.2 Influence of the solvent on solubility

10.2.2.2.1 Influence of the choice of solvent on solubility

The primary choice of the solvent for the crystallization of an API is governed by the ICH guideline Q3C [6]. The choice is mostly confined to the solvents classified as class 2 and especially class 3. Thus, the developer can choose from two to three dozen solvents. A second criterion applying to APIs and intermediates might be the availability of the solvents in the plant and the compatibility with the plant.

The influence of the solvent on solubility with respect to crystallization issues is twofold:

- the most prominent influence of solvent is the one on absolute solubility and
- a less pronounced influence is the dependence of solubility on temperature which is because the enthalpy of dissolution is mostly governed by the enthalpy of melting of the solid and to a much lesser degree by the solute–solvent interaction.

The possibility of certain solvents to dissolve a given solute can often be roughly estimated from chemical intuition. For example, the presence of polar groups or long aliphatic chains in an API or intermediate is a good criterion for choosing solvents by intuition. This is also reflected in the rule that “like dissolves like ”.

Attempts have been published over the years to use quantitative or semi-quantitative descriptors of the solvent to arrive at more rational guesses. These approaches range from solubility parameters to a principal component analysis of a variety of properties of the solvents. A closer discussion of such in-silico approaches is given in Chapter 3 of this book. More quantitative approaches lie in the estimation of the activity coefficient of the solute in different solvents. Some of the models will be discussed in part 4 of this chapter.

With respect to the choice of solvent, namely according to the precise values of solubility, the author deems

- the criteria for the choice of solvent in the pharmaceutical industry today for process development and optimization and namely for maximizing of the space–time yield of a crystallization process not as primary focus. It is more appropriate to adjust the crystallization conditions rather to choose an otherwise suboptimal solvent as the ease of a rough experimental determinations of the solubility cannot yet be beat by modelling.

10.2.2.2.2 Influence of the quality of the solvent on solubility

In general, the quality of the solvent is disregarded when measuring solubility. Other, components being present in the solvent in small quantities are generally regarded as having no influence on solubility. However, cases exist, where going from lab-grade solvent to technical grade can have a large influence. Technical-grade solvents stemming from any work-up can contain small amounts of a second solvent.

For the solubility of a steroid in dichloromethane, a substantial influence on the concentration of water was observed, even for concentrations of water well below the miscibility limit of water in dichloromethane as shown in Figure 10.4. Reworked solvent in a plant can easily contain 0.1 % water, which in turn can alter the solubility by 5–20 %;.

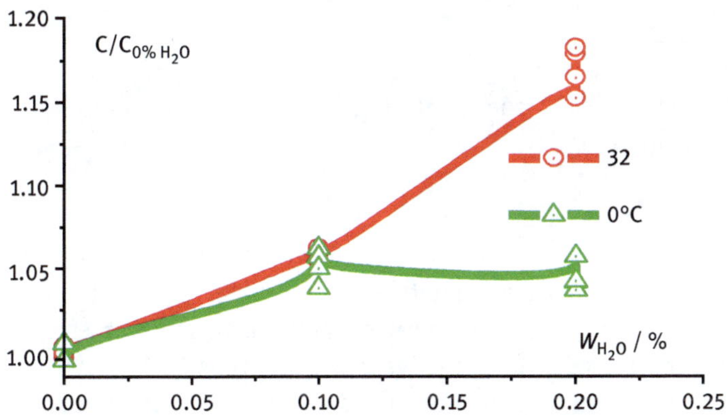

Figure 10.4: Effect of the water content in dichloromethane on the solubility of a steroid. The solubility for 0 and 32 °C is reduced with the respective solubility at 0% water. This makes the solubility trends for both temperatures comparable. Note that the non-negligible effect is for quantities of water below the low miscibility limit of water in dichloromethane of ≈0.6%.

10.2.2.2.3 Influence of the composition of solvent mixtures on solubility

The use of solvent mixtures can roughly be grouped into two categories. In both categories, the second solvent normally has a much lower solubility for the solute than the primary solvent:

- Use of the second solvent as an anti-solvent for drowning-out crystallizations
- Use of the second solvent as diluent to decrease the solubility and thus suspension density

The effect of the second solvent on the solubility can be roughly grouped into three behaviours as schematically shown in Figure 10.5. The dependencies observed are

- linear or near -to -linear dependence on the composition of the solvent mixture,

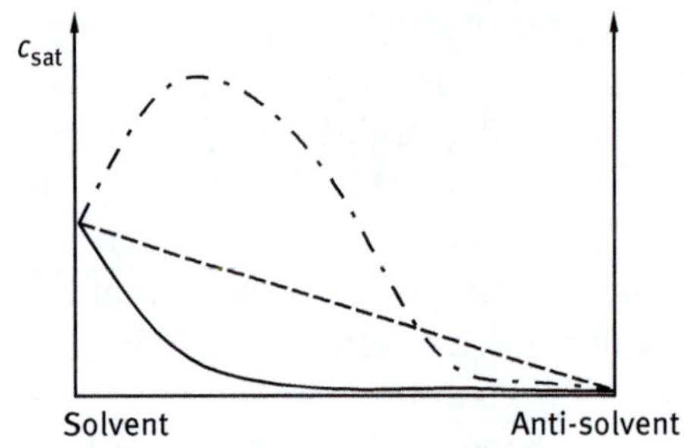

Figure 10.5: Effect of the use of a mixture of a good, primary solvent and a solvent with a low solubility for the moiety, often called anti-solvent. Besides a linear dependence on concentration, dashed line, curves with a maximum in solubility are observed, dash-dotted line. However, in most cases, a concave shape is encountered, solid line. The latter is necessary for a drowning-out crystallization.

- solubility exhibiting a maximum and
- concave curve – this is the most commonly observed case and required for drowning-out crystallizations.

With respect to crystallization processes, the most important case of the influence of solvent mixtures on solubility is the concave-shaped curves, as shown by the solid line in Figure 10.5. Only in this case, a mixture of a solute in the primary solvent with a poor solvent will yield a supersaturated solution. In most cases, the change in solubility with the fraction of anti-solvent is quite drastic. The ratio of solubilities between the solvent and anti-solvent can cover several orders of magnitude. A linear plot is not able to represent these data to check for consistency and outliers. For a check of the consistency of the data, it is helpful that many systems follow the log-linear relationship as proposed for aqueous organic systems [7]. In gross simplification, the solubility in a solvent mixture can be calculated using eq. (10.10). In eq. (10.10) x_S represents the solubility of the solute, x_{AS} is its solubility in the anti-solvent, x the solubility in the solvent mixture, and w represents the fraction of the anti-solvent:

$$\log x \approx \log x_S - w \log x_{AS} \tag{10.10}$$

Figure 10.6 shows such a plot of the solubility for a steroid in methanol as primary solvent and water as anti-solvent.

The solubility dependence on solvent composition is not perfectly linear. However, as the data are expected to lay on a smooth line, any deviation would be indicative for either errors in measurements or a change in solid-state form of the residue. In the case depicted in Figure 10.6, a change in slope is encountered. Following an analysis of the solid residue it becomes clear that this is due to a change in the solid-state form from an ansolvate to a methanol solvate. The near-to-linear relationship is obeyed for both solid-state forms.

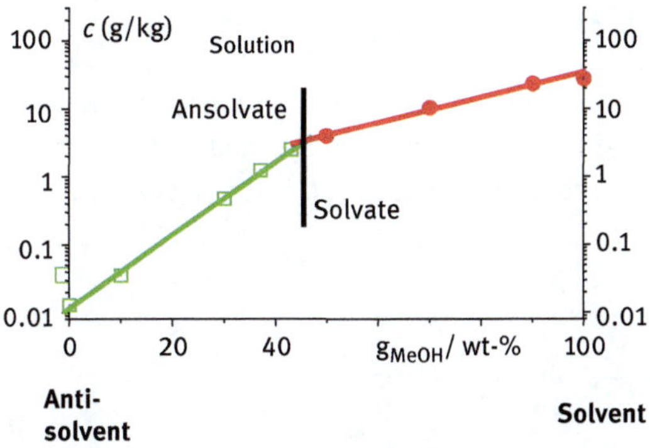

Figure 10.6: Isotherm of the solubility of a steroid as a function of solvent composition in a system forming solvates with one or both solvents using log-linear plot. The change in slope at ≈45 wt% methanol corresponds to the change in the solid-state form from an ansolvate to a methanol solvate.

10.2.2.2.4 Effect of solvent composition and temperature on solubility

The combined effects of temperature and solvent composition on solubility can best be treated by a reduction of the solubility isotherms with the solubility in the primary solvent. As is shown in Figure 10.7 for the solubility of a given moiety in acetone with hexane as anti-solvent, one master curve is obtained. This reduction has also been applied in Figure 10.4.

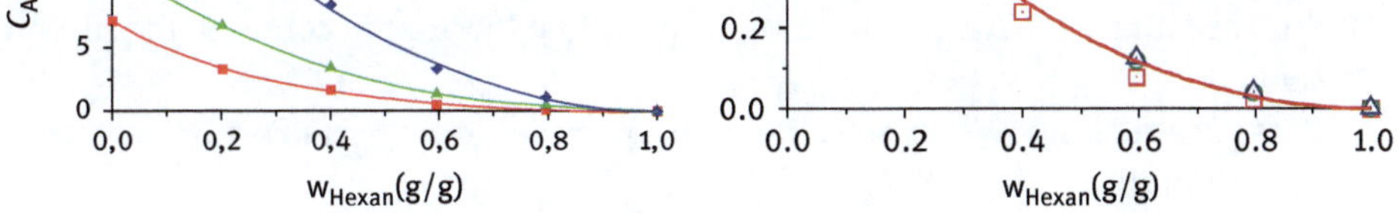

Figure 10.7: Solubility of a moiety in acetone–hexane mixtures at 20–60 °C. The isotherms of the solubility, left side, are reduced by the solubility in the pure primary solvent. Thus, a master curve of the solubility is obtained, right side.

10.2.2.3 Purity of the system – effect of second solute on solubility

The effect of a second solute can best be depicted in a ternary phase diagram as shown in Figure 10.8. This type of diagram is best known for the separation of enantiomers, where the second solute is the other enantiomer [8]. Three cases must be distinguished:

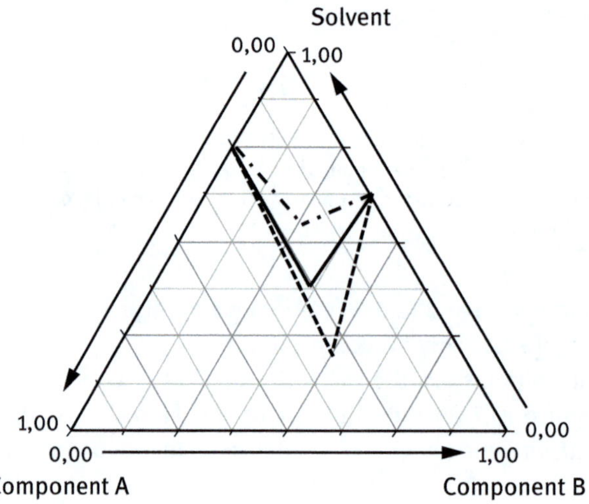

Figure 10.8: Representation of the solubility of two solutes *A* and *B* in a ternary diagram. Three cases are distinguished. The solid line indicates a case where the solubility of solute *A* is not influenced by the second solute *B*. The dashed and dot-dashed lines indicate cases where the solubility of solute *A* is increased by the second solute *B* (dashed line), and where the solubility of the solute *A* is decreased by the second solute *B* (dash-dotted line).

- If the second solute behaves indifferently, it behaves like a solvent with the molar fraction given by eq. (10.11), with *m* denoting the mass and *M* the molar mass. In this case, the solubility line is on the straight line toward the corner of the second solute.
- Cases are also known, where the second solute either increases or decreases the solubility.

$$x_2 = \frac{\frac{m_{\text{solute}}}{M_{\text{solute}}}}{\frac{m_{\text{solute}}}{M_{\text{solute}}} + \frac{m_{\text{2nd solute}}}{M_{\text{2nd solute}}} + \frac{m_{\text{solvent}}}{M_{\text{solvent}}}} \tag{10.11}$$

In technical systems, the influence of solid impurities is difficult to assess, mostly due to the fact that the solubility is influenced by a multitude of components. It is thus advisable to measure the solubility for a pure system and compare with measurements of the solubility of one or more impure systems. Figure 10.9 shows such solubility data for a pure solvent, toluene, and for two solutions in toluene, which are contaminated by a large number and a high concentration of impurities related to the solute. In this case, the impurities largely increase the solubility of the solute. For these measurements, different mother liquors are a good choice, as characteristic impurities are present to an elevated level.

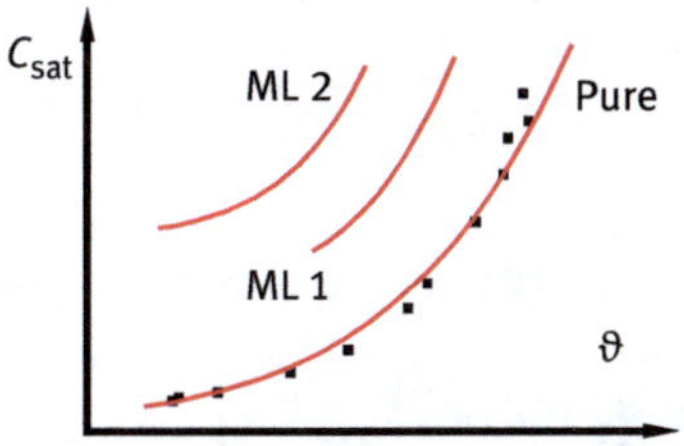

Figure 10.9: Solubility of a moiety pure solvent, i.e. toluene, and in different mother liquors (ML1, ML2) containing different types and quantities of related compounds as impurities. The level of the related compounds was on the order of 40–60%. Note the substantial influence of the impurities on solubility.

An increase in solubility by impurities is undesired for technical processes as the yield is decreased. A measurement of the true solubility in such a system can differentiate losses in yield due to

- a slow approach to the equilibrium conditions and
- an increase in saturation concentration.

It is only in the first case that an increase in process time can decrease losses.

For the case presented in Figure 10.9, the equilibrium solubility was reached for impure systems very slowly. For higher concentration of impurities, the time required to reach equilibrium was longer than 2 weeks.

In cases where the amount of other solutes is lower, their effect of solubility can be disregarded. Figure 10.10 shows the effect for a concentration of 5% and 10% of related impurities. It is evident that the solubility is not affected by the impurities. Note that the width of the metastable zone for pure and impure systems differs by more than 10 K.

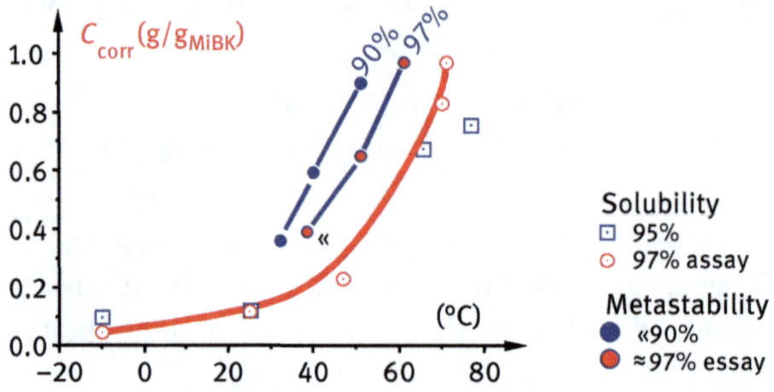

Figure 10.10: Effect of different levels of impurities on the solubility of a steroid. The sum of related compounds is 3 % and 5 % and has a negligible influence on the equilibrium solubility. Conversely, the impurities vary with the width of the metastable zone as expected.

10.2.3 Relation between solid residue and solubility

10.2.3.1 Influence of the purity of solid residue on solubility

Very few or even no published data exist on the influence of the purity of the solid residue on the solubility of the target compound such as an API. The issue here is that:

- Crystallization generally entails a purification of the target, consequently, solid residues containing considerable amounts of impurities do not exist for which the solubility could have been assessed.
- More importantly, it is hitherto unclear for most systems if the impurity forms a second solid phase or a – partial – solid solution with the target molecule of the crystallization.

It is only in the second case, the formation of a partial or complete solid solution, that the activity of the target compound is influenced by the impurity. Assessments of the influence of an impurity on the activity of the target compound in the solid state are practically non-existent. Further, experimental data exist that show a change in the enthalpy of fusion due to the inclusion of impurities into the lattice [9].

The influence of the purity of the solid residue on solubility must not be confused with the influence of the purity of the mother liquor on solubility. The latter can have a measurable influence (Section 10.2.2.3).

10.2.3.2 Influence of the solid-state form on solubility – polymorphs and solvates

The influence of the solid-state form on solubility is a well-researched and understood area and is discussed in Chapter 9 of this book. Relevant solid-state forms might be either polymorphs or solvates.

An extensive literature survey on the difference in solubilities between polymorphs and solvates has been published. For different polymorphs, the difference in solubility is on the order of 10–20% of most systems [10]. Much larger differences – up to a factor of 10 – are observed in some cases. The difference in solubility between solvates and ansolvates is generally larger.

10.2.3.2.1 Polymorphs

Thermodynamic relation between polymorphs

The relation of the solubility of two polymorphs, “1” and “2”, can be derived from eq. (10.5) by writing the equation for both polymorphs, which leads to

$$-RT\ln\frac{x_1\gamma_1}{x_2\gamma_2} = \Delta_d G_1 - \Delta_d G_2 \tag{10.12}$$

As the difference in solubility is generally small, the activity coefficients of the dissolved solute at both concentrations can be set equal, so that eq. (10.13) is obtained:

$$\frac{x_1}{x_2} \approx \exp\left(-\frac{\Delta_d G_1 - \Delta_d G_2}{RT}\right) \tag{10.13}$$

The difference in free enthalpies between two polymorphs is related to the difference in heats of fusion via eqs. (10.3) and (10.6) as follows:

$$\Delta_d G_1 - \Delta_d G_2 = \Delta_d H_1 - \Delta_d H_2 - T\,(\Delta_d S_1 - \Delta_d S_2) \tag{10.14}$$

Writing for the difference in enthalpies of fusion between two polymorphs $\Delta_d H_1 - \Delta_d H_2 = \Delta_{tr} H$, eq. (10.15) is obtained for the relationship in free enthalpies. The difference in free enthalpies and thus the difference in solubilities can be expressed by values that can be assessed by thermoanalytical methods:

$$\Delta_d G_1 - \Delta_d G_2 = \Delta_{tr} H - T\frac{\Delta_{tr} H}{T_{f,1}} + \Delta_d H_2 \frac{T_{f,2} - T_{f,1}}{T_{f,1} T_{f,2}} \tag{10.15}$$

Equations (10.13) and (10.15) show two important aspects of the relation of solubilities between polymorphs:

- The difference in solubility is a function of temperature.
- Exactly one temperature exists, for which the solubility of both polymorphs is equal. This is when the free enthalpy of dissolution of both polymorphs are equal: $\Delta_d G_1 - \Delta_d G_2 = 0$. This temperature is called transition temperature, T_{tr}, that can be derived from eq. (10.15):

$$T_{tr} = \frac{\Delta_d H_1 - \Delta_d H_2}{\Delta_d S_1 - \Delta_d S_2} \tag{10.16}$$

For illustration, stearic acid and the relation of its three known polymorphs can be used. The solubility has been determined as a function of temperature using the bracketing technique for decane, methanol, and butanone [11]. The heats of dissolution derived are summarized in Table 10.1. Also given are the enthalpies of melting as measured by differential scanning calorimetry, DSC, and the enthalpies of dissolution as derived from solution calorimetry [12]. The absolute values of the enthalpies differ due to different interactions between the solute molecule and the different solvents. However, the differences between the three forms are independent of the technique used, values behind brackets.

Table 10.1: Comparison of the heats of dissolution of stearic acid in decane, methanol, and butanone with the enthalpies of melting determined by DSC and the enthalpies of dissolution as determined by solution calorimetry, values behind brackets.

Polymorph	ΔH (kJ/mol)												
			ΔH from solubility							ΔH from solution calorimetry			
	Δ H from DSC	Δ	Decane	Δ	Δ	Methanol	Δ	Butanone	Δ	Decane	Δ	Methanol	Δ
C	64.0	6.0	64.4		4.6	84.4	4.5	73.2	4.7	62.2	6.0	75.4	5.9
B	70.0		69.9	3.3		88.9		77.9		68.6		81.3	
A			65.7										

Indicated are also the differences between the forms, see respective brackets. although the absolute values of the enthalpies vary, the differences between the polymorphs are identical.

The difference in enthalpies of fusion or dissolution between two polymorphs can be significant and thus the difference in solubility. However, although few cases are known, this difference between polymorphs can be close to negligible. Such polymorphs are called isoenergetic, and these polymorphs do have undistinguishable differences in solubility over a wide range of temperatures.

Monotropic and enantiotropic relationship between polymorphs

Except for isoenergetic polymorphs, the solubility of two polymorphs has a different dependence on temperature. A temperature exists, where both forms have equal solubility and thus equal stability. At all other temperatures, the form having the lower solubility is the more stable form. Depending on the slope of the solubility–temperature curve of the two i.e. depending on the heats of fusion of the two forms, two cases are possible. This is schematically shown in Figure 10.11 for two polymorphs having an enthalpy of dissolution of $\Delta_d H_1 > \Delta_d H_2$.
The transition temperature, the crossing of the enthalpy–temperature curves, can be either below or above the melting point of both forms leading to the two systems:

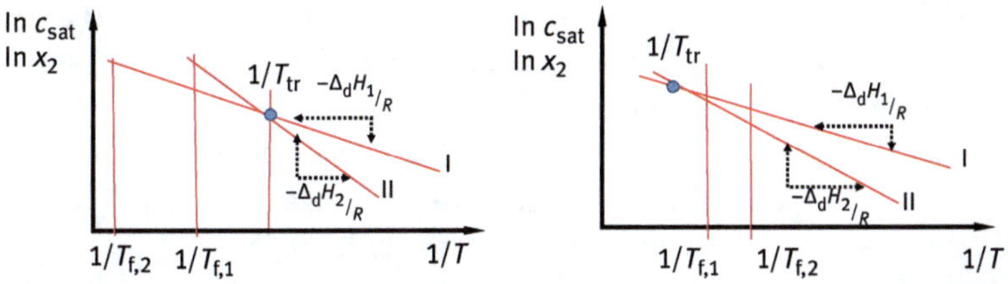

Figure 10.11: Representation of an enantiotropic, left, and a monotropic, right, system of two polymorphs 1 and 2 with an enthalpy of dissolution of $\Delta_d H_1 > \Delta_d H_2$. The crossing of the solubility curves of both forms can be either below the melting point, top, or above, bottom.

- Enantiotropic system
  The stable from changes from one to the other at the transition temperature T_{tr}.
- Monotropic system
  The polymorph with the higher enthalpy of dissolution is stable over the entire range of temperatures, that is, polymorph II in this case.

Enantiotropic and monotropic systems are discussed in more detail in Chapter 9. It is important to know the relationship between the polymorphs in the system under consideration, both for the choice of the polymorph and for process development.

As the difference in enthalpy of dissolution of two polymorphs is independent of the solvent, the transition temperature is also independent of the solvent. Thus, for the determination of the transition temperature one can resort to any solvent that is appropriate. This is different for solvates and hydrates. This will be discussed in Section 10.2.3.3.

The difference in heats of dissolution can also be determined by thermal analysis, for example by DSC. Though the values of the heats of dissolution and heat of fusion can differ, these differences are very close to both polymorphs. This can serve as an important sanity check for the experimentally determined difference in solubility, as shown in Table 10.1.

The solubility–temperature curves do not cross at all for monotropic systems and only cross once for enantiotropic systems. Thus, the stability of two polymorphs does only change once for enantiotropic systems upon heating or cooling. Finding more than one crossing indicates errors in assessment of the respective data.

The van't Hoff plots shown in Figure 10.11 are equivalent to the well-known energy–temperature diagrams. For the energy–temperature diagrams, the melting points of the forms and their enthalpy of fusion are used. For the present work, the enthalpy of the melt is subtracted, and the dependence of the enthalpy of fusion on temperature is set to zero. This makes the diagram easier to construct and easier to grasp.

Range of temperatures to be covered for process-relevant solubility data

The range of temperatures over which the solubility should be measured is to be chosen to cover at least the temperature range covered in the production process.

Preferably the range is to be extended to some 10 K outside this range. The reason being that effects that have a large influence and are otherwise difficult to observe are more easily discovered using this approach.

Figure 10.12 shows the solubility in a system with two relevant polymorphs, *B* and *E*. The dependence of the solubility on temperature is relatively low. The differences in enthalpies between the two forms is relatively large, ≈7.5 kJ/mol. This entails that the difference in solubility between both forms rapidly increases below the transition temperature of 5 °C. The implication of this with respect to process development will be discussed in Section 10.5.4.

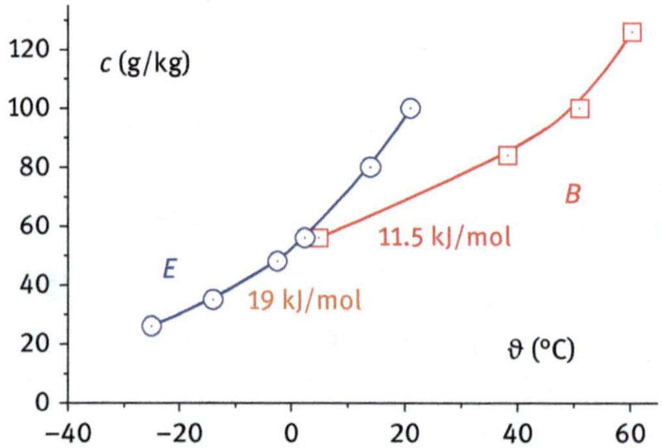

Figure 10.12: Solubility for a system with two polymorphs, *B* and *E*. The temperature dependence on the solubility is relatively low. However, the two polymorphs have a considerable enthalpy of phase transformation of 7.5 kJ/mol. Thus, the solubility of the low-temperature form *E* starts to deviate rapidly below the transition temperature of 5 °C. The determination of solubility at −20 °C easily identifies this fact red flagging a process to obtain polymorph *B* below 0 °C to increase the yield.

10.2.3.3 Solvates and hydrates

Stability of solvates and hydrates as function of temperature

For systems forming solvates or hydrates, the solubility depends on temperature in a similar way as for polymorphs. The data follow the van Laar equation. In addition, for solvates and hydrates, a decrease in the degree of solvatization is observed with increasing temperature. This is shown in Figure 10.13 assuming a system that can form two different solvates and an ansolvate. The temperatures at which the solid-state forms change are called transition temperatures and depend on the solvent. Solvates can form different polymorphs, the relationship between polymorphs discussed also above also applies to polymorphs of solvates and hydrates.

The slope of the solubility curves corresponds again to the heat of dissolution of the respective solid-state form. The difference in heats of dissolution for different solvates of a certain moiety are substantial. For hydrate forming systems, the

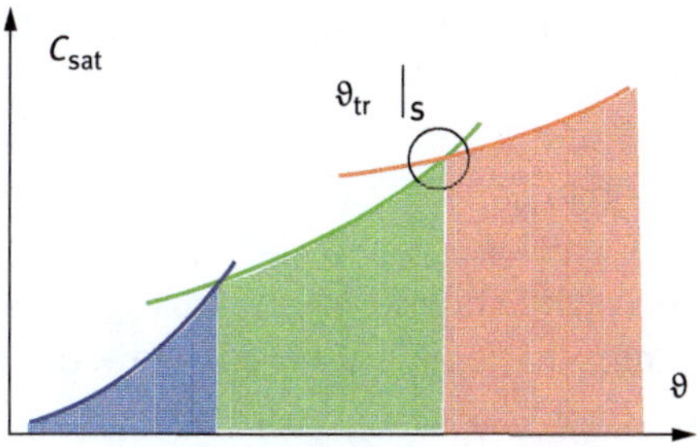

Figure 10.13: Solubility in a system forming solvates with the solvent. Three different pseudo-polymorphs are distinguished (blue = di-solvate, green = mono-solvate, red = ansolvate). Each transition has a proper transition temperature.

difference between an anhydrate and a hydrate are in the order of 10 kJ/mol and thus roughly correspond to the enthalpy of crystallization of water [13].

Solid-state form as a function of solvent composition

In a solvent mixture, the solute can either form a solvate with one or both solvents. The solvents itself can be two organic solvents or even more important aqueous mixtures with miscible organic solvents.

To discuss the solubility as a function of solvent composition and temperature, solubility isotherms can be used, as depicted in Figure 10.14. Mixtures of two solvents, s_1 and s_2, are shown, with the target moiety forming a solvate either with both or with one of the solvents. With decreasing concentration and thus decreasing activity of the solvent of solvatization, the respective solvate becomes destabilized. At a certain solvent composition, a transition between the different solvated forms occurs. The solubility of a solvated form drastically

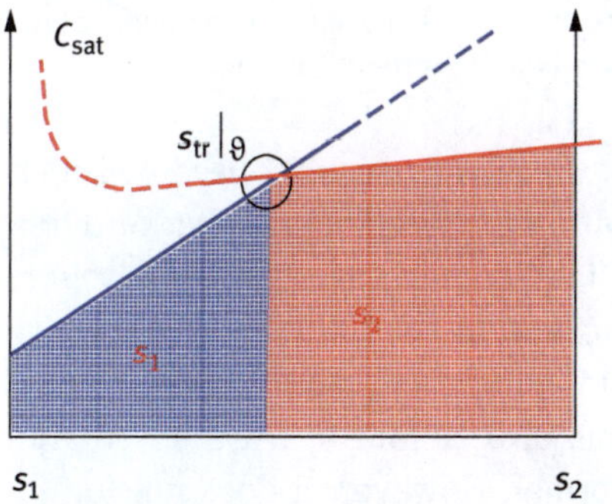

Figure 10.14: Solubility and stability isotherms of a system forming two different solvated forms, s_1 and s_2, in a solvent mixture of the two solvents. A critical concentration, $s_{tr,\vartheta}$, exists for the transition between the two solvated forms. The solubility of a solvated form drastically increases when the concentration of the solvent of solvatization approaches zero, depicted here for solvate s_2.

increases for the concentration of the solvent of solvatization approaching zero. This is depicted in Figure 10.14 for solvate s_2.

As an example of a system forming solvates and hydrates as well as a mixed solvate with an organic solvent and water, the solubility isotherms for a steroid are shown in Figure 10.15. Solubility data and the solid-state form of the residue were assessed between 20 and 60 °C. Three different solid-state forms are observed: a mono-hydrate, a mono-ethanol solvate and a mixed ethanol–water solvate. Thus, two marked changes in solubility exist that are clearly associated with a change in solid-state form of the residue. A log-linear rendering of the data is used in which the temperature dependence on solubility is hardly discernible.

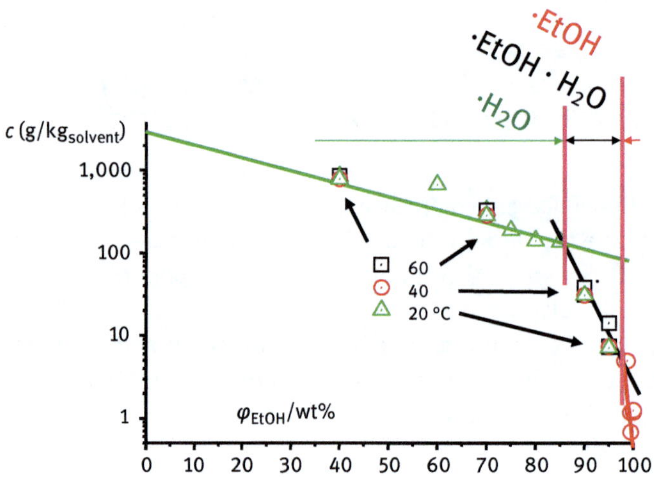

Figure 10.15: Solubility isotherms of a steroid in aqueous ethanol. The moiety can form both a mono-solvate with ethanol and a mono-hydrate, as well as mixed solvate with ethanol and water. The change in solid-state forms as a function of solvent composition is indicated by the red lines and at the same time accompanied by a marked change in slope of the solubility curve. A log-linear plot has been chosen so that the dependence of solubility on temperature is hardly distinguishable.

While in most relevant systems the number of coexistent solvates that need consideration is limited, accounts have been published with a considerable number of different solvates. For example, a determination of the stability domains of cephaloglycin in ternary mixtures of water, acetic acid, and methanol and its solubility in the respective binary mixtures revealed a larger number of solvates, single and mixed solvates [14].

Figure 10.16 shows three solubility isotherms of a steroid in a water–ethanol mixture. This steroid is a promiscuous solvate former. However, it does not form a hydrate, so that the ansolvated and anhydrated forms are directly accessible. Also indicated are the respective transition points for three temperatures. As expected, the ethanol solvate is stable above a critical concentration of ethanol. This critical concentration increases steeply with temperature covering a wide range of fractions

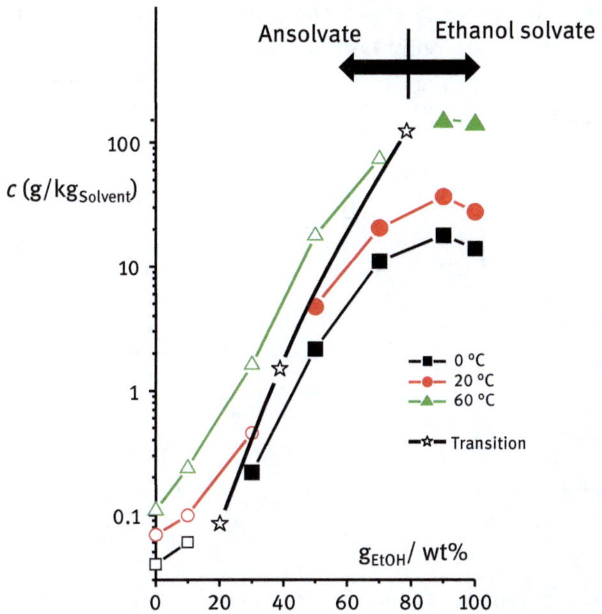

Figure 10.16: Solubility isotherms and transition points between an anhydrate, left, and an ethanol solvate, right, as a function of solvent composition in a mixture of water and ethanol. Water is a poor solvent. Solubility decreases in such a way that water can be used to drown the steroid out of a solution in ethanol. The solubility of the ethanol solvate goes through a maximum at high fractions of ethanol. This effect can be observed frequently.

for the ethanol over a small range of temperature. The enthalpy calculated from the slope of ≈12 kJ/mol representing the solvate–ansolvate transition is on the order of the expected value and thus plausible.

The development of the transition point as a function of temperature follows the same rules as discussed for a pure solvent system, that is, with increasing temperature, an increasing concentration of the solvent of solvatization is required to stabilize the solvate, as shown in Figure 10.13. The free enthalpy for the solvate–ansolvate transition is related to the activity of the solvent of solvatization:

$$-\Delta G_{\text{solvate}-\text{ansolvate}} = RT\ln a_{\text{soc}} \tag{10.17}$$

From eq. (10.17) one can derive the dependence on temperature of the transition point in mixed solvent systems as

$$\frac{d\ln a_{\text{SoC}}}{d1/T} = -\frac{\Delta_{\text{tr}}H_{\text{solvate}-\text{ansolvate}}}{R} \tag{10.18}$$

Thus, with increasing temperature a higher fraction of the solvent of solvatization is required in the solvent phase to stabilize the solvate. The slope of the line is given by eq. (10.18). Figure 10.17 schematically shows the trends in a mixture of an organic solvent with water for the transition point between a hydrate and anhydrate and for a solvate with the organic solvent and an ansolvate. The slope is in both cases towards the solvent-of-solvatization side.

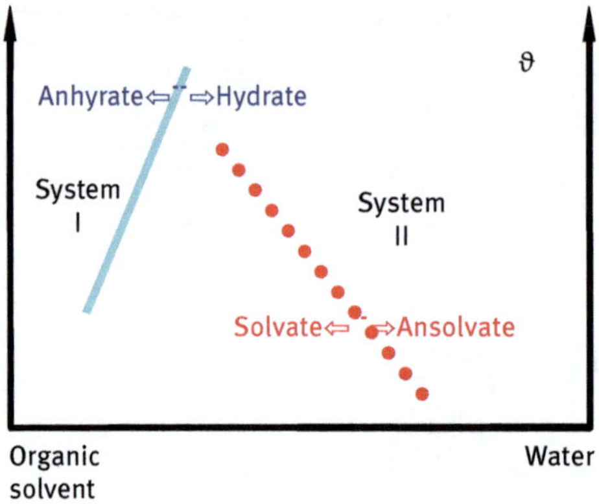

Figure 10.17: Development of the transition point as a function of temperature for the transition between an anhydrate and hydrate, system I, and a solvate and ansolvate, system II. In both cases, the slope is towards the solvent-of-solvatization side.

The transition point at constant temperature between an ansolvate and the solvate in mixed solvents apparently depends also on the nature of the second, inert solvent. The free enthalpies of the ansolvate and solvate are equal at the transition point. This is the same principle as for polymorphic phase transformations. For the solvate, the activity of the solvent of solvatization, a_{soc}, in the solution phase must also be considered, eq. (10.17).

For the hydrate-anhydrate transition of theophylline, the transition in mixtures of water with methanol and with isopropanol occurs at different concentrations of water, as shown in Table 10.2. Transforming the concentration of water to its activity, the transformations occurred at the same water activity [15]. The activity coefficient

Table 10.2: Transition points for the anhydrate–hydrate transformation of theophylline in mixtures of water with methanol and isopropanol at 25 °C.

Solvent	Anhydrate–hydrate transition observed at	
	Water fraction	Water activity
Water + MeOH	0.20	0.25
+iPrOH	0.07	0.25

Given are the concentration of water and the activity of water at the transition point. The activity of water in the solvent can be estimated from vapour pressure measurements of the respective organic–water mixture.

of water can be derived from the vapour–liquid equilibria (VLE) of the respective solvent mixtures, neglecting an influence of the solute on the VLE:

$$a_{\mathrm{tr}} = \frac{p}{p_{\mathrm{e}}}$$
$$\frac{p}{p_{\mathrm{e}}} \equiv rH|_{\mathrm{water}} \tag{10.19}$$

The transition point at constant temperature between an ansolvate and the solvate in solution and in an atmosphere of the respective solvent is inter related. This is of interest namely for hydrates. In both cases, the activity a_{tr} of the solvent of solvatization, for example, water, at the transition point is equal. In the vapour case, the activity is given by the vapour pressure (eq. (10.19)). For water as solvent of solvatization, this corresponds to the relative humidity, $rH|_{\mathrm{water}}$.

10.2.3.4 Incongruently dissolving systems

Incongruent dissolution of solids is relevant for multi-component solids such as hydrates, solvates, pharmaceutical salts, and co-crystals. A multi-component solid is a solid phase on its own. For such a system, the solubility of the compound as well as that of its other components must be considered. Two cases can be distinguished as shown in Figure 10.18 using the example of a co-crystal:

- The solubility of the co-crystal is lower than the solubility of its components, indicated by co-crystal I

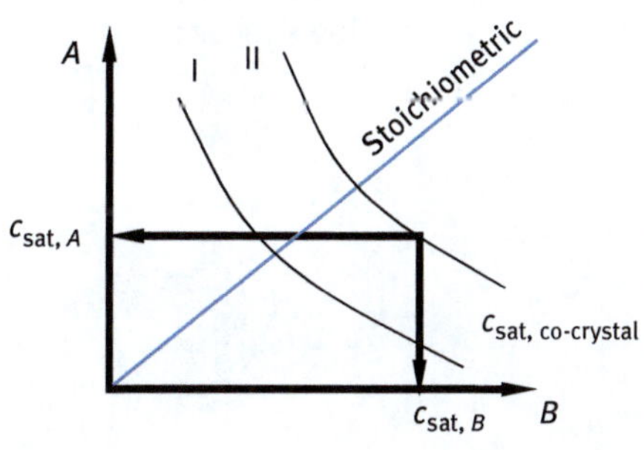

Figure 10.18: Incongruent dissolution of a co-crystal: The solubilities of the two components *A* and *B* are assumed to be independent of each other. Two cases of the solubility of the co-crystal are distinguished: (I) the solubility of the co-crystal being lower than the solubility of the single components and (II) the solubility of the co-crystal being higher than the solubility of the single components. The latter case will lead to the – partial – precipitation of one of the components when measuring the solubility of the co-crystal. In the case presented, it will be components A/B that precipitate. The partitioning will proceed until the concentration in the supernatant reaches the corner.

- The solubility of the co-crystal is higher than the solubility of its components, indicated by case co-crystal II

Thus, in thermodynamic equilibrium, the solid phase will consist of both the co-crystal and the low-soluble constituent. The degree of partitioning will depend on the ratio of the solubilities of the components of the co-crystal. It proceeds until the concentration of both components of the co-crystal have reached the corner in the supernatant, as shown in Figure 10.18.

The case of an incongruent dissolution is often a desired property of pharmaceutical salts as well as of co-crystals. It is desired that a pharmaceutical salt or a co-crystal has a higher solubility for the corresponding base, acid or active moiety itself. Thus, for a solubility-limiting concentration, the disproportionation might occur.

10.2.4 Solubility in systems with phase separation

More often than expected, a phase separation of the supernatant can occur. This phase separation can either be metastable, occurring only upon supersaturating the system, or stable. A phase separation will lead to a system with one phase having a high concentration of the solute and a second with a low concentration of it. If the high concentration phase is carried during the crystallization process to conditions where the phase split is no longer stable, crystallization will instantly occur in that phase. This can have three effects:

- Large chunks of material are generated, sometimes called spherical agglomerates or lolly pops that are difficult to handle. Figure 10.19 shows an example.
- The solidification of the phase rich in solute is not accompanied with a purification. High inclusions of solvent and mother liquor occur. The structures are not accessible to a removal of the liquid phase, i.e. a washing process.

Figure 10.19: Agglomeration of crystalline material within droplets to nearly spherical agglomerates during an oiling-out. Liquid inclusions of mother phase are clearly seen in the middle and lower right part.

- A persistent liquid phase can lead to a considerable loss in target compound during solid–liquid separation, as often the solute has a considerable concentration in one of the liquid phases.

Two cases of a stable phase separation can be distinguished:
- In a single solvent system, the dissolution of the solute in the solvent can induce a phase separation of the supernatant into a phase rich in solute and a phase with a low concertation of the solute.
- In a system comprising of two or more solvents that are completely miscible, the dissolution of solute can also induce a phase separation into a phase rich in one of the solvents and one phase poor in this solvent. The solute will mostly dissolve in the phase rich in the better solvent or solvents.

In equilibrium, the chemical potential of the solute in the two phases is equal:

$$\mu_{\mathrm{I}} = \mu_{\mathrm{II}} \tag{10.20}$$

Thus, if one of the phases is saturated with respect to the solute, the other phase is also saturated, as the chemical potential of a solute is independent of the solvent (see Figure 10.20). Two equilibria are observed, the solid-solution equilibrium, that is, solubility equilibrium, and the partitioning of the solute between the two liquid phases.

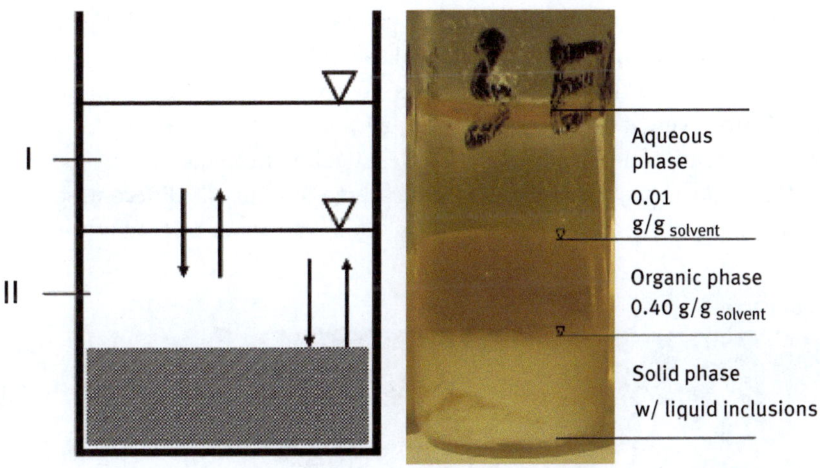

Figure 10.20: Solubility in a system with two liquid phases, I and II. The two equilibria between solid and phase II and between the two liquid phases I and II are shown on the left. A flask with a system exhibiting a liquid–liquid phase split is shown to the right. The system is a steroid dissolved in aqueous acetone. The solid is at the bottom while two liquid phases are above. The acetone-rich, lower phase dissolves considerable amounts of steroid, ≈0.4 $g/g_{solvent}$ or 400 g/kg. As a result, this organic phase, though having mainly acetone – which has a lower density compared to water – as solvent forms the bottom phase while the aqueous phase is the upper phase.

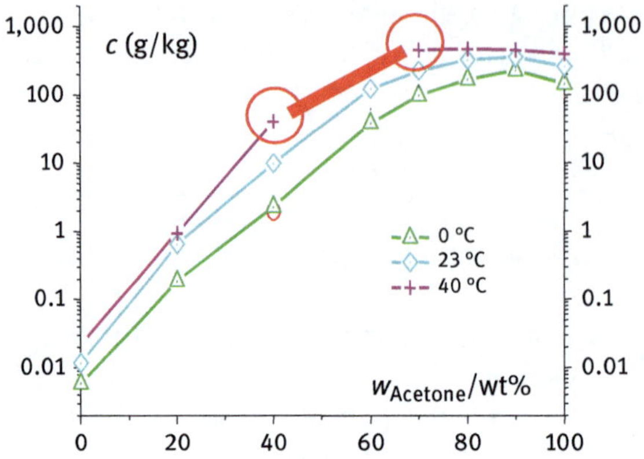

Figure 10.21: Solubility isotherms of a steroid in aqueous acetone. The system shows a phase split at and above 40 °C, see red circles indicating the two phases in equilibrium. The tie line connecting the two saturated liquid phases is also drawn, red, bold straight line connecting the circles, which is in the representation as solubility isotherms not a horizontal line.

The isotherms for the solubility of the steroid in aqueous acetone are shown in Figure 10.21. The phase split occurs at and above 40 °C. Due to the high solubility of ≈400 g/kg, The dissolution properties of the acetone for water change. The phase rich in acetone, right circle in Figure 10.21, has a concentration of ≈400 g/kg, while the aqueous phase, left circle, has a concentration of ≈30 g/kg. This explains why the acetone-rich phase is the heavy phase in Figure 10.20.

A liquid–liquid phase (LLP) separation can go unnoticed by observation with the naked eye. However, it is easily observed by microscopic inspection of the – suspected – suspension. An LLP split will manifest itself in droplets.

Figure 10.22 shows two micrographs of the same systems at different stages. First, an emulsion is formed consisting of droplets in the lower micrometre range

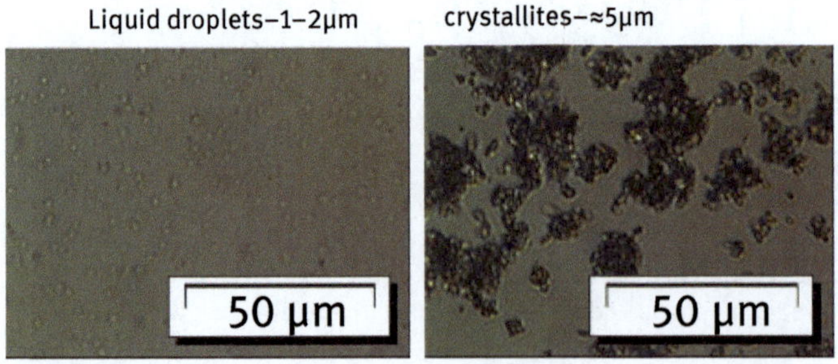

Figure 10.22: Micrographs of a system showing metastable oiling out. To the left, the emulsion is shown, while on the right crystals obtained from this emulsion are shown. Oiling out is easily detected by microscopy. The crystalline phase distinguishes itself by its birefringence.

by supersaturating a system. Out of these droplets, crystals are formed over time, as the oiling out for this system is metastable.

The phase separation leads to a steep increase in solubility as a function of temperature, dashed line (Figure 10.23). Below the temperature of demixing, T_m, the solubility is X_2, but at T_m, the solubility increases to $X_{2'}$. The phase separation is observed above a certain concentration of the solute. The immiscibility dome as shown in Figure 10.23 can be below the solubility line, which leads to the metastable phase separation. In this case, the phase separation is not observed. This fact is of relevance for process development purposes. A phase separation can be avoided by carrying out the crystallization slowly around the phase separation point.

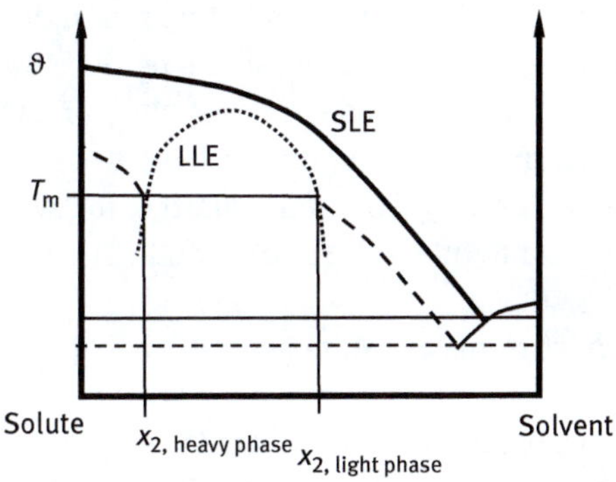

Figure 10.23: Phase separation in a composition temperature diagram. Two cases are distinguished: (a) The miscibility dome penetrates the solubility curve, dashed solubility line, and (b) the dome is below the solubility line, solid black solubility curve. In the first case, the phase split is thermodynamically stable while it is metastable in the second case. Contrary to the conventional notation, temperature and concentrations have been inverted, the y-axis showing temperature, while the *x*-axis shows concentrations. SLE and LLE denote the solid–liquid equilibrium, the solubility line, and the liquid–liquid equilibrium, the oiling out.

10.3 Determination of solubility

10.3.1 General

Several methods for solubility determination relevant for process development are available, which fit different purposes and require more or less experimental effort. In many cases, the effort is a manual one, though concepts for automation are available. Details of the methodology used are also often subject to modifications according to the requirements, such as accuracy or number of parameters to be treated. The quality and reliability of the data may depend on the technique used for determining solubility, but certainly the effort put into

- covering the correct space of parameters,
- making sure, that the solid-state form is known for which the solubility is determined and,
- in making sure that the data are correct and consistent has to be adequate.

Thus, these latter topics are treated alongside with the techniques for measuring solubility.

10.3.1.1 Units for solubility data

Units in which solubility can be expressed are manifold and should be chosen according to the specific needs. However, the data must be unique. Grams per litre do not fall into this category, as the density of the solution needs to be known, for example, when preparing the solutions. Molar or mass fractions are often used. However, for systems containing more than two components, for example, for systems with more than one solvent, these units might be misunderstood. In such cases

- concentrations of the solute can be given as grams of solute per gram of solvent on a solute free basis, or
- the composition of the solvent for a solvent mixture can be given as mass of the respective solvent per total mass of solvent on a solute free basis.

10.3.1.2 Requirements for accuracy of data

The accuracy with which solubility data is measured depends on the purpose of the data, which might vary. However tempting it might be to roughly estimate solubility first and increase accuracy later, this might lead to bad data, which is never challenged and thus never accurately assessed. Some general considerations should always be considered:

- The solid-state form for which the solubility was assessed needs to be unambiguously known, as this is one of the major influencing parameters.
- The quality of the solvent should be known.
- The saturation temperature should be known to ±1, . . ., ±2 K, for example, if the data is to be used for seeding point estimation, this corresponds to a typical error in saturation concentration of ±3% to ±7%.

The concentration in a plant vessel has also a certain error adding to the uncertainty.

A plot of data and a test of consistency can facilitate assessing the accuracy of the data. Figure 10.24 shows data of the solubility of all three modifications of stearic acid in decane and the result of a fit using the van Laar equation. It is evident that the data

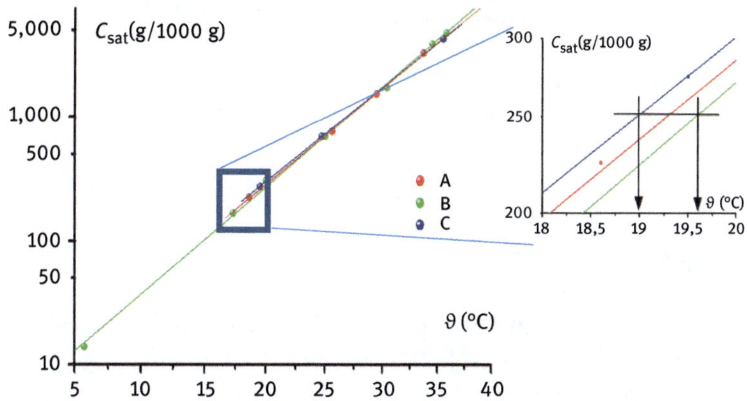

Figure 10.24: Data for the solubility of the three modifications *A*, *B*, and *C* of stearic acid in decane as a function of temperature. The data points are shown as well as a fit using the van Laar equation, and a zoom into the data, insert. The deviation of the individual measurements is ≤0.2 K. The error of 0.2 K corresponds to a considerable error in the saturation concentration of ≈5%.

can be represented over the entire range of temperatures covered by this fit. A zoom into the data reveals individual deviations from the straight line of ≤0.2 K. This is at the same time the maximum error in concentration of ≈5% which is considerable.

10.3.1.3 Reproducibility

Solubility data are thermodynamic values, which are independent of person, location, and technique and do no not vary over time. Any changes encountered, deviations from previously obtained data, any influence of the quality of the starting materials should be taken seriously and indicative of hitherto unknown variables.

The most obvious errors are sampling of the supernatant from the saturated suspension and determining the concentration of the solute. Thus, repeated samplings, at least at one temperature, and an assessment of the concentration of the solute with independent techniques are advisable. As will be discussed below, evaporation to dryness of the supernatant is prone to an inclusion of residual amounts of solvent, namely for higher solubilities.

Measurements at high temperatures are prone to an evaporation of the solvent during equilibration and namely during sampling. Evaporation of solvent during equilibration has only an influence if solvent mixtures are concerned through the preferential evaporation of one of the solvents inducing a change in solvent composition. The evaporation of solvent during sampling and work-up is a general point of concern which applies also to assessment of solubility in a pure solvent. The determination of the saturation temperature using hermetically sealed containers is a work-around, see section 10.3.4.

Thus, it is a good idea for every lab to assess the magnitude and influence of the errors, the experimental imperfectness of the processes used, at least once and for new systems.

10.3.1.4 Kinetics of equilibration

The kinetics of equilibration of a solution with the solid residue is often a question of concern and rarely answered completely. The best technique to unambiguously determine the saturation concentration is schematically shown in Figure 10.25. By starting both from a subsaturated point, the usual case when mixing solvent and solute, and from a supersaturated solution, the thermodynamic equilibrium is reached from both sides. In addition, the error in saturation concentration can be estimated.

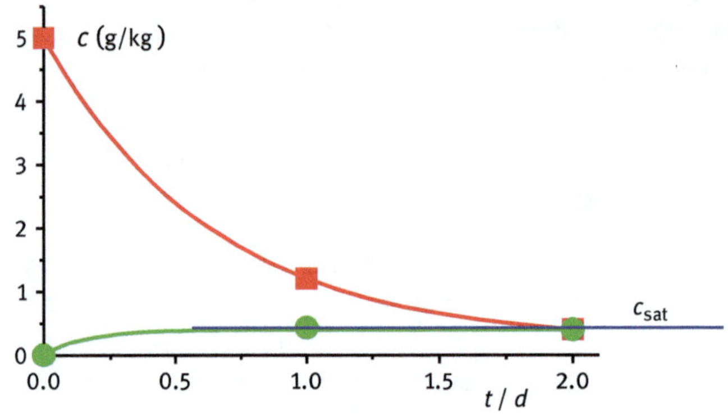

Figure 10.25: Strategy to corner the "true" solubility isotherm by observing the concentration in solution reached both from a subsaturated starting point, circles, the typical situation in a measurement of solubility, but also by starting from a supersaturated solution, squares. The latter starting point is rarely used. The supersaturated starting solution can be prepared by cooling a solution prepared at higher temperatures. From both approaches, the saturation concentration can unambiguously be stated.

The kinetics of equilibration depends on the system itself, the solute, and the solvent. The most prominent influencing parameter is the impurity content of the system. Its effect on the kinetics of equilibration is twofold:

- The dissolution kinetics can be affected by the presence of impurities; in most cases, the level of impurities is determined by the purity of the solid used in

the measurements. The influence on the kinetics is to be distinguished from the influence on the equilibrium solubility itself.
- When reaching saturation from a supersaturated solution, the level of impurities can be considerably higher, as the crystallization leading to a saturated solution is in most cases associated with a purification and thus enrichment of impurities in the mother phase.

In impure systems the kinetics of growth of the crystals can be drastically impeded. Cases are known, where fresh and pure crystals are added to a supersaturated solution of a solute in an impure system, where the crystals grew for some time and subsequently ceased to grow or grew at a much lesser rate. Equilibrium, i.e., saturation took weeks to reach.

Typically, but not always, the kinetics of dissolution is affected to a lesser extent by the level of impurities than the growth of the crystals. This might explain the different rates with which equilibrium is reached for the case shown in Figure 10.25.

A further effect that can slow the equilibration process is a change of the solid-state form of the residue from a metastable to a more stable form, see the path for cases c and c' in Figure 10.2. Of course, this effect is immediately evident if the solid-state form is assessed.

If solubility as obtained from the subsaturated and supersaturated starting point is not the same, this might reflect that in both approaches different solid-state forms have been obtained.

10.3.1.5 Range covered and number density of points

The range of conditions that the solubility measurements should cover depend on the properties of the system. In general, the range of temperature should at least cover the range of temperatures envisaged for the process. This range should consider the current as well as possible extension of conditions at later stages of the development process. Preferably, the range should cover somewhat of 10 K outside this range.

The extension of data points outside the range envisaged for the process is instrumental in detecting changes in the solubility behaviour that might otherwise go unnoticed. The driving force for possible changes in the solid-state form are increased, thus increasing the possibility of detecting changes.

The number of data points to be measured depends on the accuracy required. For measurements of solubility as a function of temperature, at least three data points at different temperatures are necessary to assess the slope and thus the temperature dependences. If this number is set to five, this enables both the detection of errors in individual measurements and in detecting possible changes in solubility behaviour by detecting deviations in the trends. These changes in solubility behaviour can both be changes in the solid-state form or other effects, for example, oiling-out processes.

For the measurements of solubility in solvent mixtures, the influences of solvent composition should always be determined over the entire range of relevant solvent compositions. Additionally, the inclusion of pure solvents, even if not envisaged for the process, allows for the possible finding of hitherto undiscovered solvates that might otherwise appear later.

10.3.1.6 Accuracy of solvent composition in solvent mixtures

For solvent mixtures, two aspects are of relevance. The solvent composition can change due to the preferential evaporation of one solvent during the measurement, in most case of the lower boiling solvent and the consumption or production of solvent in case the solvation state of the solute changes. This is namely important, if the solid residue has a high fraction, or if the solvent consumed or produced has a low fraction in the solvent mixture.

It is thus advisable, to check if a significant change in solvent composition has occurred.

10.3.1.7 Programme for data measurement

It is advisable to assess the solubility variable by variable as it allows for the determination of influencing parameters as well as a consistency check of the data. Thus, solubility should be determined as a function of temperature for a certain solvent or solvent mixture at fixed solvent composition, and at constant temperatures (at least for two temperatures) as a function of solvent composition.

The assessment of any influence of impurities such as a second solute can follow at selected temperatures and solvent compositions and using the trends observed for pure systems.

10.3.2 Rendering of data

Rendering of solubility data is an important process as it can serve several purposes such as a supporting sanity check, detecting outliers, interpolating data, and to a certain degree help to extrapolating data. Additionally, it is useful to understand phase diagrams.

Several types of plots are used for this purpose:
- Solubility as a function of temperature for a given solvent composition; see Figures 10.3, 10.12, and 10.13 for example.
- The influence of the composition of the solvent can be plotted in isotherms as shown in Figures 10.5, 10.6, and 10.16.
- For three-component systems, triangular diagrams are in use; see for example Figure 10.8.

In all cases, attention should be paid to the solid-state form of the residue.

For the development and optimization of crystallization processes, it is often advantageous to follow or to depict the course of the crystallization in the phase diagram. By this, a change in the domain of the stable solid-state form can be identified, see chapter 10.5.4.

10.3.3 Flask method

The flask method is a simple yet very versatile method to measure solubility. The basic idea is to equilibrate a solid with solvent, for which a flask is used. To accelerate the equilibration, the flask is usually shaken, or its content is stirred. Two strains will be discussed in detail, a semi-quantitative and a quantitative flask method.

10.3.3.1 Semi-quantitative flask method

Rough estimations of solubility are often performed for solvent screening by a semi-quantitative flask method. Two types of this method are known:
- A certain amount of solute, for example, 50 or 100 mg of solute are weighted into a flask, and increasing amounts of solvents are successively introduced into the flask.
- The interval between a fully dissolved system and the system with a certain amount of undissolved solute present gives the estimate of solubility.
- A certain amount of solvent, for example 2 g, is introduced into a flask and successively small amounts of solute are added.
  Here the solubility is bracketed between a system still fully dissolved and with a small amount of undissolved solute present.

After each addition of solute or solvent, the system should be shaken or stirred intensely to accelerate equilibration. If measurements at higher temperatures make shaking or stirring difficult, one can resort to a longer equilibration time.

This methodology is very much limited to systems reaching equilibrium consistently and quickly, as after each addition of either solvent or solute, the system must be given enough time to attain equilibrium. This method gives only an interval of the solubility.

With respect to phase behaviour, the major drawback of this method is that no control and no reliable information is available on the solid-state form of the residue for which the solubility is estimated.

10.3.3.2 Quantitative flask method

The quantitative flask method can be used to determine the solubility for both polymorphs as well as solvates. Typically, it is employed to assess the data for the stable form. However, depending on the system, it is also possible to determine, or to estimate the solubility of metastable forms.

10.3.3.2.1 Flask method – stable polymorphs or solvates

For the quantitative flask method, solid and solvent are weighed into a sealed flask and incubated including stirring or shaking to equilibrate at chosen constant temperature. The saturation concentration for this temperature is determined by sampling the supernatant of the suspension (Figure 10.26).

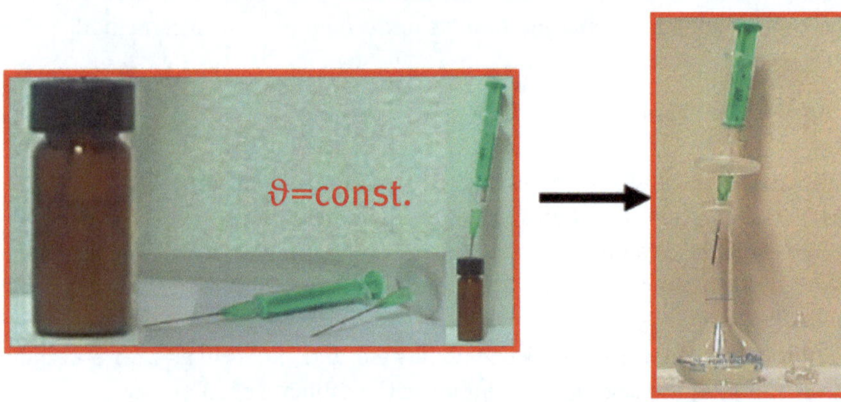

Figure 10.26: Equilibration of a suspension using the isothermal flask method, left, and procedure for sampling of supernatant, right.

For the sampling process the amount of solute dissolved and the mass of the sample withdrawn are relevant. Thus, the sampling consists of several steps:

- Sampling of the supernatant

This can be achieved by pulling a larger quantity of supernatant into a syringe. To avoid evaporation of the solvent, sampling should not (!) be through a needle but directly into the syringe. To avoid crystallization in the syringe, it should be at the temperature under investigation or higher.

- Separation of solid residue from the sample

This can be achieved by pressure filtering through a 0.4–0.2 µm membrane filter, with the filter material being chosen according to the solvent under investigation, either hydrophobic or hydrophilic. The filters area itself should be chosen large enough for easy filtration.

To avoid crystallization in the filter, it is mandatory that the filter is at a temperature equal or higher than the temperature under investigation. As the concentration of solute is typically high, a pre-saturation of the filter with solute is not necessary.

- Assessment of the amount of solution taken

This can be achieved by filtering the sample into a flask containing a weighed amount of solvent at room temperature. This solvent serves both as a dilatant but namely as a quench in the case that hot solutions are filtered entailing the danger of a loss of solvent and thus weight due to evaporation. The size of the sample filtered can be assessed by weighing the flask before and after filtration.

Once the samples are diluted and subsaturated, with the quantity of sample known, the amount of solute present is to be determined. Several methods are in use:

- The most versatile method is the use of HPLC or UPLC

The technique has the advantage to cover a very wide range of concentrations for which the concentration can be assayed. Relatively small concentrations for solubility are accessible. A further advantage being that it can distinguish and quantify impurities present in the supernatant.

- Quantitative UV spectroscopy is sometimes used.

The method can cover a limited range of concentrations. Its results are obscured by low specificity as not only the solute but other moieties, for example, related substances, might also absorb in the UV range chosen.

- Evaporating the solvent in the sample, by a "rotavap" or any other device

The use of a "rotavap" is advisable, as the surface area of the solid in the round bottomed flask can be kept large, minimizing the risk of solvent inclusions in the solid. A drawback of this technique is that the mass of solute and solid impurities present in the supernatant are not differentiated, which might lead to erroneous results.

Thus, only the chromatographic method can unambiguously quantify the amount of dissolved target compound, while the latter two cannot distinguish impurities stemming from the dissolution of the starting material.

As already mentioned, the solid-state form of the solid residue during equilibration with the supernatant must be assessed. This requires the sampling of the residue and assessment of the actual solid-state form. Sampling can be achieved by either a quick filtration or a decanting of a suspension that has settled. For the settling, centrifuges with a temperature control can be used. For the assessment of the solid-state form, several techniques such as PXRD, DSC, IR, or Raman spectroscopy can be used. Care must be taken during drying of the solid residue to avoid conversion of solvates to ansolvates or lower solvates due to loss of solvent of solvatization. The same holds true for hydrates. A comparison of the solid-state form of wet and dried samples is recommendable.

Finally, a microscopy of the crystals might be of help, as changes in the solid-state form during sampling and measurement will most probably be a solid-solid phase transition, as the time would not suffice for a solution mediated phase transformation. Such a phase transformation is generally accompanied with a destruction of the homogeneity in the crystal appearance as shown in Figure 10.27.

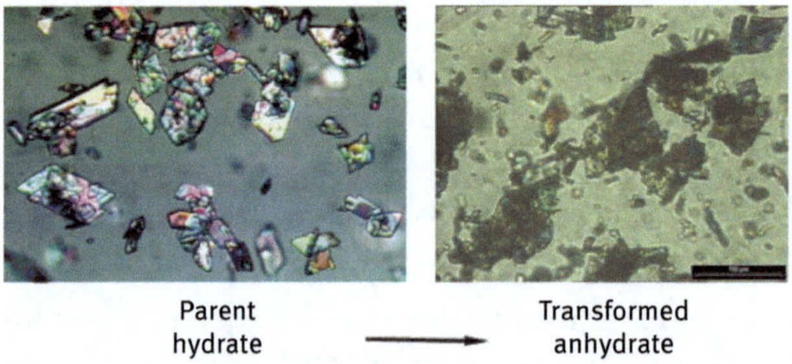

Figure 10.27: Microscopic images of a solvate form of a crystal, left, undergoing a desolvatization in a solid-solid phase transformation during drying, right. The neat and homogeneous birefringence is lost while the outer shape and symmetry of the crystals remains.

The quantitative flask method has its advantages but also some drawbacks. Advantages are:
- The method is simple and cheap to set up.
- The solid-state form for which the solubility is determined can be unambiguously assessed by sampling the residue.

Some of the advantages are also directly a drawback:
- The equilibration takes time, where the actual time required is not known. A solution of this issue is a repeated sampling of the supernatant and a determination of the concentration in solution. Figure 10.25 shows data for a solution being saturated. In addition, data for the desaturation of a supersaturated solution is shown.
- Data for temperatures close to the boiling point of the solvent are difficult to assess, as the effects of solvent evaporation during sampling become important. In such cases, techniques discussed in Section 10.3.4 are better suited.

10.3.3.2.2 Flask method – metastable forms

The equilibration of solid and solution takes time, so that a metastable solid-state form can transform to a more stable, and thus less soluble one before equilibrium has been reached. Thus, the flask method is mainly used to assess solubility of the stable form. However, if the nucleation of and transformation into a more stable from is slow, the flask method can used at least to estimate the solubility of a metastable solid-state form.

A suspension with the form under investigation is prepared. Both the concentration of the solute in the supernatant and the solid-state form of the residue are assessed as a function of time. If the concentration has reached a plateau and stays at this plateau for at least some hours, while the solid-state form of the residue remains – mainly – of the form under investigation, the concentration at the plateau can be taken for its solubility (Figure 10.28). As the concentration in solution roughly corresponds to the solubility of the dominant phase of the solid residue, small amounts of a more stable form do not significantly influence the results.

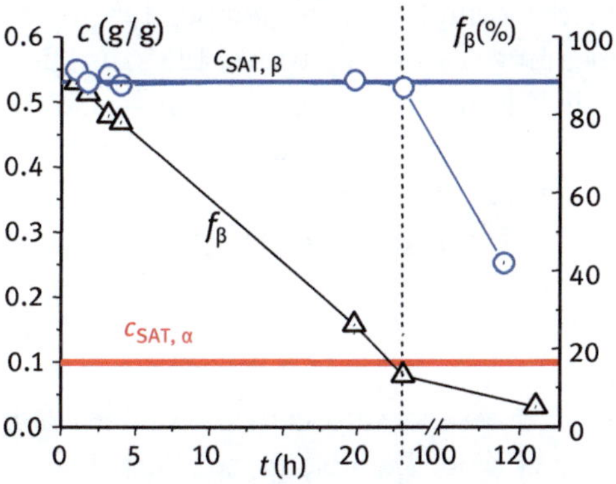

Figure 10.28: Concentration of a solute in its supernatant using a metastable form, either a polymorph or a solvate, circles. Both the concentration and the solid-state form are followed as a function of time. The solid state form is expressed as fraction of the unstable ß (editor: sign for beta that is) form, triangles. If the concentration reaches a plateau for some time while the solid-state form of the residue remains – mainly – of the form under investigation, the concentration at the plateau can be estimated for the solubility of the metastable form used for preparation.

The data can be checked by plotting the solubility of both the metastable form and the stable form as a function of temperature in a van't Hoff- plot. The difference in enthalpies of dissolution should be comparable to the difference in fusion as assessable by thermal analysis.

10.3.4 Saturation temperature methods

The drawbacks of solubility determination via measurement of the concentration of solute in the supernatant, namely during sampling and workup of the supernatant, can be avoided by the determination of the saturation temperature for a solution with known concentration. Several techniques for the determination exist. All have certain advantages and disadvantages.

10.3.4.1 Slow heating –manual procedure

A suspension with a known concentration is prepared, for example, by weighing solvent and solute into a tightly sealed container. The amount of solid weighed must be such that the system still contains an amount of undissolved solute at the starting temperature. This suspension is slowly heated, for example, in 5 K or better 1 K steps. At each step, the development of the residue is observed visually, over one day. If the solid material is not dissolved completely, the temperature is increased to the next temperature level. Depending on the step width, the saturation temperature is accessed to a certain interval. The suspension must be observed for a time long enough that the dissolution process has attained equilibrium. However, this is only necessary for the last interval before the saturation temperature is reached, where the amount of solid residue is low. This considerably shortens the time for complete equilibration.

The container must be tightly sealed to avoid any evaporation. As a check, the container can be weighed after each temperature step.

This method has the advantage of being accurate with respect to the solid residue, as the eye can easily detect even tiny amounts of solid. The time for equilibration can be chosen over a wide range.

The solid residue can be controlled to be the stable form by preparing the initial suspension with this solid-state form.

If no solid is present after preparation or after bringing the system to complete dissolution, crystalline material can be generated by a cooling step. No active control over the solid-state form of the solid residue is possible during this procedure. One way to tackle this issue is to analyse the solid residue generated. However, after sampling the concentration is no longer known, so that this determination is possible only retrospectively.

10.3.4.2 Crystal16 method

An automated version of the saturation temperature measurement was first published by Mettler [16] while commercial equipment is available under the name

"Crystal16" [17]. The basic idea is the same as described above for the determination of the saturation temperature, i.e. the temperature of complete dissolution: A suspension with known composition is slowly heated until all solid is dissolved. The process is automated using a turbidity probe to monitor the dissolution process (Figure 10.29).

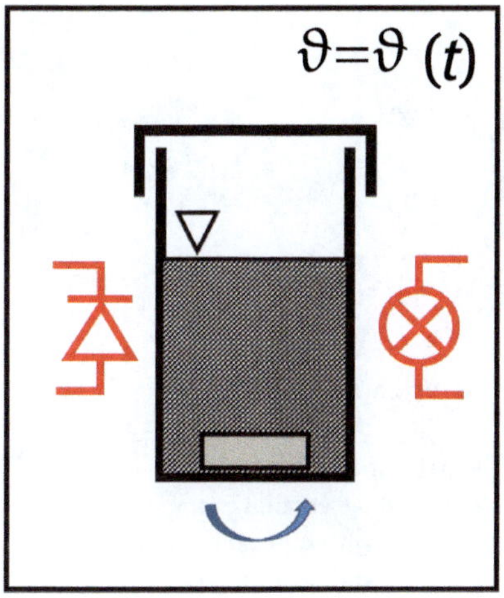

Figure 10.29: Experimental set-up for the "Crystal16" automated measurement of solubility. The sample is in a 2 mL vial equipped with a magnetic stirrer. This vial is placed inside a temperature-controlled heating-block. The dissolution of crystals during heating and the appearance of crystals after cooling is followed by a transmission turbidity probe.

Afterwards, each vial is cooled to generate a suspension. Different heating and cooling rates can be employed, thereby extrapolating the temperature for infinitely slow heating.

The apparatus accommodates in total 16 vials grouped into four blocks. The temperature of each block can be controlled independently. Thus, four independent trials with four repeats each can be carried out simultaneously.

Figure 10.30 shows a scheme of heating and cooling cycles and typical curves for turbidity as measured by the probe. This technique allows the determination of solubility, but also the determination of the metastable zone limit.

By plotting the turbidity versus temperature (Figure 10.31), the saturation temperatures and their variability as well as the temperatures required for nucleation become evident.

The second and all following cycles need the nucleation of crystals, which can be difficult for lower solubility at lower temperatures as nucleation in organic systems typically requires a substantial subcooling of typically 10–15 K.

Figure 10.32 shows the results for measurements at elevated temperatures. The heating and cooling rates varied between 0.25 and 1 K/min. The clear points, i.e. the apparent saturation temperatures do not depend significantly on the heating rates. However, as expected, the nucleation temperatures or cloud points are a steep function of the cooling rate.

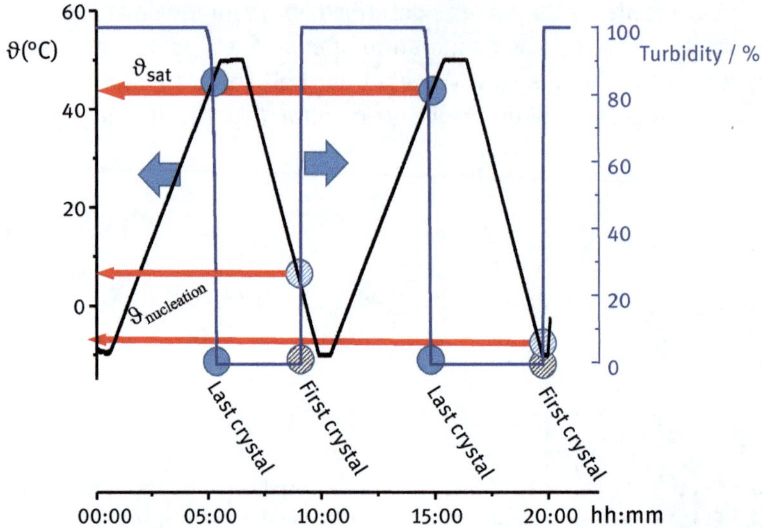

Figure 10.30: Temperature cycles for the determination of the saturation temperature of a suspension with known composition. Each heating ramp is followed by a cooling ramp until nucleation occurs. Repeated saturation temperature determinations with different heating and cooling rates can be carried out. This technique also yields the metastable zone width.

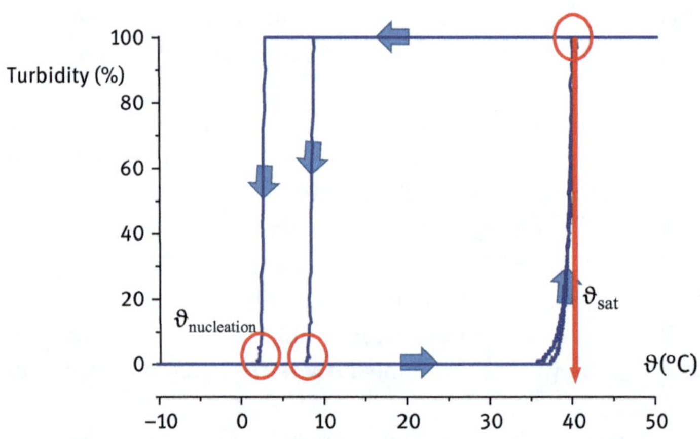

Figure 10.31: Plot of turbidity versus temperature for the measurements presented in Figure 10.30. A considerable variability between the cycles can be seen, namely for the temperature of nucleation.

Solubilities as determined by the Crystal16 method are compared with values obtained by the static flask method for the same system (Figure 10.33). The agreement is satisfactory.

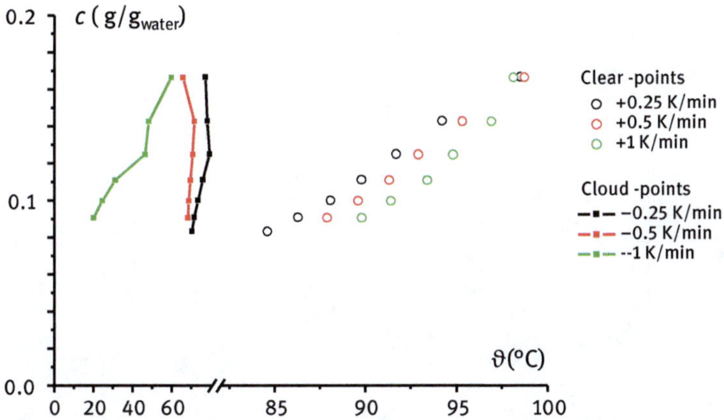

Figure 10.32: Solubilities determined by the Crystal16 method. The clear points, that is, the apparent solubilities, do not vary significantly with the heating rates, while the cloud points significantly depend on the cooling rate. Note the different scales for temperatures 0 to 70 and 85 to 100°C.

Using the synthetic solutions and the Crystal16-method is especially advantageous at higher temperatures, where sampling of the supernatant might be difficult due to excessive evaporation of the solvent.

There are several issues that need to be addressed for the Crystal16 method to yield reliable data:

- The solid-state form of the crystals in suspension is not unambiguously known. Namely the spontaneous nucleation during the cooling cycle is prone to yield a metastable modification.
  Oiling-out might occur and be mistaken as crystallization by the turbidity probe. Additionally, disappearing of the emulsion might be mistaken as saturation temperature of the crystals.
- Conversely, any transformation in solid-state form might occur during heating but go unnoticed.
- Cases are known where the crystals tend to agglomerate during the heating cycle or even stick to the walls and thus disappear from the field of view of the turbidity probe. This yields erroneous results for the saturation temperature.

Awareness of these issues and addressing them by visual inspection of the suspension or by sampling the suspension can solve these problems.

A further improvement is the automatic addition of known quantities of solvent after heating/cooling cycles, allowing for the true automated determination of the solubility diagram.

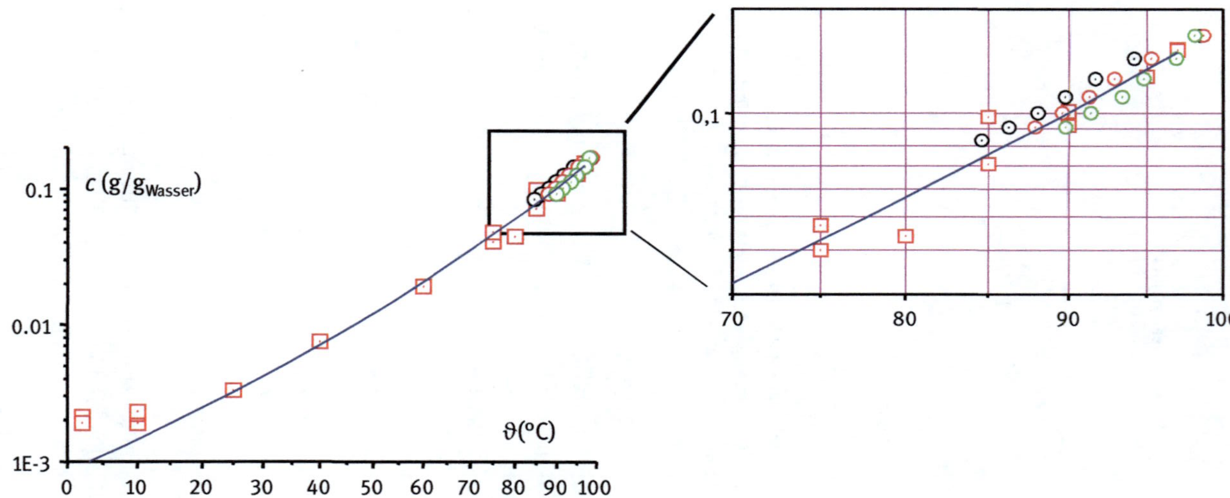

Figure 10.33: Comparison of the solubility as determined by the static flask method, squares, and the Crystal16 method, circles. The values for the Crystal16 method are taken from Figure 10.32.

10.3.4.3 Bracketing technique

One of the few techniques that can unambiguously assess the solubility of all solid-state forms is the bracketing technique [11, 18, 19]. The basic idea is to place one crystal of the respective solid-state form under consideration into a solution with known composition and to observe the partial dissolution and partial growth of that specific crystal of known modification under a microscope in a temperature interval around the saturation temperature of this form (Figure 10.34).

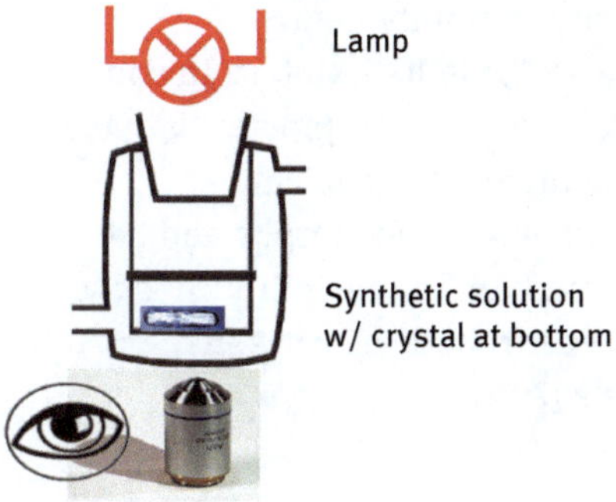

Figure 10.34: Set-up for the bracketing technique to assess the solubility of a crystal of known solid-state form. The crystal is placed into a solution of known composition. Its dissolution and growth are observed under a microscope at certain constant temperatures. For the choice of temperature points see text and the sequence of temperatures as indicated in Figure 10.35.

The temperature of the solution is first set to a temperature where the crystal slowly and slightly dissolves. Afterwards, this temperature is lowered to a temperature where the crystal grows. This cycle in repeated with a decreasing temperature span until the saturation temperature is cornered by a dissolution and a growth temperature to a degree small enough for the purpose (Figure 10.35). It can be important to start the cycle with a dissolution step to clean and activate the surface. Otherwise, growth might be impeded at low supersaturations.

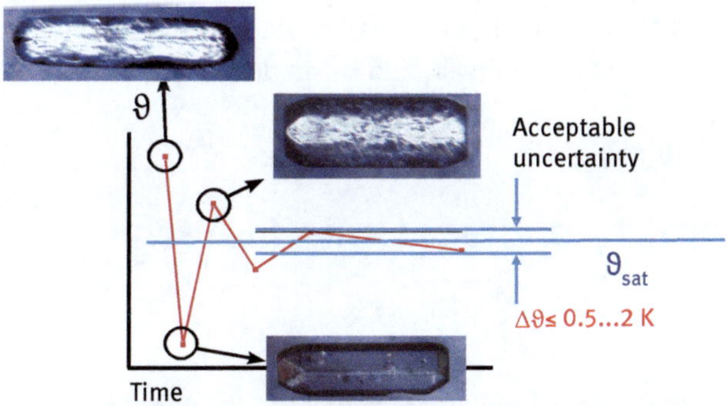

Figure 10.35: Temperature programme to determine the saturation temperature of a desired solid-state form via the bracketing technique. The saturation temperature is cornered between dissolution and growth until the saturation temperature is cornered within an acceptable limit for the saturation temperature.

The bracketing technique is very tedious to employ. Consequently, it has been used only sporadically.

10.3.5 Plausibility checks

Solubility data are unique and do not depend on person, place or time. However, solubility data can be prone to depend on parameters that the experimenter may not be aware of, and solubility data can be prone to errors due to experimental conditions. Some simple plausibility tests have been summarized in Table 10.3. These are easily applied and can prevent from using faulty solubility data.

Plotting the data according to the suggestions in Table 10.3, that is, in a manner that the data points form a straight line and making some good practice judgement can help improve the quality of the solubility measurements drastically.

Some few rules apply to the transition points between polymorphs and between differently solvated forms. These should also be used for a plausibility check of the data (Table 10.4). However, there are also exceptions from these rules, like hydrates that convert to anhydrates in aqueous suspensions [20].

10.3.6 Interpolation and extrapolation

The interpolation of solubility data is best made from a plot of solubility that renders the data in a straight line, i.e., ln c versus $1/T$ for the dependence of solubility on temperature and ln c versus solvent composition for isotherms as a function of solvent composition. This should be done for the solid-state form under consideration.

An extrapolation of solubility data bears the risk that a change in solid-state form is not considered as this is a first order phase transition occurring at a defined point, which cannot be extrapolated. As in any extrapolation, small deviations from the trend might become too large. Determination of solubility should be made in such a way that an extrapolation is not required.

10.4 Modelling solubility of crystalline material

10.4.1 General remarks

In this chapter, modelling solubility is understood as predicting the solubility either ab initio or from a very limited set of information. This topic is also discussed in chapter 3 of this book. For process development and optimization purposes, this means treating fully crystalline solids. Attempts are made to model the increase in solubility by going

Table 10.3: Plausibility checks for solubility data that can be inferred from plots of solubility data.

Type of measurement	Plotting	Tests
Solubility as a function of temperature	Plot ln x, the logarithm of the molar fraction as a function of $1/T$, if the solubility is not too high, concentration can be plotted instead of the molar fraction	– The data should be on a straight line, or a line with a small curvature. Outliers are easily detected, If outliers are observed, a redetermination of solubility will either yield the "true" data or reveal a dependence meriting further investigation. – The slope of the line should be in most case close to or somewhat lower than the heat of fusion. – Check for changes in slope and assess if sudden changes in the slope are accompanied by changes in the solid-state form A change in solid-state form in the range of temperatures covered is accompanied by a change in slope of the solubility curve. The change in slope between solvates or between a solvate and an ansolvate is usually quite high, with the solid-state form of the higher solvated form having the lower heat of dissolution and consequently a lower slope. – In enantiotropic systems, there is only one crossing between the forms. – If a change in solid-state form occurs and if at least one of the solid-state residues is a solvate, the degree of solvatization decreases with increasing temperature.
Solubility isotherms as a function of solvent composition	Linear plot	Curves can be either concave, convex, or even exhibit a maximum.
	Plot log of concentration as a function of solvent composition	– For a solubility isotherm, a change of the solid-state form of the solid residue can only concern a change in the degree of solvatization or in the solvent of solvatization. – The higher the concentration of the solvent of solvatization, the more stable is the respective solvate and the higher is the degree of solvatization with this solvent. – Plot solubility isotherms reduced with the solubility in pure primary solvent

Table 10.4: Rules for transition points between polymorphs and between differently solvated solid-state forms that should be checked for phase diagram assessments.

Transition between	Rules
Polymorphs	– Only one or no transition between two forms possible
Solvates	– Degree of solvatization decreases with temperature
Solvates and polymorphs	– In solvent mixtures, concentration of solvent necessary to stabilize solvate increases with increasing temperature

from a crystalline to amorphous solid-state forms, mostly for solubilization of low solubility moieties in water, but will not be discussed here.

Modelling of solubility of crystalline solids is an evolving field. One of the underlying goals of the development of solubility models lies in a reduction of experimental efforts to accelerate process development, namely for solvent (pre-) selection or even in a pre-selection of development candidates from their solubility behaviour.

10.4.2 Thermodynamic basis of models

As detailed in part 2 of this chapter and summarized by eq. (10.5), solubility is given by the free enthalpy of dissolution, $\Delta_d G$, and the activity coefficient, γ_2, of the moiety in the specific solvent. The two terms can be discussed as follows:

- $\Delta_d G$
  The free enthalpy of dissolution influences solubility exponentially. Therefore, it must be known with some accuracy. Up to now, no models are known to calculate the free enthalpy of dissolution ab initio. However, in first approximation, the free enthalpy of dissolution is independent of the solvent if the interaction between solvent and solute are comparable and can be estimated from the heat and temperature of fusion (eq. (10.7)). Thus, the free enthalpy of dissolution can either:
- be estimated experimentally from DSC data. Again, a high accuracy is required, for which the sample must be pure enough, namely free of residual solvent, and it must be of high crystallinity. These conditions are often not fulfilled, at least in early stages of development when a modelling would be most helpful.

or

- it can be estimated from the solubility of the moiety in one solvent.
  This of course requires the activity coefficient to be known for the first and all other solvents.

- γ_2
  The activity coefficient plays an important role in solubility, its prediction, and transfer between solvents, cf. eq. (10.5).

The discussion of solubility modelling is thus concentrating on modelling the activity coefficient of the solute in the given solvent.

Thus, ab initio modelling of solubility from structural information, including the crystal structure and accordingly the solid-state form, so far appears to be impossible to any necessary accuracy for process development. However, the influence of different solvents on solubility can be modelled, if the solubility in one or a small number of solvents is known.

10.4.3 Modelling solubility via activity coefficients

As discussed in the last paragraph, modelling solubility at least partially boils down to the task of modelling the activity coefficients of the solute in solution. Several approaches are known:
- Solubility parameter models [21, 22]
- Wilson, NRTL and polymer NRTL models [21]
- UNIQUAC model [21]
- UNIFAC group contribution models, namely those specialized for pharmaceutical compounds [23]
- COSMO-RS [24]
- PC- SAFT [25]

10.4.3.1 Example of group contribution modelling

To model the solubility of a compound, more specifically to model the activity coefficient of the solute in a solvent, the group interaction parameters must be known. For the group interaction parameters, both the solute and solvent molecule are broken into basic groups, which are assumed to have universally applying interactions. Thus, these interaction parameters can be derived from model systems and then transferred to the system in question. However, given the molecular complexity of modern research and development compounds, splitting the molecules into groups is difficult, and the interaction parameters of these groups are not all known.

For a simple molecule such as long-chain aliphatic carboxylic acid, the solubility in methanol, acetone, and ethylacetate has been measured and modelled using UNIFAC. From the solubility measurements, the activity of the solute has been calculated backward using the experimental heat and temperature of fusion. These experimental activity coefficients are compared with the modelled ones (Figure 10.36). The agreement for this simple molecule is acceptable, that is, within an error of 30%.

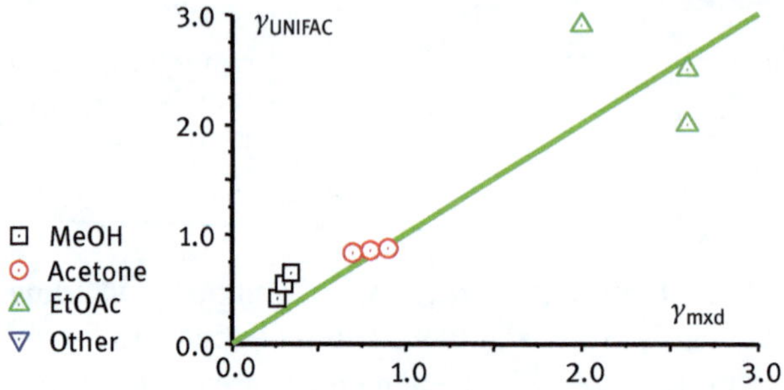

Figure 10.36: Comparison of experimental activity coefficients of long-chain aliphatic carboxylic acids in three organic solvents (methanol (MeOH), acetone and ethylacetate (EtOAc)) as derived from solubility data and experimental data for fusion. Activity coefficients were calculated using UNIFAC. Solubility increases with decreasing activity coefficient.

10.4.4 Modelling solubility via transfer – SAC-NRTL

The NRTL model has been extended to use experimentally determined solubility data to improve the modelling. A small number of five solvents has been proposed in which the solubility has to be measured. These solvents cover the spectrum of four types of "conceptual surface segments"

- Hydrophobicity
- Hydrophilicity
- Hydrogen-bond donors and acceptors
- Electrostatics: electron donors and acceptors
- Solvation: attractive to hydrophilic molecular surface
- Dipolar: repulsive or attractive to hydrophilic molecular surface

These data first yield values for the enthalpy of dissolution as well as NRTL interaction parameters.

The solubility of Lovastatin was modelled for 13 solvents using SAC-NRTL with the experimentally determined solubility in water, methanol, acetone, toluene, and heptane as calibration. The comparison between calculated and measured values is shown in Figure 10.37. The agreement between measured and predicted values appears sufficient. However, in some cases, deviations of half an order of magnitude are also observed.

The dependence of the solubility on temperature can also be predicted using SAC-NRTL. Figure 10.38 shows the comparison between measurement and prediction for the solubility of Lovastatin in five solvents. Again, the deviation of the prediction from the measured values is reasonable and on the order of ½ magnitude.

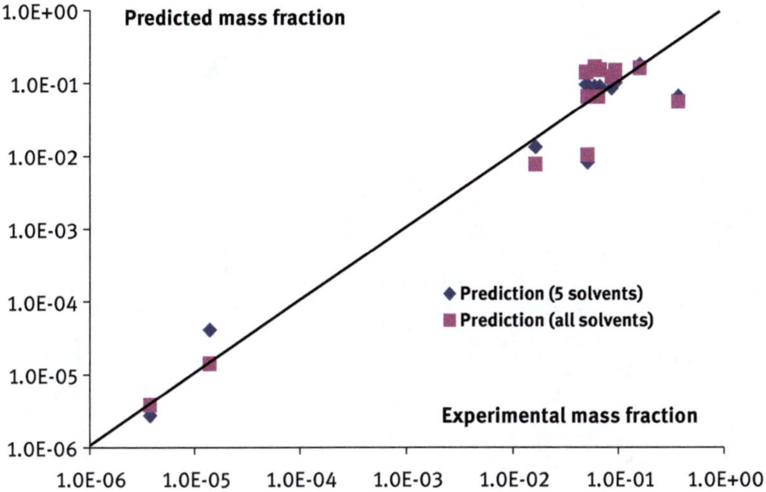

Figure 10.37: Comparison of the solubility of Lovastatin in thirteen solvents as calculated using SAC-NRTL based on a calibration with experimental solubility data in water, methanol, acetone, toluene, and heptane. All data are for 45 °C. Graph redrawn from [26].

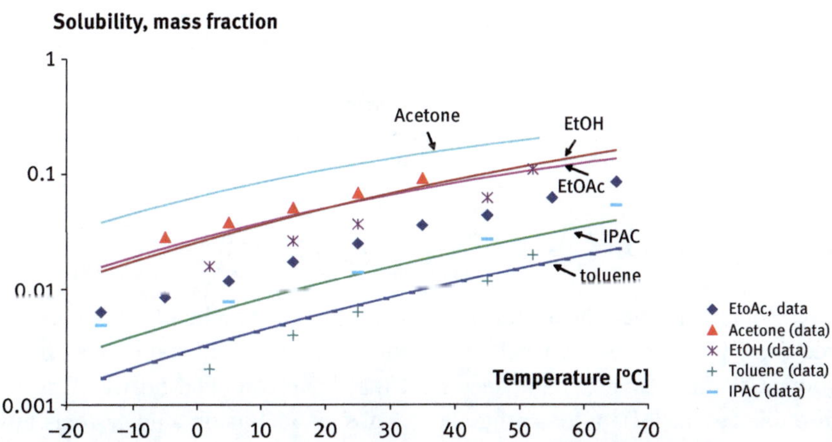

Figure 10.38: Solubility of Lovastatin in five solvents as a function of temperature: Comparison of measured data and data predicted by SAC-NRTL. Graph adapted from [26].

The prediction of solubility isotherms in aqueous mixtures with organic solvent yields the results shown in Figure 10.39. The absolute values of solubility are predicted in an acceptable way. However, the curvature itself and namely the position of the maxima in the solubility curves are not always met.

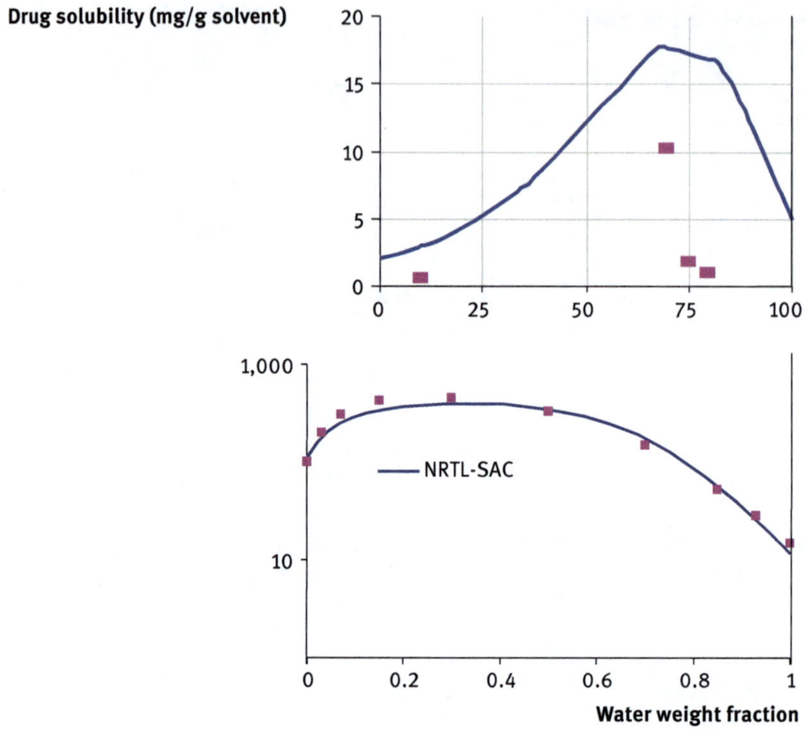

Figure 10.39: Comparison of solubility isotherms in mixed solvents. Predicted and measured solubilities are compared for two moieties dissolved either in an acetone–water mixture, top and an *n*-propanol–water mixture, bottom. Graph redrawn from [26].

10.4.5 Literature data of solubility

For most systems in the healthcare but also in the life science industry, solubility data is required for moieties, for which hitherto no solubility data have been assessed nor published. Several databases exist that have compiled nearly all solubility data published [27]. Data for inorganic systems as well as organic systems are included in the database. Many of the data have been checked for consistency.

10.5 Solubility and crystallization development

The development, trouble-shooting, and optimization of a crystallization process is very much facilitated by the knowledge of solubility data or better, by knowledge of the phase diagram. Some examples of conclusions derived from solubility data and phase diagrams are presented below.

10.5.1 Solubility and crystallization conditions

A crystallization process of a drug substance, an intermediate or any other species requires the substance to be dissolved in a given solvent. Starting from the homogeneous solution, the solubility of the moiety is decreased. The solution becomes supersaturated, and the substance is crystallized and harvested. With this crystallization process, the substance is not only yielded in a solid form, but also the solid-state form, the polymorph is defined. Additionally, the particle size distribution and crystal shape are determined. Finally, but equally important, a purification is achieved. A parameter beyond quality of the drug substance is the yield of the crystallization process. All these parameters are governed by solubility of the substance, and with regard to purification of respective impurities.

Depending on the solubility behaviour, the crystallization technique is chosen from the following techniques:

– Cooling crystallization

For organic moieties, the cooling crystallization is by far the most frequently applied crystallization technique. It is easiest to control via a temperature programme. Cooling crystallizations require the solubility to change by a factor of 10–20 over a temperature range from close to boiling point of the solvent to the lowest temperature practically used, which is around −10 °C.

– Evaporative crystallization

An evaporative crystallization can be carried out for any kind of temperature dependence of the solubility. However, control of such a crystallization process is more advanced, namely the control of local supersaturation heavily depends on the mixing conditions.

Drowning-out crystallization

Drowning-out crystallizations are rarely used. This is since drowning-out crystallizations are usually associated with high (local) supersaturations that do not allow for an easy control of crystallization conditions. They entail no or a reduced purification effect.

Cases when a drowning-out is advised are crystallizations requiring high yields such as for the recovery of intermediates where the purification does not matter much and crystallizations of moieties that are promiscuous solvate formers, for which no primary solvent can be found with which no solvate is formed.

Knowledge of solubility as a function of solvent composition will facilitate finding appropriate crystallization conditions.

When considering the crystallization process, it is not only the crystallization itself, but also the solid–liquid separation including the washing that is affected by solubility:

- Suspension density, that is, the concentration of solids in the suspension:

The suspension density in a crystallization process is determined by the starting concentration and yield desired. As the desired yield is typically around or above 90 %, the suspension density is equivalent to the starting concentration. Depending on the equipment, the suspension density that can be used in a batch crystallization is typically on the order of 20 % with maximum values up to 40 %, which limits the starting concentration to these values.

- Solubility at the end of the crystallization process:

Solubility at the end of the crystallization should be low to minimize losses during washing. In most cases, the washing liquor will reach concentrations close to saturation. As the washing liquor is removed, this would amount to substantial losses in yield. For a typical cake porosity of 25 %, a solubility of 2 % would result in losses of at least 1 %. Finally, a high solubility in the mother liquor will most probably lead to more agglomeration during drying when the material dissolved in the residual moisture crystallizes out.

In cases that no solvent with appropriate solubility of the substance can be found, the solubility can be decreased by adding an anti-solvent. A close to linear relationship between solvent composition and solubility as depicted in Figure 10.5 and discussed in Section 10.5.2 is needed for facile and robust processes.

10.5.2 Choice of solvent–solvent mixture to reduce suspension density

Basic considerations for the choice of solvent are the ICH classification of solvents and the quantity of solvent remaining in the crop after drying, the availability of the solvent in the plant, the solvents used on previous stages, and finally, the solubility and its dependence on temperature.

For the crystallization of a steroid, a small number of solvents came into focus. The solubilities of the steroid in these solvents at room temperature are listed in Table 10.5. A high solubility in various solvents, such as ethylacetate (EtOAc), methyl-ethyl-ketone (MIBK), and isopropanol (iPrOH), is found. An exemplary determination of the solubility as a function of temperature for MIBK showed sufficient dependence of solubility on temperature (Figure 10.40). As can be deduced from Table 10.5, non-polar solvents such as diisopropylether (DIPE) or hexane are bad solvents.

Opting for MIBK would have resulted in a suspension density of >800 g/kg, a way too high value.

Testing mixtures of MIBK and DIPE, it was found that these two solvents do not behave like a solvent–anti-solvent pair. Rather a close to linear dependence on solvent composition was found (Figure 10.41).

Table 10.5: Solubility of a steroid in few solvents estimated for 25 °C.

Solvent	Solubility at 25 °C in g/kg of solvent
EtOAc	130
MIBK	120
iPrOH	97
DIPE	1
n-Hexane	0.6

It is evident that the steroid has a very high solubility in the first three solvents, while the solubility in the latter two is negligible.

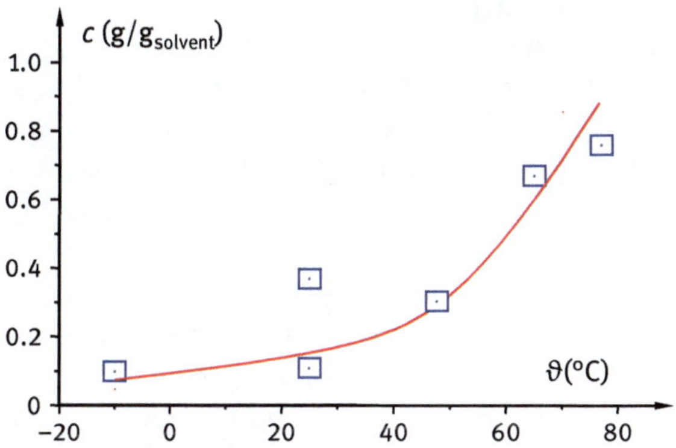

Figure 10.40: Solubility as a function of temperature of the steroid also discussed in Table 10.5 in MIBK. Data give only the trend but show that a cooling crystallization with a high yield appears feasible. The high solubility of up to 800 g/kg of solvent at 80 °C would result in a suspension density that cannot be handled, neither in the lab nor in the plant.

It was thus decided to crystallize this steroid from a 1:3 mixture of MIBK and DIPE using a cooling crystallization, the preferred crystallization process. The suspension density for these conditions was ≈20 %, a value easily within the process conditions. Finally, the crystallization afforded a yield of 90–95 %.

10.5.3 Seeding – point of addition

Common practice for seeding is the addition of seeds shortly after passing the saturation line and well before reaching the metastable zone, that is, typically one-third into the region between saturation and spontaneous nucleation. Thus, the measurement of solubility will give valuable information on the point of seeding. The point

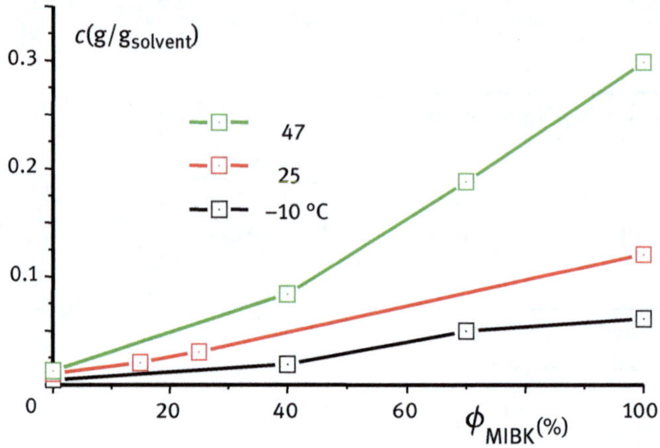

Figure 10.41: Solubility isotherms of a steroid in MIBK–DIPE mixtures. The solubility has a close to linear dependence on composition. DIPE does not act as an anti-solvent but can rather be used to decrease the solubility in pure MIBK and to arrive at a suspension density that can be handled.

of seeding can be equally well determined by a titration, that is, by a subsequent addition of seeds to a solution slowly driven from subsaturation to spontaneous nucleation. A comparison of the point of seeding with solubility measurements might reveal hitherto undetected parameters influencing solubility.

For a solution containing 37 g of solute per 1 kg of solvent, dichloromethane in this case, the temperatures for seeding were titrated. The interval of temperatures between dissolution of the seeds and their growth are shown in Figure 10.42. Contrarily, the

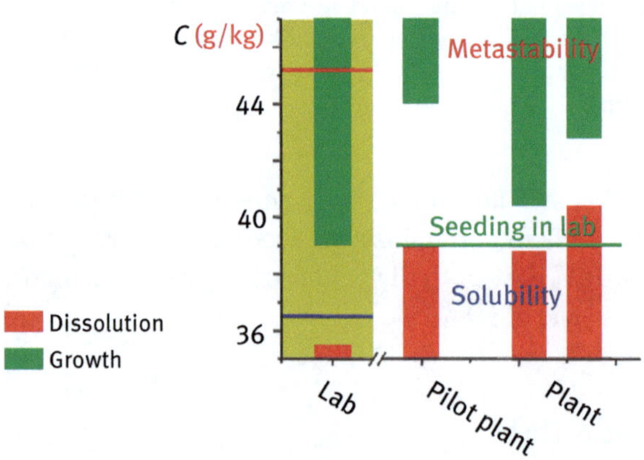

Figure 10.42: Temperatures at which seeds introduced as a suspension to a solution containing 37 g of solute per 1 kg of solvent dissolved or grew, depending on the lieu of the process. Contrarily to laboratory conditions, the seeds dissolved in the pilot plant and plant under conditions under which growth was occurred in the laboratory.

seeds dissolved in the pilot plant and plant conditions under which growth already occurred under laboratory conditions. In effect, the point of addition of the seeds was increased by more than 5 K, indicating an increased solubility.

It was shown by solubility measurements that the effect was due to the presence of small amounts of water in the dichloromethane which was used as solvent resulting from the work-up of the solvent under pilot plant and plant conditions. As discussed in Section 10.2.2 and quantified in Figure 10.4, small levels of water below the miscibility gap increase the solubility considerably. Consequently, the parameter water is to be monitored and taken into account.

10.5.4 Representation of the crystallization process in phase diagrams

The course of crystallization processes can be followed in the respective phase diagram. This is not a prerequisite in process development but can facilitate process optimization.

10.5.4.1 Polymorphs

A representation of the crystallization process in the solubility phase diagram is advisable for systems forming more than one polymorph and namely for polymorphs with an enantiotropic relation. The system presented in Figure 10.12 will be used to elaborate this. For the process development and optimization, the low dependence of the solubility on temperature lead to the choice of an evaporative crystallization close to the boiling point of the solvent, followed by a cooling crystallization to maximize yield.

The evaporative crystallization at elevated temperatures definitely leads to the crystallization of form *B* . This is confirmed by sampling of the suspension and determination of the concentration in the supernatant and the polymorph (Figure 10.43). For the cooling part of the crystallization, it becomes obvious that the cooling is carried out too fast. Consequently the concentration in solution deviates considerably from the equilibrium solubility. A cooling below the transition temperature between forms *B* and *E* of 5 °C quickly supersaturates the system. Form *E* is nucleated. The solubility diagram indicates the rapid increase in the supersaturation of form *E* below 5 °C.

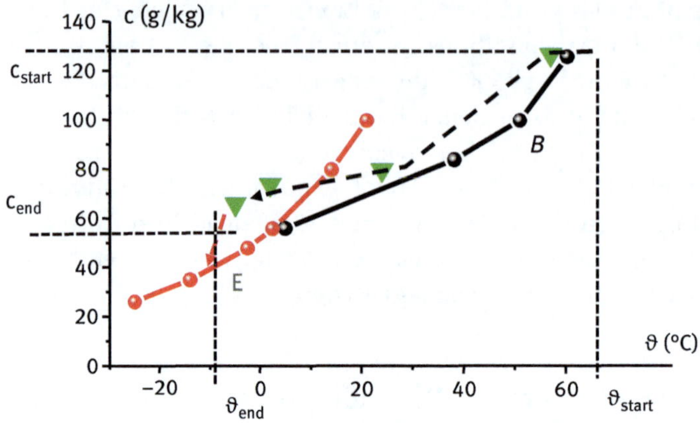

Figure 10.43: Course of a cooling crystallization of a system with two polymorphs – forms *B* and *E* – that are enantiotropically related. The enantiotropic transition temperature is above the final- temperature of the crystallization process but below the starting temperature. For a run on production scale, the concentration in the supernatant has been followed, green triangles. A considerable supersaturation in the supernatant is necessary for the crystallization to proceed. From the solubility curves for both polymorphs, the supersaturation of modification *E* can be estimated, if the primary cooling results in the high-temperature modification *B*.

10.5.4.2 Solvate-forming systems

For a system forming an anhydrate and a hydrate, the solubility phase diagram has been determined for mixtures of water and ethanol. Solubility was determined as isotherms as a function of solvent composition (Figure 10.44, left). These values were consequently converted to solubility at constant solvent content as a function of temperature for three characteristic water contents, 1 %, 4.5 %, and 9 % water (Figure 10.44, right). The transition points between the anhydrate and hydrate are shown.

The moiety is a hydrochloride salt. Water is the good solvent while ethanol is a mediocre solvent. Salt formation is initially carried out at a temperature of 25 °C by the addition of hydrochloric acid followed by cooling to increase yield.

The phase diagram shows that the starting concentration of water plays a determining role in the hydrate form formed and its stability throughout the process. The path of cooling crystallizations after the addition of hydrochloric acid starting at c_{start} are shown in Figure 10.44 for three different fractions of water. For the fractions < 8 % water, the solid-state form initially stable is the anhydrate. Depending on the degree of cooling and the fraction of water present, a change in stable solid-state form from the anhydrate to the hydrate might occur. In cases ① and ③, either the anhydrate or hydrate remain stable, while the cooling at intermediate fractions of water, case ②, the stable solid-state form changes upon cooling. In effect, the anhydrate is nucleating first

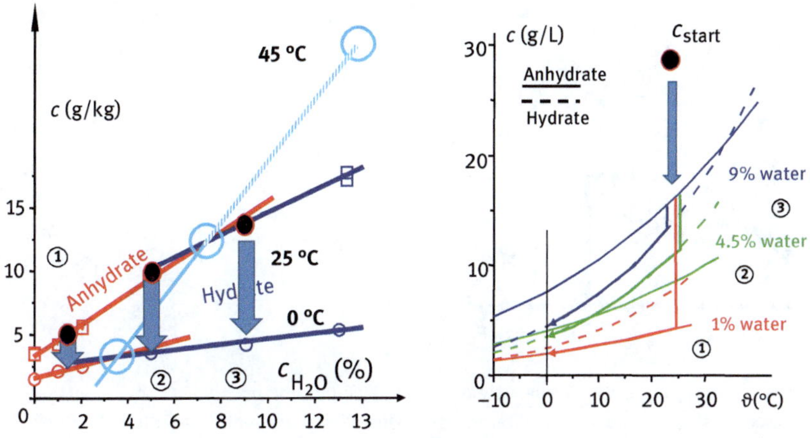

Figure 10.44: Solubility isotherms for a hydrochloride salt in aqueous ethanol. The moiety forms both a hydrate and an anhydrate. The data were measured for 0 and 25 °C. Transition are shown for these temperatures. For 45 °C, only the transition point was assessed. Measurements were done by assessing the isotherms, left, and then used to calculate the solubility at constant solvent composition, right. The path of cooling crystallizations starting at c_{start} are shown for two different fractions of water. For the fractions shown, that is, < 8 % water, the solid-state form initially stable is the anhydrate. Depending on the degree of cooling and the fraction of water, a change in stable solid-state form from the anhydrate to the hydrate might occur. In cases ① and ③, the anhydrate or hydrate, respectively, remains stable, while the cooling at intermediate fractions of water will stabilize and possibly yield the hydrate upon cooling, case ②.

but the nucleation of the more stable hydrate upon cooling below the stability temperature occurs within ≤1 h. This is far too short for any stable process.

Starting from such a phase diagram, the developer can make an informed choice for the process conditions.

10.5.4.3 Incongruent dissolution of solid-state forms

In case of a compound that is incongruently dissolving, the crystallization process must make sure that the solid-state form crystallized is the desired one and possesses the desired purity. In case of incongruent dissolution, the path of the crystallization is shown in Figure 10.45.

The crystallization is started in a subsaturated state. A removal of solvent will drive the concentration along the diagonal. Two cases can be distinguished:

- In case of a congruent melting, the saturation line of the compound will be reached, and the desired compound will crystallize out, full line in Figure 10.45.
- In case of an incongruent melting, the concentration in solution will first reach the saturation of one of the components that will crystallize.

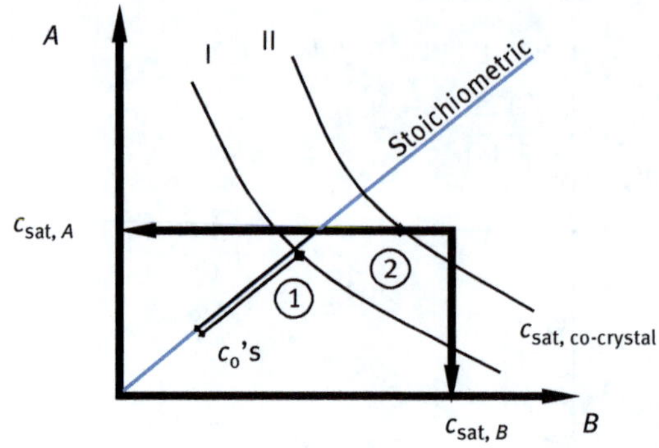

Figure 10.45: Course of a crystallization for a system forming a compound of the two components in solution, that is, a co-crystal. Two cases are differentiated: One with congruent dissolution, II and one with incongruent dissolution I. For the latter case, the crystallization follows the blue line.

Via this crystallization of one of the components, component A in Figure 10.45, the concentration in solution will tend towards the solubility line of the compound, case II.

10.5.5 Liquid–liquid phase separation

An LLP separation must be avoided in a crystallization process, as it might lead to a decrease in purification, high residual solvent content, and particles difficult to handle. Thus, it is of importance to detect such a phase separation and to take counter-measures. In case of a:

– Stable oiling-out

The only remedy is to change the solvent system or for a given solvent system to start the crystallization below the LLP separation. However, due to a possible insufficient solubility at this temperature, this might be impracticable as it might lead to low yields. In essence, a change in solvent is recommended.

– Metastable oiling out

The crystallization can be carried out at supersaturations low enough so that the system stays between the saturation line and the immiscibility dome: Frequently this means to carry out the crystallization very slowly. In case of a cooling crystallization the temperature difference between the suspension and the wall of the vessel must be kept small.

An example of a metastable oiling out and its mediation is shown in Figure 10.46. The top of the miscibility dome is only ≈5 K below the solubility line and follows the solubility line closely.

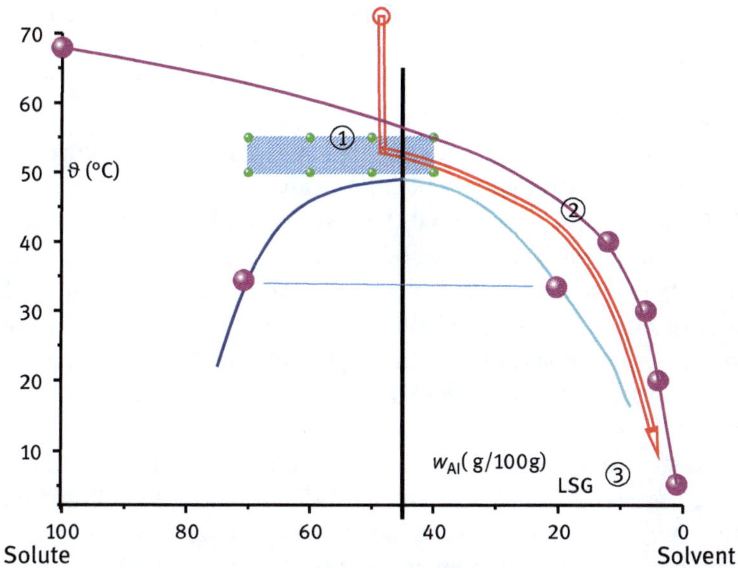

Figure 10.46: Solubility line with immiscibility dome in a single component solvent/API system. Also indicated is the path of the crystallization, as shown by the red arrow. The immiscibility and the solubility range are close. The tip of the miscibility dome in only ≈5 K below the solubility line. At low temperatures, the liquid–liquid equilibrium and solubility line are close. Three critical points are noted. First, the point where the system is first supersaturated by cooling, ①, and where the cooling must be slow enough to reduce the supersaturation generated by the cooling sufficiently. Second and third, the cooling itself must be slow enough for the system to follow the critical path between solubility and liquid–liquid phase split, ② and ③.

In this case, crystallization needs to be carried out with a low supersaturation, including a heavy seeding some few degrees below saturation, point ① in Figure 10.46, and a slow cooling to the final temperature, line ②, without any acceleration at lower temperatures, point ③. The latter point also limits the temperature difference between the suspension and the coolant to some few degrees at the expense of a decreased cooling rate in the plant.

10.6 Summary

The development, optimization, as well as the trouble-shooting of crystallization processes are vastly facilitated by a sound knowledge of solubility data and phase diagrams. These are especially important for the crystallization of moieties that can form polymorphs or pseudo-polymorphs such as hydrates and solvates. Especially in the case of moieties with multiple solid-state forms, following the crystallization process in the phase diagram can greatly improve the process understanding and help in assessing the limits within which the crystallization process can be safely carried out, namely with respect to the desired solid-state form.

Assessing solubility data and the respective phase diagrams is straightforward and can be carried out in a structured approach that can be tailored according to the polymorphic behaviour of the system and the solvent system under consideration. Equally important and helpful are thermodynamic considerations for the solubility behaviour that can serve both to structure and guide the measurements and to corroborate and cross-check the results obtained. The experimental effort for the determination of solubility and phase behaviour is relatively small. The equipment necessary is simple, an automation of the measurements is possible, but not necessarily advantageous.

Optimization strategies for crystallizations based on approaches such as quality by design and design of experiment should be based on measured solubility data and by following the path of the crystallization in the phase diagram.

The models available to predict solubility for process development are evolving. For the time being, the most promising are based on a small subset to extrapolate to other solvents. However, given the current accuracy of the predictions and the ease of the experimental determination of solubility data, the latter approach might still be the best choice.

References

[1] Grant, D.J.W., & Higushi, T. Solubility Behavior of Organic Compounds, Wiley, 1990.

[2] Predel, B., Hoch, M., & Pool, M. Phase diagrams and heterogeneous equilibria – A practical introduction, Springer, 2004.

[3] Ricci, J.E. The Phase Rule and Heterogeneous Equilibrium, Dover Publications, Inc., New York, 1966.

[4] Lorenz, Heike. Solubility and Solution Equilibria in Crystallization, in Crystallization Basic Concepts and Industrial Applications, Wolfgang Beckmann, ed., Wiley, 2013.

[5] Yu, L. Inferring thermodynamic stability relationship of polymorphs from melting data. Pharm. Sci. 1995, 84, 966.

[6] ICH International Conference on Harmonisation, Quality Guidelines Q3C. http://www.ich.org.

[7] Yalkowsky, S. H. Solubility and solubilization in aqueous media, Oxford University Press, 1999, S.H. Yalkowsky, G.L. Flynn and G.L. Amidon, J Pharm Sci, 61(1972)983.

[8] Jacques, J., Collet, A., & Wilen, S.H. Enantiomers, Racemates, and Resolutions, Krieger, 1994.

[9] York, P., & Grant, D.J.W. A disruption index for quantifying the solid state disorder induced by additives or impurities. I. Definition and evaluation from heat of fusion. Int J Pharm 1985, 25, 57–72. D.J.W. Grant und P. York, A disruption index for quantifying the solid state disorder induced by additives or impurities. II. Evaluation from heat of solution. Int J Pharm, 28 (1986) 103–112.

[10] Pudipeddi, M., & Serajuddin, A.T.M. Trends in solubility of polymorphs. J Pharm Sci. 2005, 94, 929–939.

[11] Beckmann, W., Boistelle, R., & Sato., K. Solubility of the A, B, and C polymorphs of stearic acid in decane, methanol, and butanone. J Chem Eng Data 1984, 29, 211–214.

[12] Beckmann, W., Rouquerol, F., & Beckmann, H. Calorimetric determination of the heats of dissolution of stearic acid in some organic solvents. Thermochimica Acta, 1983, 66, 295

[13] Nordhoff, S., Bechtloff, B., Ulrich, J., Wagner, G., & Dumpelmann, R. Properties of pseudopolymorphs in industrial use; sodium-2-keto-L-gulonate monohydrate as an example. 14th International Symposium on Industrial Crystallization IChem^{E}, 12. – 16.9.1999, Cambridge, Bechtloff, B., Ulrich, J., Wagner, G., Dumpelmann, R., Nordhoff, S., Thermodynamic principles and prediction of the transition points in dehydration processes. 14th International Symposium on Industrial Crystallization IChem^{E}, 12. – 16.9.1999, Cambridge.

[14] Pfeiffer et al. Crystal pseudopolymorphism of cephaloglycin and cephalexin. J Pharm. Sci. 1970, 59, 1809–1814.

[15] Grant, D.J.W., Zhu, H., & Yuen., C. Influence of water activity in organic solvent + water mixtures on the nature of the crystallizing drug phase. 1. Theophylline Int J Pharm 1996, 135 (1–2), 151–160.

[16] Riesen, Rudolf. Industrielle Kristallisationsprozesse, Verfahrensoptimierung durch Einsatz eines automatischen Laborreaktors, Mettler-Toledo GmbH. Schwerzenbach, Schweiz: s.n., 1992.

[17] Reus, M.A. et al. Solubility: importance, measurements and applications, white paper by technobis

[18] Taylor, M. The solubility of pure sucrose in water at high temperatures. Int Sugar J 1948, 50, 292.

[19] Madsen, H.E.L. Boistelle, R. Solubility of long-chain n-paraffins in pentane and heptan. JCS Faraday I 1976, 1078–1081.

[20] Saal, C. et al. Cilengitide – exceptional pseudo-polymorphism of a cyclic-pentapeptide". Eur. J. Pharm. Sci. 2015, 71, 1–11.

[21] Gmehling, J., & Kolbe, B. Thermodynamik, Wiley-VCH Verlag GmbH, Weinheim, 1993.

[22] Prausnitz, J.M., Lichtenthaler, R.N., & de Azevedo, E.G. Molecular Thermodynamics of Fluid-Phase Equilibria, Prentice Hall, Englewood-Cliffs, NJ, 1986.

[23] Hahnenkamp, I., Graubner, G., & Gmehling, Measurement and prediction of solubilities of active pharmaceutical ingredients. J. Int. J. Pharm. 2010, 388, 73–81.

[24] Klamt, A. COSMO-RS (COnductor like Screening Model for Real Solvents). Encyclopedia of Computational Chemistry, Wiley, 1998.

[25] M. Kleiner, F. Tumakaka, G. Sadowski: In: X. Lu a. Y. Hu (Eds.): Molecular Thermodynamics of Complex Systems, Springer, Berlin, 2009.

[26] Chen, C-C., & Song, Y. Solubility modeling with a nonrandom two-liquid segment activity coefficient model. Ind. Eng. Chem. Res. 2004, 43, 8354–8362, .

[27] detherm data base, Dechema Frankfurt am Main, Germany.

David Elder

11 Biowaivers

11.1 Introduction

The International Conference on Harmonization (ICH) guideline M9 titled "Biopharmaceutics Classification System (BCS)-based Biowaivers" [1, 2] was first adopted in June 2016 and reached the public consultation stage by June 2018. A biowaiver provides regulatory support for in vitro testing, to be used in lieu of in vivo bioavailability and/or bioequivalence studies to enable product approval, based on biopharmaceutical considerations. In vitro testing covers aspects such as solubility, permeability, and dissolution. A biowaiver is therefore predicated on the assumption that two solid oral immediate release products (test and reference) will behave similarly within the gastrointestinal (GI) tract due to a combination of high solubility, rapid in vitro dissolution and an absence of drug precipitation once the drug is solubilized, such that the in vivo dissolution profiles will be similar. This allied with high passive permeability should ensure that the formulations exhibit the same rate and extent of absorption resulting in bioequivalence [3].

It is therefore the expectation that ICH M9 will consequently minimize unnecessary in vivo studies in man and allow greater public access to vital medicines. However, it should be recognized that this approach has not always universally aligned or internationally accepted. Indeed, the sparsity of successful biowaivers during clinical development may be a result of a lack of harmonization [4], which it is hoped will be addressed by the ICH M9 initiative.

BCS and biowaiver considerations play an important role in approval of generic drug products [5]. Over the 10-year period (2000 to 2011), the FDA approved 263 ANDA (Abbreviated New Drug Application) submissions. Prospective assessment of these ANDA submissions using BCS criteria showed that 42% were BCS class I drugs, 21% were BCS class II drugs, and the remaining 37% were for BCS class III drugs. Not surprisingly, there were no BCS class IV drugs approved.

The most common deficiencies with respect to non-approval of proposed BCS-based biowaivers were:
- Absence of pH-solubility profiles
- Inappropriate methodology for solubility determinations
- Absence of dissolution data across all marketed strengths
- Absence of critical aspects of quality management systems, for example, SOPs (standard operating procedures) to support analytical methodologies
- Absence of data that support GI stability

David Elder, Hertford, Hertfordshire, SG14 2DE, UK

https://doi.org/10.1515/9783110559835-011

- Absence of any data assessing impact of efflux transporters on the cell line, for example, Caco-2, MDCK (Madin-Darby Canine Kidney), etc., used in the in vitro permeability assessments
- Absence of bidirectional data on the in vitro permeability assessments on the compounds

11.2 The role of biowaivers

In certain cases, regulatory applications using biowaivers are likely to be permissible. As such, submissions can be based on either BCS considerations, in vitro–in vivo correlations (IVIVC), in vitro–in vivo relationships (IVIVR) or simply based on in vitro dissolution profile comparisons at different physiologically relevant pHs, that is, pH 1.2, 4.5, and 6.8. Current examples of BCS, IVIVC, or IVIVR biowaiver approaches [6] include:
- supporting generic product entry versus the existing innovator product,
- supporting early-stage regulatory submission, that is, changing the product type during early clinical development, for example, capsules to tablets, etc.,
- supporting the use of over-encapsulated comparator products, which are frequently used in pivotal clinical studies, that is, Phase III,
- supporting bridging between the product used in pivotal clinical studies, that is, Phase III and the 'to be marketed' commercial product,
- supporting product line extensions (PLEs), for example, oral liquids, for different clinical populations, for example, age-appropriate formulations for vulnerable populations, that is, paediatric or geriatric products,
- supporting various post-approval CMC (chemistry, manufacturing and control) changes.

11.3 History and evolution of BCS

11.3.1 BCS definitions

Solubility can be simplistically defined as the amount of a drug substance that will dissolve in a given amount of solvent (typically water or a simple aqueous buffer) at a designated pH, temperature, and pressure (see chapter 1). From a BCS perspective, "a drug substance is classified as highly soluble if the highest single therapeutic dose is completely soluble in 250 mL or less of aqueous media over the pH range of 1.2–6.8 at 37 ± 1°C" [2].

Permeability can be defined as the movement or flux of a molecule across a biological membrane. BCS is typically concerned with passive transcellular permeability.

From a BCS perspective "a drug substance is considered to be highly permeable when the systemic bioavailability or the extent of absorption in humans is determined to be ≥85% of an administered dose based on a mass balance determination (along with evidence showing stability of the drug in the GI tract) or in comparison to an intravenous reference dose"[7].

Dissolution can be defined as the in vitro rate of solubilization (or release) of drug from a solid oral dosage form at a designated pH, temperature, apparatus, agitation speed, and using a specified media. To "qualify for a BCS-based biowaiver for BCS Class I drug substances both the test product and reference product should display either very rapid (≥85 for the mean % dissolved in ≤15 min) or rapid (≥85 for the mean % dissolved in ≤30 min) and similar in vitro dissolution characteristics under all of the defined conditions" [2].

11.3.2 Regulatory history on the use of BCS-based biowaivers

The BCS approach was introduced over two-decades ago in the mid-90s [3] to facilitate the introduction of biowaivers into the regulatory lexicon. The BCS approach comprises of a four-box model for drug product evaluation based on an assessment of drug solubility and permeability of the API and related in vitro dissolution profiles of the drug products.

There are four BCS classifications: class I (high solubility and high permeability), class II (low solubility and highly permeability), class III (high solubility and low permeability), and class IV (low solubility and low permeability). In the current regulatory climate, biowaivers are typically granted for highly soluble compounds, that is, BCS class I or class III products. The BCS is summarized in Table 11.1.

Table 11.1: Summary of BCS 4-box model [3].

BCS Class I High Solubility/High Permeability	**BCS Class II** Low Solubility/High Permeability
BCS Class III High Solubility/Low Permeability	**BCS Class IV** Low Solubility/Low Permeability

The US Food and Drug Administration (FDA) subsequently published regulatory guidance for BCS Class I drugs describing the essential regulatory requirements in which biowaivers would be applicable [8]. The Japanese National Institute Health Sciences also published comparable guidance [9, 10]. Subsequently, the World Health Organization (WHO) [11] and European Medicines Agency (EMA) [12, 13] followed suite and additionally they also addressed the potential for the inclusion of

BCS class III compounds. This encouraged updates in Japan [14] and the US [7] to attempt to align international biowaiver guidance, which culminated in this topic being recommended for inclusion in ICH discussions [1, 2].

11.3.3 Application of BCS-based biowaivers to essential drugs

A very comprehensive series of BCS biowaiver assessments of essential drugs is being undertaken and concurrently published by the International Pharmaceutical Federation (FIP) [15]. FIP have currently published 49 monographs covering all four BCS classes and many drugs that cannot be easily assigned unambiguously into a single BCS class. This is typical because either literature data for solubility or permeability is equivocal or inadequate, that is, insufficient data to accurately permit BCS categorization (see Section 11.1 for typical deficiencies) or in those cases where the solubility classification changes with respect to the definition of dose. That is, the drug may be BCS class I for lower doses, which could be the highest market dose strength of the product, but this subsequently changes to BCS class II for higher doses which maybe the maximum single dose administered. Each monograph summarizes the drugs' physicochemical and pharmacokinetic performance, including importantly the recommended dose in various international territories. The monograph then summarizes dosage form performance including typical excipients used, in vitro dissolution performance, and details of any bioavailability studies that have been published. The discussion then focusses on solubility, permeability, BCS classification, evidence of bio-inequivalence from the literature, and any patient risks associated with potential bio-inequivalence. A full summary of the findings from the FIP initiative (to date) is provided in Table 11A.1 in Appendix 1, and condensed summaries of those drugs that can be unambiguously assigned to BCS categorization are provided in Tables 11.2. The BCS classification of the API is based on the specifications of the guidances that were valid at the time of publication.

The data fully exemplify the difficulties in applying BCS-based approaches for the assessments of biowaivers. There are 49 monographs (accurate as of January 2019), some of which cover more than one pharmaceutical salt, that is, Chloroquine (phosphate, sulfate, and hydrochloride) and diclofenac (sodium and potassium). Of these only two-thirds of the monographs (i.e., 29 compounds) can be unambiguously assigned into one of the four BCS classes. The remainder cannot be unambiguously assigned to a single category. The lack of harmonization in the guidance and lack of adequate data, especially arising from definition of dose and pH specifications, are the reasons for most of the ambiguities in BCS classification.

From a biowaiver perspective, the following findings hold true for compounds listed in Table 11.2. All BCS class I compounds (i.e., twelve) and all BCS class III

Table 11.2: Condensed Summary of FIP Biowaiver Monographs for Essential Drugs (BCS Classes I-IV) [15].

Essential Drugs Assigned into Various Biopharmaceutical Classification System (BCS) classes (Biowaiver (Y/N) in brackets)			
I	II	III	**IV**
Acetylsalicylic Acid (Y)	Diclofenac (Y[1])	Atenolol (Y)	Ciprofloxacin (N)
Bisoprolol Fumarate (Y)	Fluconazole (Y[1])	Cimetidine (Y[2])	Folic Acid (N)
Chloroquine (Y)	Ibuprofen (Y[1])	Enalapril (Y[2])	Furosemide (N)
Codeine (Y)	Ketoprofen (Y[1])	Lamuvidine (Y[2])	
Doxycycline Hyclate (Y)	Nifedipine (N)	Metoclopramide (Y[2])	
Levetiracetam (Y)	Piroxicam (N)	Pyrazinamide (Y[2])	
Levofloxacin (Y)	Rifampicin (N)	Ranitidine HCL (Y[2])	
Metronidazole (Y)			
Propanolol (Y)			
Primaquine Phosphate (Y)			
Stavudine (Y)			
Zidovudine (Y)			

[1]These so-called BCS IIa compounds (weak acids with good intestinal solubility) were initially granted biowaivers under the original WHO guidance. However, the most recent WHO guidance only covers BCS class I and class III compounds.
[2]As long as the composition of the test product is qualitatively the same and quantitatively very similar to the reference product, with demonstrated usage in marketed products, then biowaivers can be considered.

compounds (i.e., seven) will support biowaivers; whereas, all BCS class IV compounds (i.e., three) do not support biowaivers.

11.3.4 Evolution of BCS concepts

Tsume et al. [16] proposed a modification of the existing BCS approach for poorly soluble BCS II/IV drugs, sub-dividing them into BCS II/IVa, II/IVb, and II/IVc. BCS class IIa and IIb exhibit pH-dependent solubility; the former are weakly acidic drugs which are poorly soluble at gastric pHs, for example, ibuprofen, ketoprofen, but show good solubility at intestinal pHs. In contrast, class IIb drugs, for example, carvedilol, ketoconazole, are weakly basic drugs, which show the inverse solubility relationship. Class IIb drugs are prone to supersaturation and precipitation as they move from the gastric into the intestinal compartments [15], see chapter 2.

In contrast, BCS class IIc drugs, for example, danazol, fenofibrate, are neutral and exhibit low, pH-independent solubility. The BCS class IV drugs can similarly be classified into IVa (high intestinal solubility), IVb (high gastric solubility), and IVc (low solubility in both gastric and intestinal compartments). This is summarized in Table 11.3.

Table 11.3: Proposed Re-classification of BCS Class II and Class IV (Adapted from Tsume et al., 2014 [16]).

Proposed BCS Sub-Classifications	Permeability	Solubility at pH 2 (H/L)	Solubility at pH 6.5 (H/L)
I	High	High	High
IIa	High	Low	High
IIb	High	High	Low
IIc	High	Low	Low
III	Low	High	High
IVa	Low	Low	High
IVb	Low	High	Low
IVc	Low	Low	Low

There have also been some suggestions of modifying the BCS system by including routes of drug elimination and assessing the effects of efflux and absorptive transporters on oral drug absorption. Wu and Benet [17] identified that for those drugs exhibiting high intestinal permeability rates, that is, BCS class I and II compounds, the main route of drug elimination in man was via urinary metabolites. In contrast, those drugs showing poor intestinal permeability rates in man, that is, BCS class III and IV compounds, were principally eliminated unchanged in the faeces [18]. They proposed that this new classification system, that is, Biopharmaceutics Drug Disposition Classification System (BDDCS), was complimentary to the BCS approach, particularly in simplifying and accelerating drug development [18]. The authors indicated that the BDDCS [17] was useful in assessing the (a) type, mechanism, and relative importance of food in the absorptive process, (b) likely transporter effects on systemic drug concentrations following both oral and intravenous dosing, and (c) overall drug disposition in those cases when transporters or enzymatic interactions can produce clinically significant effects [19]. Wu and Benet [17] suggested that there would be an increase in the number of compounds that could fall into the BCS class I biowaiver category if the BDDCS approach was utilized. However, many of these transporter and efflux systems are dose-dependent and can be saturated at higher doses. Benet et al. [20] then applied the BDDCS approach to classify nearly 1,000 drugs, as well as trying to predict the BDDCS classification for new chemical entities (NCEs) using computational approaches [21]. Based on these studies the authors demonstrated the importance of the solubility/dose ratio in the BDDCS approach.

The impact of dose was explored further in the dose-dependent BCS (DDBCS) [22]. The authors used in silico simulations based on mathematical modelling of F (fraction of dose absorbed) versus drug dose. It showed a constant value for F across a wide range of doses for BCS classes I-III drugs thereby justifying biowaiver

Table 11.4: The Biopharmaceutics Drug Disposition Classification System (BDDCS) [17].

BDDCS Class I **High Solubility/Extensive Metabolism**[1]	**BDDCS Class II** **Low Solubility/Extensive Metabolism**
BDDCS Class III High Solubility/Limited Metabolism Renal and/or Biliary Clearance of unchanged drug	**BDDCS Class III** Low Solubility/ Limited Metabolism Renal and/or Biliary Clearance of unchanged drug

[1]Extensive metabolism is defined as metabolism >70%.

claims. The impact of increasing dose was then assessed. The authors showed that even for high-solubility drugs, that is, BCS Class I and III, F will change above a critical dose ($Dose_{cr}$), after which the quantity of drug absorbed is independent of the dose, that is, non-linearity of PK-dose response. Thus, for doses higher than $Dose_{cr,}$ BCS Class I drugs become solubility limited and change to BCS class II and similarly, BCS class III drugs become BCS Class IV. Therefore, $Dose_{cr}$ was utilized to define an in vivo effective solubility, that is, $S_{eff} = Dose_{cr}/250$ mL, based on an assessment of literature data.

The authors showed that it is possible to classify drugs with $F \geq 0.90$ as class I drugs; similarly, drugs with limited absorption ($F \leq 0.20$) as class IV drugs. Finally, those drugs with $0.20 < F < 0.90$ can be classified as either class II or class III drugs (dependent on administered dose) with $F \geq 0.90$ would be appropriate for biowaiver consideration. This is a similar approach to that adopted by world-wide regulatory authorities that the human F should be ≥0.85.

Table 11.5: The Dose-Dependent Biopharmaceutics Classification System (DDBCS) [22].

Class I ($F \geq 0.90$)	**Class II ($0.20 < F < 0.90$)** **Dose > Dose $_{CR\ I\text{-}II}$**	**Class IV ($F \leq 0.20$)**
Class II ($0.20 < F < 0.90$) **Dose < Dose $_{CR\ I\text{-}II}$**	**Class III ($0.20 < F < 0.90$)** **Dose < Dose $_{CR\ III\text{-}IV}$**	**Class III ($0.20 < F < 0.90$)** **Dose > Dose $_{CR\ III\text{-}IV}$**

Shaded areas indicate those classes where biowaivers should be achievable

In parallel with this, Butler and Dressman also proposed the Developability Classification System (DCS) [23], which was intended to have a greater focus on drug development criteria, rather than merely biowaivers of marketed drugs. The DCS system is a four-box model based on a dose/solubility ratio (rather than solubility per se) and permeability. Dose/solubility ratio is used because if the dose is low, for example, ondansetron (4–8 mg), biperiden (2 mg), dexamethasone (2-4 mg) [24], then even for those drugs that could be classified as poorly soluble according to standard pharmacopoeial criteria, they could still have adequate solubility within the GI tract. The DCS system also stresses the importance of intestinal solubility in bio-relevant media, that is, FaSSIF (Fasted State Simulated Intestinal Fluid) and FeSSIF (Fed State Simulated Intestinal Fluid). This is because bio-relevant media have significantly more solubilizing power than conventional aqueous buffers, due to the presence of naturally occurring surfactants, for example, bile acid salts, which can emulsify hydrophobic drugs [25] (see chapter 6). The ratio of FeSSIF/FaSSIF solubilities can be used to predict the likelihood of a food effect in man. Consequently, the DCS system also has significant utility in predicting the appropriate formulation strategy based on DCS categorization.

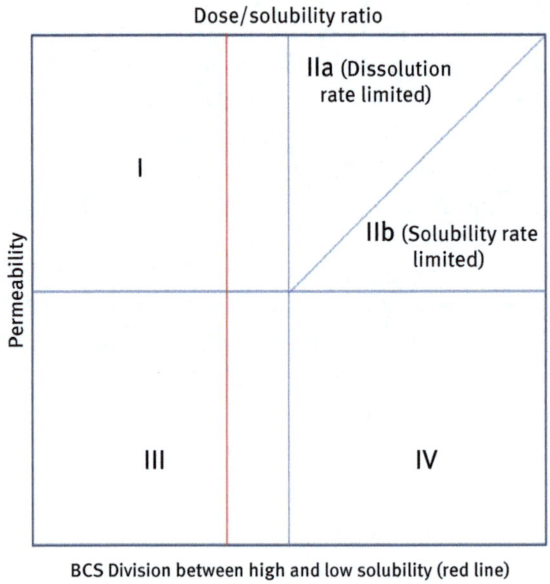

Figure 11.1: Developability Classification System (DCS) [23].

DCS utilizes a value of 500 mL for the volume of available GI fluids (based on both gastric and intestinal fluid volumes); whereas, BCS recommends using only the gastric volume, that is, 250 mL. Consequently, some BCS II compounds would become class I using the DCS criteria and would likely benefit from being included in any future biowaiver programme. The DCS approach also takes into account the synergistic interplay between solubility and permeability, that is, this is a dynamic equilibrium between these two parameters and if dissolved drug is removed from the intestinal compartment as a result of absorption, then more GI fluid is free to solubilize additional drug. DCS defines a solubility limited absorbable dose (SLAD), which equates to the dose above which absorption is constrained by solubility [23], which is a similar concept to $Dose_{CR}$ [22]. This then leads to a sub-division of the DCS class II compartment into DCS IIa (dissolution rate limited absorption) and DCS IIb (solubility limited absorption); not to be confused with BCS classes IIa and IIb by Tsume et al. [16], which refers to ionization behaviour of drugs and consequently pH dependence of solubility. The DCS also includes an assessment of the drug substance particle size and its role in helping to address "dissolution limited" absorption. Reduction in particle size increases the rate of solubility, that is, the dissolution rate, but does not enhance the intrinsic solubility of the API. In contrast, for "solubility limited" compounds, DCS can be used to identify which formulation strategies are appropriate. In all cases, these formulation approaches will endeavour to enhance the intrinsic solubility, either by using lipid solutions or by using amorphous stabilized formulation strategies. DCS has a greater capability than the BCS system in predicting those factors that are critical to in vivo performance.

11.4 The role of in vitro dissolution testing

In vitro dissolution testing is always used in support of any biowaiver proposal. The concept of "rapidly dissolving" is also important for biowaivers, and it is defined as not less than (NLT) 85% of the dose dissolves in 30 min using either (i) USP apparatus 1 (or equivalent) at 100 rpm or (ii) USP apparatus 2 (or equivalent) typically at 50 rpm (or at 75 rpm when appropriately justified – but see later discussions on coning), using either 900 mL of media (for EU, Japan, WHO) or 500 mL or less (or the higher volume of 900 mL when appropriately justified) (for US) in pH 1.2, pH 4.5, and pH 6.8 buffer at 37°C [26]. A similar concept of "very rapidly dissolving" is used to support biowaivers of BCS class III compounds (highly soluble/poorly permeable), this is defined as NLT 85% of the dose dissolves in 15 min with the other conditions as before. Quite why a more aggressive dissolution specification should be applied to highly soluble drugs that also happen to be poorly permeable (that is, BCS class III), when dissolution per se cannot discern any differences in permeability is somewhat of a conundrum. There is also the significant concern that the hydrodynamics within the dissolution vessel will result in increased variability for any time point less than 30 min. For example, Sanofi Aventis performed a world-wide comparative assessment of the dissolution testing of 5 mg glibencamide instant release tablets across its global network of 29 centres. All sites used a USP paddle apparatus with a rotation speed of 75 rpm and pH 7.4 phosphate buffer (900 mL). Sampling intervals were 10, 20, 30, 45, 60, 90, and 120 min. The overall variability (both method and product) as a function of time point is shown in Table 11.6. It can be readily seen that the method variability at 10 min is nearly 2.5 x greater than that seen at 30 min [27].

Table 11.6: Variability in Dissolution Testing as a Function of Time Point: Results from a Collaborative Study across 29 Centres using 5 mg glibencamide immediate release tablets. (adapted from Weinandy 2005) [27].

Time point (Minutes)	Variability (% RSD)
10	16.1
20	9.3
30	7.1
45	5.6
60	4.9
90	4.6
120	4.5

Indeed, 30 min is often seen as the best compromise between variability and discrimination [28]. At earlier time points, variability is too high; whereas at later time points, discrimination is too low.

If the dissolution testing conditions need to be changed to better reproduce rapid in vivo dissolution (e.g., use of a different non-compendial rotation speed), such modifications need to be justified by matching the in vitro dissolution with in vivo absorption data (e.g., performing a relative BA study), that will allow a robust IVIVC to be developed [26].

Comparative dissolution testing of 12 units of both the test and reference compounds are typically assessed at three different pH values, that is, pH 1.2, 4.5, and 6.8. Applicants can use standard aqueous media or buffers, or use simulated gastric fluid (SGF) or simulated intestinal fluid (SIF), both without enzymes. The dissolution profiles are then compared for equivalency or similarity using the f2 test, or other similarity test [7, 11–14, 26]. The similarity test (f2) permits the use of mean data ($mean_{12}$) in the comparison of dissolution profiles, but specifies that the coefficient of variation (CV) should not be more than 20% at the earlier time points (e.g., ≤10 min), and should not be more than 10% at later time points; again re-emphasizing that the intrinsic variability at time points less than 30 min are significant.

There are several advantages inherent in using these pharmacopoeial methodologies; they are easy to operate, robust (as long as the time point is ≥30 min), universally available, extensive experience exists (but only when using standard media volumes, that is, 900 mL – see next section), and they are accepted by academia, industry, and the regulatory bodies [25]. However, although aqueous buffers reflect the typical pH conditions experienced within the GI tract, that is, pH 1.2–6.8, they are not reflective of other key attributes of bio-relevant media, that is, ionic strength, osmolality, viscosity, wetting, and surface tension properties [25]. However, at this time in the ICH M9 proceedings (1Q 2019), there appears to be little appetite to use these bio-relevant media as part of the biowaiver strategy.

Interestingly, the recent FDA guidance on dissolution requirements for biowaivers [7, 26] recommends 500 mL of media (or less) should be used in the comparative dissolution tests. However, this proposed change is not aligned with EU [12], Japanese [14], or WHO [11] requirements for dissolution media volume. Historically, the media volume for USP apparatus 1 and 2 was 900 mL, even though volumes between 500 and 900 mL are still deemed to be acceptable [29, 30]. It would appear that this modification is to make the dissolution media volume more "bio-relevant", as under fasting conditions. The typical total volume of gastric (250 mL) and intestinal media (250 mL) compartments are 500 mL [23, 31], but this is not explicitly stated. Whilst changing media volume is commendable in its objectives, there is not huge amount of historical scientific data in support of the routine use of reduced media volume for dissolution in a production setting. The concern would be that the hydrodynamics within the dissolution vessel may be adversely impacted as it will be working at the bottom end of its design range, so issues such as coning may become more prevalent [32]. Coning was observed in doxycycline formulations at 50 rpm [33]. EMA and WHO consequently suggest 75 rpm with paddle apparatus. However, the current ICH M9 guidance [2] does not support a default increase in paddle speed from 50 rpm to 75 rpm to address coning, rather, it states that "When high variability or coning is observed in the paddle apparatus at 50 rpm, the use of the basket apparatus at 100 rpm is recommended. In addition, the use of sinkers in the paddle apparatus, to reduce method variability, may be considered with justification" [2].

Interestingly, dissolution may be a better arbiter of bio-relevant "solubility" than solubility itself. Yazdanian et al. [34] highlighted that, "an inherent limitation of the solubility classification is that it relies on equilibrium solubility determination, which is static and does not take into account the dynamic nature of absorption" (see also Solubility – Definition and basic physicochemical considerations and discussions on Equilibrium Solubility). It should also be noted that the defined acceptance criterion for "rapidly dissolving" dosage forms, that is, NLT 85% dissolved in 30 min, is independent of, and not correlated with the drug's solubility/dose ratio, which is a critical factor controlling the rate of dissolution [35]. As such dissolution could in principle be used as the "sole indicator for biopharmaceutic drug classification" [36]. However, dissolution is inherently more complex than solubility measurements, with significantly more critical parameters; that is, apparatus, agitation speed, media (stirring rate, pH, ionic strength, sink conditions, surfactants, bio-relevance), Q value (where Q is a pre-defined value that is established for each drug product in its monograph at a pre-specified time, that is, Q = 80% after 30 min), time point, etc., and whether this will ever be a meaningful strategy is open to doubt.

For biowaiver applications, FDA reviews dissolution data from different perspectives dependent on whether this is a new chemical entity (NCE) or a generic product. For new drug applications (NDA) bioavailability data always assumes primacy and in vitro dissolution data is supportive in nature, (ii) For abbreviated NDA (ANDA) the reverse holds true and in vitro dissolution data is likely to be pivotal [37].

11.5 ICH M9: potential discussion points

As the new ICH M9 guidance progresses towards step 4 ratification, several areas of debate/contention are becoming apparent.

11.5.1 Supportive data for BCS assessment

11.5.1.1 Solubility

Solubility should be assessed using the shake-flask (or other justified) methodology, within the pH range of 1.2–6.8 and if the drug is ionizable then the pH conditions should also include the pKa and at pH 1.2, 4.5, and 6.8 [2, 7, 12–14]. The pH should be measured both before and after the addition of the drug substance and adjusted if required. The volume is typically 250 mL (to reflect fasted state gastric volumes), and the temperature is 37 ± 1 °C. The lowest measured solubility value over the biologically relevant pH range, that is, 1.2–6.8, defines the solubility criteria for the BCS assessment. In addition, adequate stability of the API in the media should be demonstrated using an appropriately validated stability-indication method, for example, HPLC. See acetylsalicylic acid entry in Table 11A.1 in the Appendix (Summary of FIP Biowaiver Monographs for Essential Drugs) for additional commentary on the stability of the API.

The major area of contention focusses on whether the solubility should be based on either the highest therapeutic dose (EMA [13]) or conversely the highest strength (FDA [7]) of the medicinal product. For example, the anti-malarial drug quinine sulfate [38] has a maximum therapeutic dose from the product label of 648 mg; whereas, the highest dose strength is 324 mg/capsule. The guidance indicates that the target dose/solubility (D/S) ratio is based on the gastric volume, that is ≤250 mL [3]. Quinine sulfate shows greater solubility at acidic pHs, that is, BCS class IIb compound. A summary of the solubility, D/S, and BCS assignment in various media of differing pHs is given in Table 11.7.

Therefore, quinine sulfate is a BCS class I compound across all physiologically relevant pHs based on the highest capsule strength criterion. In contrast, quinine sulfate is a BCS class I compound at acidic pHs, but a BCS class II compound at neutral pHs, based on the highest dose criterion.

Table 11.7: Solubility, D/S and BCS assignment of quinine sulfate in various media (derived from Strauch et al., 2011 [31]).

Media	Solubility (mg/mL)	D/S[c] mL (based on 324 mg- highest strength)	BCS[d] Assignment	D/S mL (based on 648 mg – highest dose)	BCS Assignment
SGF[a]	12.0	27	1	54	1
SIF[b]	1.3	249	1	498	2

[a]SGF simulated gastric fluid, pH 1.2;
[b]SIF simulated intestinal fluid, pH 6.8;
[c]D/S dose/solubility ratio;
[d]BCS Biopharmaceutics Classification System

However, if the D/S ratio was based on the total GI volume, that is, ≤500 mL [23], then quinine sulfate would be a DCS class I compound – based on both highest strength and highest dose calculations, although in the latter case, the drug is borderline class I at pH 6.8.

There are other examples of commercial drugs, particularly weak anionic drugs, for example, NSAIDs (non-steroidal anti-inflammatory drugs) where BCS classification changes from BCS I to II at higher doses. For example, the $Dose_{CR}$ for ibuprofen is 1,200 mg; for diclofenac, it is 100 mg; and for flurbiprofen and fenoprofen, it is >300 mg [22]. It is also increasingly common for new chemical entities (NCE's) in Phase I to straddle DCS I to IIb [23] across the range of doses intended for dose escalation.

11.5.1.2 Permeability

ICH M9 aims to harmonize different approaches to assessing permeability, that is, in vitro or in vivo assessments. Currently, data from man is preferred using either absolute bioavailability or mass balance studies [39]. In addition, the cut-off between high and low permeability often prompts debate and is currently " ≥85% or more of an administered dose based on a mass balance determination (along with evidence showing stability of the drug in the GI tract) or in comparison to an intravenous reference dose" [2, 13]. Guidance is provided to applicants on how to include metabolites in the discussion and only oxidative and conjugative metabolites should be considered [2].

It has been widely recognized that differences do exist in assessing permeability between the various in vivo and in vitro models in Ussing chambers and cell monolayers, that is, MDCK, Caco-2, etc [39].

As such it can be extremely difficult to unambiguously define a permeability classification. The cell-based permeability data should be "discussed in the context of available data on human pharmacokinetics" and the methodology is required to be appropriately validated" [2]. In addition, if high permeability is inferred by using cell-based methodology, then "permeability independent of active transport should be proven".

Quinine sulfate is a well-characterized drug and it is often reported as being highly permeable [38]. However, the absolute bioavailability after oral dosing was reported to be between 76% [40] and 88% [41], which would make this drug a borderline case. However, because quinine sulfate is extensively hepatically metabolized [42], the absolute bioavailability represents only a minimum estimate of the fraction of drug absorbed and as <5% of the oral dose is found unchanged in the faeces, this provides auxiliary evidence of high permeability [43, 44].

In general, there exists a reasonable correlation between log P values and small intestinal jejunal permeability. However, there are significant numbers of "false-negative" cases that have been reported, in which drugs with lower log P than metoprolol, that is, 2.3, exhibit complete absorption. These include codeine phosphate, antipyrine, cephalexin, d-glucose, levodopa, l-leucine, phenylalanine, piroxicam, valacyclovir, pseudoephedrine, and sotalol. This shows the challenges inherent in using limited physicochemical characteristics in trying to assign BCS classification [45].

For biowaiver applications, FDA is looking for a study report, use of appropriate reference compounds to define that the system would be able to discriminate between high and low permeabilities and characterization of active transporters and any efflux mechanisms that might be present [36]. Frequently missing from these reports are drug stability data in the GI tract and data supporting the appropriateness of the in vitro or in situ permeability methods. FDA is concerned with issues of bio-inequivalence at very low permeability values [37].

11.5.1.3 Published literature

Literature data are typically not acceptable for regulatory purposes, as the underlying data cannot be reviewed [2, 37]. However, data from FIP monographs are likely to be acceptable as it is a very complete review of available data for solubility, permeability, and dissolution [15].

11.5.2 Scope of biowaivers

Biowaivers are constrained to instant release solid oral dosage forms. Specifically excluded for consideration in the US are any product designed to be released/absorbed in the oral cavity, for example, sublingual or buccal tablets [7]. In the EU, buccal, sublingual, and oro-dispersible tablets designed to be released/absorbed in the oral cavity are excluded [13]. The WHO provides similar guidance to EU but supports biowaivers for oro-dispersible products if there is no absorption in the oral cavity [11]. The ICH M9 [2] guidance is supportive of the EU position (no biowaivers) and states further that "an oro-dispersible product is eligible for a biowaiver application only if there is no buccal or sublingual absorption and the product is labelled to be taken with water only".

All territories (and ICH M9) currently exclude narrow therapeutic index (NTI) drugs. They can be characterized as "drugs where small differences in dose or blood concentration may lead to serious therapeutic failures and adverse drug reactions that are life-threatening or result in persistent or significant disability or incapacity" [7]. FDA (and other agencies) recommends tighter quality and bioequivalence standards to ensure the safety and efficacy of generic NTI drugs.

11.5.3 Setting the requirements for in vitro dissolution testing

It is interesting to note that the ICH M9 guidance links the dissolution testing with a "representative" batch, that is, "using one batch representative of the proposed commercial manufacturing process for the test product relative to one batch of the reference product. The test product should originate from a batch of at least 1/10 of production scale or 100,000 units, whichever is greater, unless otherwise justified". ICH M9 also identifies that smaller batch sizes may be acceptable (if justified) during the clinical development phase [2].

It is recommended to determine the dose/solubility ratio of the NCE prior to initiating dissolution studies, as a ratio of <250 mL in aqueous media of various different pHs, that is, pH 1.2–6.8 is indicative that "dissolution is very unlikely to limit drug absorption" [25]. Good guidance for dissolution method development is provided by Klein et al. [46]. In vitro dissolution testing will be a pre-requisite for any biowaiver. The ICH M9 guidance [2], in terms of apparatus (basket or paddle), agitation rate (100 or 50 rpm, respectively), media volume (900 mL), pH of media (pH 1.2, 4.5 or 6.8), the use of the f_2 similarity factor for comparison of profiles and the number of units to be tested (12) are fairly standard and well established. The only new areas in the guidance are how to address the coning of tablet dosage forms in a 50 rpm paddle method (see earlier Section 11.4).

Currently, there are two different criteria for high-solubility compounds, either Q = 85% after 30 min (BCS class I) or Q = 85% after 15 min (BCS III). However, given the well-established and poor hydrodynamic mixing capabilities [28] of pharmacopoeial dissolution apparatus (particularly USP 2), it makes little sense to apply different specification limits, particular as both BCS class I and class III compounds are highly soluble. Since it is practically impossible to perform any meaningful comparability assessment between two dissolution profiles with Q values centred at 15 min, Q = 85% after 30 min may be preferred for both.

Biowaivers are likely to continue to use pharmacopoeial buffers at various different physiologically relevant pHs. However, bio-relevant media, particularly FaSSIF and FeSSIF media, may be more appropriate and may enjoy greater use in the future (see chapter 6). In particular, there may be a role for biowaivers in lieu of relative bioavailability studies in man to assess the effects of food [25], which are not part of the current ICH M9 biowaiver proposals. Galia et al. [47] compared the dissolution performance of 200 mg danazol in FeSSIF and FaSSIF media using 500 mL media and USP apparatus 2 at 100 rpm. They demonstrated that there was a threefold to fourfold increase in the dissolution rate of the drug in FeSSIF compared to FaSSIF media. These in vitro results were confirmed in man. Charman et al. [48] performed a relative bioavailability study in 11 healthy female volunteers. The data are provided in Table 11.8 and show a sixfold increase in both AUC (area under the curve) and C_{max} for the drug dosed in the fed state.

Table 11.8: Relative bioavailability study for danazol in fed and fasted state (derived from Charman et al. [48]).

PK Parameter	Fed State	Fasted State
$AUC_{0\text{-}36h}$ (ng h mL^{-1})	639 ± 259	101 ± 42
C_{max} (ng/mL)	204 ± 125	37 ± 16

11.5.4 Pharmaceutical equivalents

Thus far biowaivers have been restricted to pharmaceutical equivalents and primarily to BCS class I and III compounds. Pharmaceutical equivalence implies the same amount of the same active substance(s), in the same dosage form, for the same route of administration and meeting the same or comparable standards. In contrast, pharmaceutical alternatives contain the same therapeutic moiety, but are different pharmaceutical salts, esters, or complexes of that moiety, or are different dosage forms, that is, tablets versus capsules [49]. In Europe, under Essential Similarity rulings, alternative pharmaceutical salts have historically being considered "similar". In contrast, in the US pharmaceutical alternatives are not permitted. ICH M9 indicates that "a biowaiver is not applicable when the drug substance in the test product is a different pharmaceutical salt, ester, isomer, or mixture of isomers from that in the reference product. Pro-drugs may be considered for a BCS-based biowaiver when absorbed as the pro-drug" [2].

An interesting example of the differences in perspective between the EU and US is seen with different salts of paroxetine. The innovator salt is the hydrochloride, and this has a solubility of 5.4 mg/mL in water, high permeability, and a typical dose of 20 mg/day (as the free base equivalent) [50]. In contrast, paroxetine mesylate has significantly higher aqueous solubility (>1,000 mg/mL) in water, high permeability, and a typical dose of 20 mg/day (as the free base equivalent) [51]. Both salts are BCS class I, the significantly higher solubility of the mesylate is not clinically relevant because of the low dose (i.e., 20 mg) and both salts would be inter-changeable and be appropriate candidates for a biowaiver in the EU, prior to the implementation of ICH M9. In contrast, in the US, no biowaiver could be approved despite the fact that generic versions, that is, Paxeva [51] which contain the mesylate salt as API have been approved based on successful bioequivalence studies comparing the mesylate salt and the hydrochloride salt [50].

11.5.5 The role of excipients in biowaivers and bioequivalence studies

There has been widespread concern regarding the effect of different excipients on the permeability and thereby the bioavailability of different formulations. For example, FDA [7] guidance states, "Unlike for BCS class I products, for a biowaiver to be scientifically justified, BCS class III test drug product must contain the same excipients as the reference product. This is due to the concern that excipients can have a greater impact on absorption of low permeability drugs." ICH M9 [2] indicates that, "BCS Class III drug substances are considered to be more susceptible to the effects of excipients. These drugs are poorly permeable and may have site-specific absorption, so there are a greater number of mechanisms through which excipients can affect their absorption than for BCS Class I drugs".

As such, the composition of the test product is defined as being qualitatively the same (except for those excipients, for example, colorants, flavorants, or preservatives that will not affect the bioavailability), quantitatively very similar to the reference product" and demonstrate well-established usage in products containing that particular API, as well as no effect on GI motility, no interference with transporter processes, and finally no effect on PK of drug substance [7, 11]. In order to assess whether a new formulation of a BCS class III drug is suitable for a biowaiver, any potential effects that the excipients have on the permeability of the drug need to be assessed (see risk assessment). The ICH M9 guidance [2] provides a useful decision tree, which is reproduced below for ease of review.

If no effects are observed in a fully validated in vitro model, for example, Caco-2, MDCK, or an in situ rat perfusion model, then it would be logical to assume that there would be no effect on in vivo permeability [52].

The following are allowable differences in excipients articulated in ICH M9 [2] based principally on SUPAC IR [53] considerations:
- Changes in the technical grade of an excipient
- Changes in excipients, less than or equal to,
  - Filler (±10 %)
  - Disintegrant, starch (±6 %), or disintegrant, other (±2 %)
  - Binder (±1 %)
  - Lubricant, calcium, or magnesium stearate (±0.5 %) or lubricant, other (±2 %)
  - Glidant, talc (±2 %), or glidant, other (±0.2 %)
  - Film coat (±2 %)

Although most of the focus on excipients has rightly centred on BCS class III compounds, certain excipients, such as surfactants (e.g., polysorbate 80) and sweeteners (e.g., mannitol or sorbitol) may be problematic, and can influence the rate and extent of drug absorption of BCS class I compounds. ICH M9 [2] indicates that, "BCS Class I drugs are highly absorbed, and have neither solubility nor permeability limited absorption. Therefore they generally represent a low risk group of compounds in terms of the potential for excipients to affect absorption, compared to other BCS classes". As such any excipient effects that could affect BCS Class I drug products need to focus on any potential for changes in the rate or extent of absorption. Interestingly, the ICH M9 [2] guidance cites the example of drugs that have "high permeability due to active uptake", and hence "excipients that can inhibit uptake transporters are likely to be of concern". However, such drugs would be typically classified as BCS class III (not BCS class I) as their passive permeability is low, but they are substrates for active transport. In addition, "for BCS Class I drugs that exhibit slow absorption, the potential for a given excipient to increase absorption rate should also be considered". Therefore, for BCS Class I drugs, "qualitative and quantitative differences in excipients are permitted, except for excipients that may affect absorption, which should be qualitatively the same and quantitatively similar, i.e., within ±10.0% of the amount of excipient in the reference product". The ICH M9 [2] guidance again provides a useful decision tree, which is reproduced below for ease of review.

The case of risperidone has become something of a cause celebre amongst certain researchers. Risperidone is a BCS class I [54] and also a BDDCS class I compound [17]. It is often formulated with SLS (sodium lauryl sulfate). However, a 1 mg generic product formulated with SLS (3.64 mg) showed bio-inequivalence in a 34 subject 2 × 2 cross-over study; for both AUC (74.7–91.7 μg h^2/mL) and C_{max} (70.0–86.8 ng/mL). In contrast, in vitro dissolution data showed very rapid dissolution (<15 min) at both pH 1.2 and 6.8. There was concern that small levels of this surfactant, that is, SLS, could cause bio-inequivalence that could not be detected by in vitro tools and therefore biowaivers should not be applied to products containing these critical excipients [54].

However, we should be cautious about over-interpreting the outcomes of this study, as failed bioequivalence studies are fairly common and are often attributed to various factors: (a) an under-powered study, (b) inappropriate study designs, (c) an outlier response from one or more subjects, (d) assay issues, (e) a wrong reference, (f) baseline-corrected versus not corrected, (g) incorrect statistical analysis, (h) compliance issues, and finally (i) a formulation that is not truly bioequivalent to the reference [55]. As the reasons for the failed BE study are not fully known and risperidone known to show high PK variability [56], primarily because of CYP 2D6 polymorphism in the general population (i.e., 20% of population are poor metabolizers), it could be that the study was under-powered, or one of the other myriad factors that could have caused the study failure.

However, it is by no means true that all excipients can adversely influence absorption [57]. A recent permeability study using Caco-2 and in situ rat intestinal perfusion studies of five common excipients (lactose, povidone, HPMC, SLS, and PEG 400) with five model BCS class III drugs, that is, acyclovir, atenolol, antipyrine, ganciclovir, and nadolol, showed that these excipients caused no changes to drug permeability; with the exception of SLS at concentrations of 0.1 mg/mL or higher, where the cell monolayer integrity was affected [57]. The authors concluded that biowaivers could be supported for this combination of excipients and BCS class III drug substances. It is likely that the various precepts defined in the "scale up and post approval change" (SUPAC IR) [53] guidance documents will be useful in assessing the likely impact of any changes in excipient quality or quantity.

11.5.6 Biowaiver risk assessment

As indicated in the FIP biowaiver initiative, the "biowaiver monographs are not aimed at merely applying these guidelines, but rather critically evaluating the properties of the API and applying risk analysis to complement the biopharmaceutics classification system (BCS)" [58]. There has been some risk assessments reported mainly based on failure mode and effects analysis (FMEA) approaches [59]. The biggest risk appears to be based on the presence of excipients in the test formulation that could subsequently influence absorption, and this risk is of course highest for BCS class III compounds. The latest draft of the ICH M9 guidance [2] goes a long way towards minimizing this risk, with decision trees for both BCS class III (Figure 11.2) and BCS class I (see Figure 11.3) molecules discussing allowable qualitative and quantitative excipient changes. Other risks that are typically encountered usually arise due to omissions in the regulatory submission. The more common omissions have been previously identified in Section 11.1.

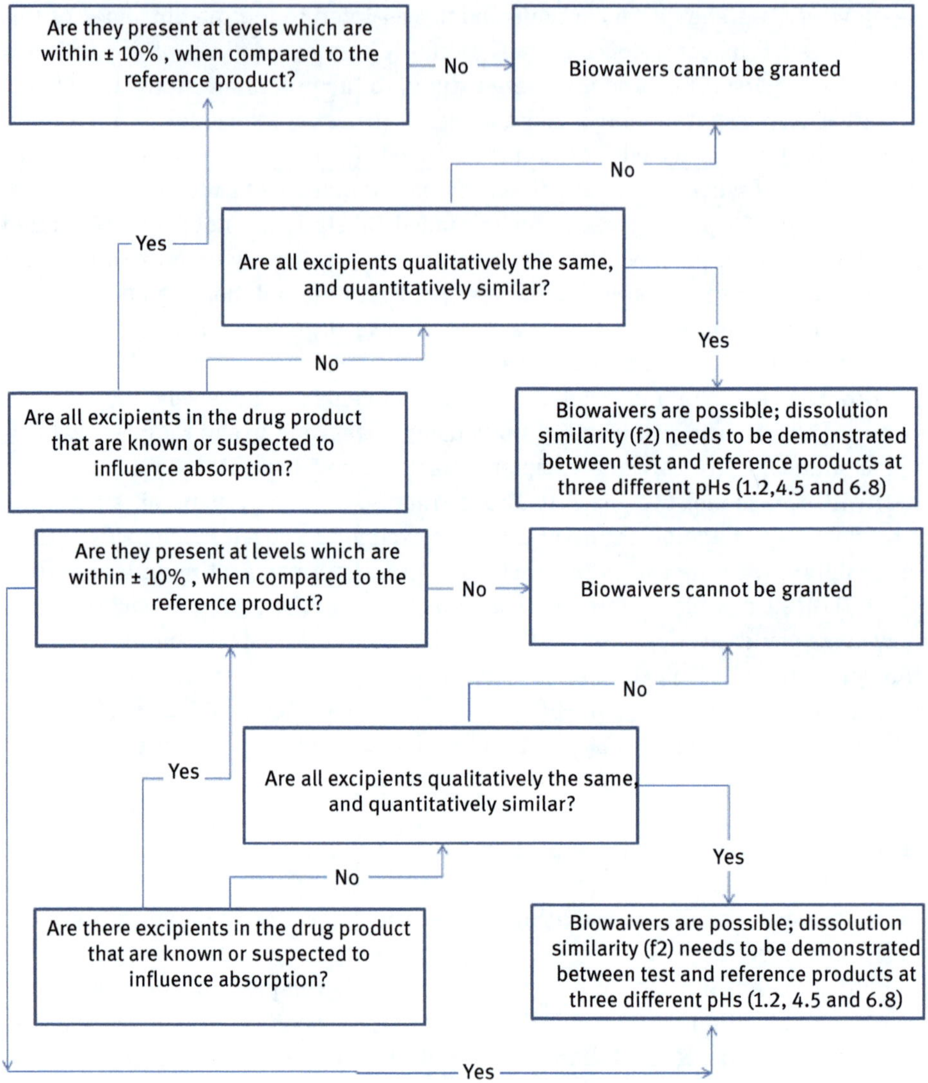

Figure 11.2: Decision Tree for the Allowable Levels of Excipient Differences for BCS Class III Drug Products (modified from [2]).

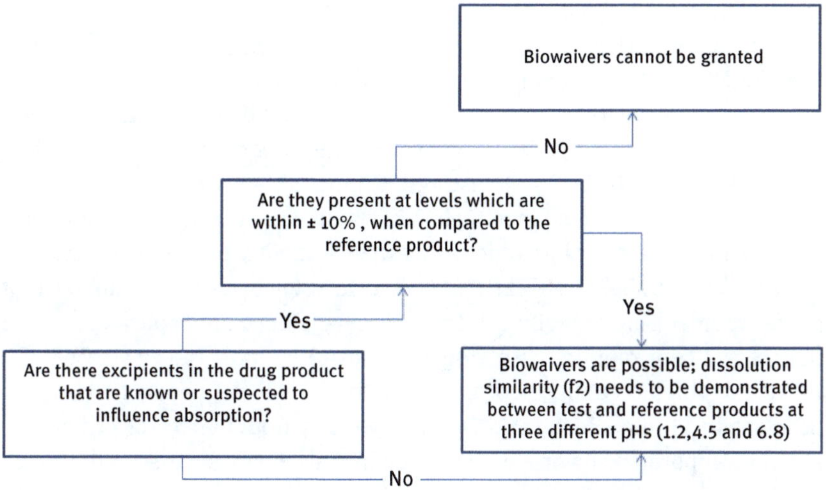

Figure 11.3: Decision Tree for the Allowable Levels of Excipient Differences for BCS Class I Drug Products (modified from [2]).

11.6 Conclusion

ICH M9 will provide comprehensive recommendations to support a BCS assessment of drug products and the potential for biowaiver relief. The objective of ICH M9 is to prevent unnecessary in vivo bioequivalence studies being performed as a result of conflicting regional recommendations on the acceptability of current BCS based biowaivers. This of course often results in increased drug development costs, as well as unnecessary human volunteer studies. However, the challenges facing ICH M9 are still significant and may be difficult to achieve.

The FIP biowaiver monograph initiative fully exemplifies the difficulties in applying BCS-based approaches for the assessments of biowaivers. There are currently 49 monographs (accurate as of January 2019), some monographs cover more than one pharmaceutical salt, that is, chloroquine (phosphate, sulfate and hydrochloride) and diclofenac (sodium and potassium). Of these, only two-thirds of the monographs can be unambiguously assigned into one of the four BCS classes. The Biopharmaceutics Drug Disposition Classification System (BDDCS) is also complimentary to the BCS approach, particularly in simplifying and accelerating drug development.

Drug products with different dose strengths are not clearly covered by the existing BCS based biowaiver initiative and this raises many questions. Where a medicinal product is marketed in several strengths, that is, amitriptyline, has four strengths: 10, 25, 50, and 75 mg, does this mean that the BCS-based biowaiver has to be performed on all four strengths? Or can the applicant bracket the strengths, that is, 10 and 75 mg or if all four strengths are compositionally similar and there is evidence of linear pharmacokinetics in the proposed dosing range, can the applicant use one strength, that is, 75 mg? Current experience would suggest that biowaivers will need to be evaluated on all marketed doses. For example, amoxicillin trihydrate straddles several BCS classes based on dose considerations and biowaivers can only be recommended based on the lowest dose of this drug, that is, ≤875 mg. The impact of dose on BCS classification is also addressed in the DDBCS system and this approach shows that above a critical dose ($Dose_{CR}$) many drugs do change their BCS classification. Similarly, Butler and Dressman [23] use the dose/solubility ratio rather than solubility per se to address this problem. Nevertheless, it should still be noted that for biowaiver applications, BCS-based biowaiver approaches as specified in the ICH M9 guidance are applicable.

The concept of "very rapidly dissolving" is used to support biowaivers of BCS class III compounds (high solublity/poor permeability), this is defined as NLT 85% of the dose dissolves in 15 min. However, given the poor hydrodynamics of pharmacopoeial dissolution apparatus, having a specification time point at less than 30 min makes limited sense, nor indeed does trying to use dissolution (which is a measure of rate of solubility) to control permeability (which is a measure of drug flux across biological membranes). Similarly, the adoption by the FDA of a default media volume of 500 mL or less for biowaivers without any meaningful discussion with interested parties (including other regulatory agencies) makes limited sense and issues such as coning may become more a prevalent topic in dissolution experiments in the future. Also, the classical approach of addressing coning issues in the paddle apparatus (USP 2), by increasing paddle speed from 50 to 75 rpm, is not supported in the ICH M9 guidance. This guidance advocates the use of the basket apparatus (USP 1) at 100 rpm in those cases where coning is seen with the paddle apparatus at 50 rpm.

One major area of contention focusses on whether the solubility should be based on either the highest single therapeutic dose (EMA) or conversely the highest strength (FDA) of the medicinal product and differences in perspective can lead to dose-related differences in BCS classifications.

There has been widespread concern regarding the effect of different excipients on the permeability and thereby the bioavailability of different formulations, particular for biowaivers for BCS class III compounds. The majority of Health Agencies have taken a pragmatic view on this issue noting that "certain excipients, such as surfactants (e.g., polysorbate 80) and sweeteners (e.g., mannitol or sorbitol) may be problematic" but that as long as the composition of the test product is qualitatively the same and quantitatively very similar to the reference product, with demonstrated usage in marketed products, then biowaivers can be considered. Others have strongly articulated that formulations containing these problematic excipients should be exempt from biowaiver consideration based on a small number of failed BE studies; without fully demonstrating that the failure was indeed a result of dissimilar in vivo performance of the product, rather than any one of the other innumerable reasons for BE failure that have been identified over the years. However, it is clear that more needs to be done to fully understand the interplay of these problematic excipients with in vivo permeability. The presence of two decision trees (see Figures 11.2 and 11.3) covering qualitative and quantitative aspects of excipient selection is to be welcomed, as it de-risks this key area of biowaiver assessment.

Finally, looking to the future, biowaivers may also have a role in paediatric and food interaction assessments, and there may also be a role for more bio-relevant media, for example, FeSSIF, FaSSIF in future biowaiver considerations.

Acknowledgements: This chapter is loosely based on an earlier article: Elder DP, Tindal S. ICH M9: Biopharmaceutics Classification System-Biowaivers, American Pharmaceutical Review, 20(3), 20–25, 2017.

References

[1] Final endorsed Business Plan M9: Biopharmaceutics Classification System-based Biowaivers, 7 October 2016.

[2] Biopharmaceutics Classification System-based Biowaivers, ICH M9, draft version. Endorsed 7 June 2018. Currently under public consultation.

[3] Amidon, GL., Lennernäs, H., Shah, VP., & Crison, JR. A theoretical basis for a biopharmaceutic drug classification: The correlation of in vitro drug product dissolution and in vivo bioavailability. Pharm. Res. 1995, 12, 413–420.

[4] Cook, J. Impact of discordance amongst regulations for biopharmaceutical classification system based waivers of clinical bioequivalence studies. Dissolution Technol. 2015, 22, 58–65.

[5] Nair, AK., Anand, O., Chun, N., Conner, DP., Mehta, MU., Nhu, DT., Polli, JE., Yu, LX., & Davit, BM. Statistics on BCS classification of generic drug products approved between 2000 and 2011 in the USA. AAPS J. 2012, 14(4), 664–666.

[6] Van Buskirk, GA., Asotra, S., Balducci, C., Basu, P. et al. Best practices for the development, scale-up, and post approval change control of IR and MR dosage forms in the current quality-by-design paradigm. AAPS Pharm. Sci. Tech. 2014, 15(3), 665–693.

[7] US Food and Drug Administration. Waiver of In Vivo Bioavailability and Bioequivalence Studies for Immediate-Release Solid Oral Dosage Forms Based on a Biopharmaceutics Classification System. December 2017.

[8] US Food and Drug Administration Guidance for Industry: Waiver of In vivo Bioavailability and bioequivalence studies for immediate release dosage forms based on biopharmaceutical classification system, US Food and drug administration center for evaluation of drug and research, 2000.

[9] NIHS Guideline for Bioequivalence Studies for Different Strengths of Oral Solid Dosage Forms, National Institute of Health Sciences, Japan, 2000.

[10] NIHS Guideline for Bioequivalence Studies for Formulation Changes of Oral Solid Dosage Forms, National Institute of Health Sciences, Japan, 2000.

[11] World Health Organization. WHO technical report series, No. 992 annex 7. Multisource (Generic) pharmaceutical products: guidelines on registration requirements to establish interchangeability, 2015. http://apps.who.int/medicinedocs/documents/s23245en/s23245en.pdf (Accessed 14 Jan 2019).

[12] European Medicines Agency, Committee for Medicinal Products for Human Use (CHMP). Concept paper on BCS-based biowaivers, 2007, London, 24 May 2007, http://www.ema.europa.eu/docs/en_GB/document_library/Scientific_guideline/2009/09/WC500003010.pdf. Accessed on 27 Jan. 2017.

[13] European Medicines Agency. Committee for Medicinal Products for Human Use (CHMP). (2010) Guideline on the investigation of bioequivalence. 2010. http://www.ema.europa.eu/docs/en_GB/document_library/Scientific_guideline/2010/01/WC500070039.pdf. (Accessed 27 Jan 2017).
[14] Pharmaceutical and Food Safety Bureau, Japan, 2012, Guideline for Bioequivalence Studies of Generic Products, Attachment 1 of Division-Notification 0229 No. 10, February 29, 2012.
[15] International Pharmaceutical Federation (FIP). (2017) Biowaiver monographs. 2016. http://www.fip.org/bcs_monographs. (Accessed 27 Jan 2017).
[16] Tsume, Y., Mudie, DM., Langguth, P., Amidon, GE., & Amidon, GL. The biopharmaceutics classification system: Sub-classes for in vivo predictive dissolution (IPD) and IVIVC. Eur. J. Pharm. Sci. 2014, 57, 152–163.
[17] Wu, C-Y., & Benet, LZ. Predicting drug disposition via application of BCS: Transport/absorption/elimination interplay and development of a Biopharmaceutics Drug Disposition Classification System. Pharm. Res. 2005, 22, 11–23.
[18] Benet, LZ. The role of BCS (Biopharmaceutics Classification System) and BDDCS (Biopharmaceutics Drug Disposition Classification System) in drug development. J. Pharm. Sci. 2013, 102(1), 34–42.
[19] Macheras, P., Karalis, V., & Valsami, G. Keeping a critical eye on the science and regulation of oral drug absorption: A review. J. Pharm. Sci. 2013, 102(9), 3018–3036.
[20] Benet, LZ., Broccatelli, F., & Oprea, TI. BDDCS applied to over 900 drugs. AAPS J. 2011, 13, 519–547.
[21] Broccatelli, F., Cruciani, G., Benet, LZ., & Oprea, TI. BDDCS predictions for new molecular entities. Mol. Pharm. 2012, 9, 570–580.
[22] Charkoftaki, G., Dokoumetzidis, A., Valsami, G., & Macheras, P. Elucidating the role of dose in the biopharmaceutics classification of drugs: The concept of critical dose, effective in vivo solubility, and dose dependent BCS. Pharm. Res. 2012, 29(11), 3188–3198.
[23] Butler, JM., & Dressman, JB. The developability classification system: Application of biopharmaceutics concepts to formulation development. J. Pharm. Sci. 2010, 99(12), 4940–4954.
[24] WHO Model List of Essential Medicines, 20th list, March 2017, Amended August 2017 https://apps.who.int/iris/bitstream/handle/10665/273826/EML-20-eng.pdf?ua=1 (Accessed on 06 April 2019).
[25] Klein, S. The Use of Biorelevant Dissolution Media to Forecast the In Vivo Performance of a Drug. AAPS 2010, 12(3), 397–406.
[26] US Food and Drug Administration Dissolution Testing and Specification Criteria for Immediate-Release Solid Oral Dosage Forms Containing Biopharmaceutics Classification System Class 1 and 3 Drugs Guidance for Industry, U.S. Department of Health and Human Services Food and Drug Administration Center for Drug Evaluation and Research (CDER), August, 2015.
[27] Weinandy, L. Variability in Dissolution Testing Results from a Collaborative Study. Mainz, 05 October 2005. https://www.tu-braunschweig.de/Medien-DB/pharmchem/dphg_2005_weinandy.pdf. (Accessed on 09 January 2018).
[28] Grady, H., Elder, DP., Webster, GK., Mao, Y., Lin, Y., Flanagan, T., Mann, J., Blanchard, A., Cohen, MJ., Lin, J., Kesisoglou, F., Hermans, A., Abend, A., Zhang, L., & Curran, D. Industry's View on Using Quality Control, Biorelevant and Clinically Relevant Dissolution Tests for Pharmaceutical Development, Registration and Commercialization. J. Pharm. Sci. 2018. http://dx.doi.org/10.1016/j.xphs.2017.10.019.
[29] United States Pharmacopeia. <1092>. The dissolution procedure: Development and validation, 2015, 1090–1096.
[30] United States Pharmacopeia. <711>. Dissolution, 2015, 486–495.
[31] Symonds, J., & Elder, DP. Dissolution testing in the modern world. Eur. Pharm. Rev. 2016, 21(1), 40–43.

[32] Higuchi, M., Yoshihashi, Y., Tarada, K., & Sugano, K. Minimum rotation speed to prevent coning phenomena in compendium paddle dissolution apparatus. Eur. J. Pharm. Sci. 2014, 65, 74–78.
[33] Strauch, S., Jantratid, E., Stahl, M., Rägo, L., & Dressman, JB. The Biowaiver Procedure: Its Application to Antituberculosis Products in the WHO Prequalification Programme. J. Pharm. Sci. 2011, 100, 822–830.
[34] Yazdanian, M., Briggs, K., Jankovsky, C., & Hawi, A. The "high solubility" definition of the current FDA guidance on Biopharmaceutical Classification System may be too strict for acidic drugs. Pharm. Res. 2014, 21, 293–299.
[35] Rinaki, E., Dokoumetzidis, A., & Macheras, P. The mean dissolution time depends on the dose/solubility ratio. Pharm. Res. 2003, 20, 406–408.
[36] Dokoumetzidis, A., & Macheras, P. A century of dissolution research: From Noyes and Whitney to the biopharmaceutics classification system. Int. J. Pharm. 2006, 321, 1–11.
[37] Polli, JE., Cook, JA., Davit, BM., Dickinson, PA., Argenti, D., Barbour, N., García-Arieta, A., Geoffroy, JM., Hartauer, K., Li, S., Mitra, A., Muller, FX., Purohit, V., Sanchez-Felix, M., Skoug, JW., & Tang, K. Summary workshop report: Facilitating oral product development and reducing regulatory burden through novel approaches to assess bioavailability/bioequivalence. AAPS J. 2012, 14(3), 627–638.
[38] Strauch, S., Dressman, JB., Shah, VP., Kopp, S., Polli, JE. et al. Biowaiver Monographs for Immediate-Release Solid Oral Dosage Forms: Quinine Sulfate. J. Pharm. Sci. 2001, 101(2), 499–508.
[39] Davit, BM., Kanfer, I., Tsang, YC., & Cardot, J-M. BCS biowaivers. Similarities and differences among EMA, FDA and WHO requirements. AAPS J. 2016, 18(3), 612–618.
[40] Kasim, NA., Whitehouse, M., Ramachandran, C., Bermejo, M., Lennernas, H. et al. Molecular properties of WHO essential drugs and provisional biopharmaceutical classification. Mol. Pharm. 2004, 1(1), 85–96.
[41] Salako, LA., & Sowunmi, A. Disposition of quinine in plasma, red blood cells and saliva after oral and intravenous administration to healthy adult Africans. Eur. J. Clin. Pharmacol. 1992, 42(2), 171–174.
[42] Paintaud, G., Alvan, G., & Ericsson, O. The reproducibility of quinine bioavailability. Br. J. Clin. Pharmacol. 1993, 35(3), 305–307.
[43] White, NJ. Clinical pharmacokinetics of antimalarial drugs. Clin. Pharmacokinet 1985, 10(3), 187–215.
[44] Moffat, AC., Osselton, MD., Widdop, B., & Galichet, LY. Clarke's analysis of drugs and poisons, 3rd ed, Great Britain: Pharmaceutical Press, London, 2004.
[45] Dahan, A., Wolk, O., Zur, M., Amidon, GL., Abrahamsson, B., Cristofoletti, R., Groot, DW., Kopp, S., Langguth, P., Polli, JE., Shah, VP., & Dressman, JB. Biowaiver Monographs for Immediate-Release Solid Oral Dosage Forms: Codeine Phosphate. J. Pharm. Sci. 2014, 103(6), 1592–1600.
[46] Klein, S., Stippler, E., Wunderlich, M., & Dressman, J. Development of dissolution tests on the basis of gastrointestinal physiology, Dressman J, Kraemer J, Eds., Pharmaceutical Dissolution Testing, Taylor & Francis, Boca Raton, Fl., 2005, 193–228.
[47] Galia, E., Horton, J., & Dressman, JB. Albendazole generics – A comparative in vitro study. Pharm. Res. 1999, 16(12), 1871–1875.
[48] Charman, WN., Rogge, MC., Boddy, AW., & Berger, BM. Effect of food on a monoglyceride emulsion formulation on danazol bioavailability. J. Clin. Pharmacol. 1993, 33(4), 381–386.
[49] Nomenclature (as excerpted from the Orange Book). https://www.fda.gov/ohrms/dockets/ac/05/briefing/2005-4137B1_07_Nomenclature.pdf. (Accessed on 13 January 2018).
[50] Prescribing Information. PAXIL® (paroxetine hydrochloride). Tablets and Oral Suspension. https://www.accessdata.fda.gov/drugsatfda_docs/label/201/020031s067,020710s031.pdf. (Accessed on 13 January 2018).

[51] Prescribing Information. PEXEVA® Brand of (paroxetine mesylate) tablets. https://www.accessdata.fda.gov/drugsatfda_docs/label/201021299s024lbl.pdf. (Accessed on 13 January 2018).
[52] Yang, Y., Faustino, PJ., Volpe, DA., Ellison, CD., Lyon, RC., & Yu, LX. Biopharmaceutical classification of selected β-blockers: Solubility and permeability class membership. Mol. Pharm. 2007, 4(4), 608–614.
[53] SUPAC IR Immediate release solid oral dosage form scale-up and post-approval changes: chemistry, manufacturing and controls, in vitro dissolution testing, and in vivo bioequivalence documentation, Center for Drug Evaluation and Research (CDER), November, 1995.
[54] Garcia-Aríeta, A. Interactions between active pharmaceutical ingredients and excipients affecting bioavailability: Impact on bioequivalence. Eur. J. Pharm. Sci. 2014, 65, 89–97.
[55] Conner, DP. ANDFA bioequivalence studies that fail to meet FDA's current bioequivalence criteria. https://www.fda.gov/ohrms/dockets/ac/00/slides/3657s2_08.ppt. (Accessed on 13 January 2018).
[56] Cabaleiro, T., Ochoa, D., Roman, M., Moreno, I., Lopez-Rodrıguez, R., Jesus, Novalbos., & Francisco, Abad-Santos. Polymorphisms in CYP2D6 have a Greater Effect on Variability of Risperidone Pharmacokinetics than Gender. BCPT 2015, 116, 124–128.
[57] Parr, A., Hidaldo, IJ., Bode, C., Brown, W. et al. The Effect of Excipients on the Permeability of BCS Class III Compounds and Implications for Biowaivers. Pharm. Res. 2016, 33(1), 167–176.
[58] Hofsäss, MA., E de Souza, J., Silva-Barcellos, NM., Bellavinha, KR., Abrahamsson, B., Cristofoletti, R., Groot, DW., Parr, A., Langguth, P., Polli, JE., Shah, VP., Tajiri, T., Mehta, MU., & Dressman, JB. Biowaiver Monographs for Immediate-Release Solid Oral Dosage Forms: Folic Acid. J. Pharm. Sci. 2017, 106(12), 3421–3430.
[59] Kubbinga, M., Langguth, P., & Barends, D. Risk analysis in bioequivalence and biowaiver decisions. Biopharm. Drug Dispos. 2013, 34, 254–261.

Appendix 1

Table 11A.1: Summary of FIP Biowaiver Monographs for Essential Drugs.The classifications mentioned in the biowaiver monographs are based on the regulatory specifications relevant at the time of monograph publications.

Drug	BCS Assessment	References
Acetaminophen (Paracetamol)	BCS class III (borderline BCS Class I) Acetaminophen is a suitable candidate for biowaiver with certain caveats (see [a]).	[a]
Acetazolamide	Ambiguous, no BCS classification possible. Available data on solubility, on oral absorption and permeability are not sufficiently conclusive for classification. Acetazolamide is not a suitable candidate for biowaiver.	[b]
Acetylsalicylic Acid	BCS class I. Hydrolytic degradation over the timeframe of the solubility determination is the probable reason behind the inconsistency among the solubility data reported in literature [c]. Acetylsalicylic acid is a suitable candidate for biowaiver.	[c]

Table 11A.1 (continued)

Drug	BCS Assessment	References
Aciclovir	Ambiguous, no BCS classification possible. Different territories have different highest tablet strengths on their markets, which would lead to different BCS classifications in those countries. The risk of an aciclovir drug product being bioinequivalent appears to be very low. Aciclovir is a suitable candidate for biowaiver for doses ≤ 800 mg (under new guidelines).	[d]
Amitriptyline HCl	Ambiguous, BCS I/II. Solubility data is lacking to unambiguously assign BCS class. Amitriptyline HCl is a suitable candidate for biowaiver with certain caveats (see [e]).	[e]
Amodiaquine HCl	Ambiguous BCS class III (WHO, EMA) and BCS class IV (FDA) Applying the principles of BDDCS, its extensive metabolism in vivo (2% excretion of unchanged drug through urine) and a positive food effect would likely support a high permeability classification. Amodiaquine HCl products (as single API) can be considered for a biowaiver only if both test and comparator products state the absence of a food effect.	[f]
Amoxicillin trihydrate	Ambiguous BCS I, II, IV. Amoxicillin doses up to 875 mg (BCS class I), 1000 mg (BCS class II) and doses > 1000 (BCS class IV). Amoxicillin trihydrate biowaivers are applicable for doses of ≤ 875 mg.	[g]
Atenolol	BCS Class III. Atenolol is a suitable candidate for granting a biowaiver (under new guidelines).	[h]
Bisoprolol fumarate	BCS I. Bisoprolol fumarate is a suitable candidate for granting a biowaiver. Other salts would be applicable for a biowaiver with certain caveats (see [i]).	[i]
Chloroquine HCl, Chloroquine phosphate Chloroquine sulfate	BCS class I. All designated salts of chloroquine (HCl, phosphate, sulfate) are suitable candidates for granting a biowaiver.	[j]
Cimetidine	BCS class III. Cimetidine is a suitable candidate for granting a biowaiver (under new guidelines).	[k]
Ciprofloxacin HCl	BCS class IV. Ciprofloxacin HCl is not a suitable candidate for biowaiver.	[l]

Table 11A.1 (continued)

Drug	BCS Assessment	References
Codeine phosphate	BCS class I. Codeine phosphate is a suitable candidate for biowaiver.	[m]
Diclofenac sodium Diclofenac potassium	BCS Class II. Diclofenac salts (Na, K) are typically not suitable candidates for biowaivers. However, these so called BCS IIa compounds (weak anions with good intestinal solubility) may be granted biowaivers under WHO guidance, with certain caveats (see [n]).	[n]
Doxycycline hyclate	BCS class I. Doxycycline hyclate is a suitable candidate for biowaiver.	[o]
Efavirenz	Ambiguous BCS Class II/IV. There is inconclusive permeability data. Efavirenz has been classified as BCS Class II on the basis of its metabolic and solubility behaviour using BDDCS approach [q]. Efavirenz is a suitable candidate for biowaiver.	[p]
Enalapril maleate	BCS Class III. Enalapril is a prodrug and is hydrolysed by carboxylesterases to enalaprilat. Only 60%-70% of an orally administered dose of enalapril is absorbed from the GI tract. Enalapril is a suitable candidate for biowaiver, with appropriate caveats (see [r]). Other enapril salts, that is, Na must be evaluated on a case-by-case basis.	[r]
Ethambutol Dihydrochloride	BCS Class I/III Ethambutol dihydrochloride is classified as BCS Class I using the BDDCS approach [q]. Ethambutol dihydrochloride is a suitable candidate for biowaiver, with appropriate caveats (see [s]).	[s]
Fluconazole	BCS Class II (Forms II/III). The risk of bio-inequivalence because of formulation or manufacturing considerations is deemed to be low. Fluconazole is a suitable candidate for biowaiver, with appropriate caveats (see [t]).	[t]
Folic acid	BCS IV. Small doses of folic acid (≤320 µg) are absorbed completely via active transport; however, permeability data for higher doses of 1-5 mg are inconclusive. Folic acid is not a suitable candidate for biowaiver.	[u]
Furosemide	BCS IV. Furosemide is not a suitable candidate for biowaiver.	[v]

Table 11A.1 (continued)

Drug	BCS Assessment	References
Ibuprofen	BCS Class II. BCS class II drugs are typically not suitable candidates for biowaivers. However, these so called BCS IIa compounds (weak anions with good intestinal solubility) may be granted biowaivers under WHO guidance. Ibuprofen is a suitable candidate for biowaiver, with certain caveats (see [w]).	[w]
Isoniazid	Ambiguous BCS Class I/III. Dependent on the definition of permeability, isoniazid is borderline. Isoniazid is classified as BCS 1 using the BDDCS approach, as it is extensively metabolized [q]. Isoniazid is a suitable candidate for biowaiver, with appropriate caveats (see [x]).	[x]
Ketoprofen	BCS Class II. BCS class II drugs are typically not suitable candidates for biowaivers. However, these so called BCS IIa compounds (weak anions with good intestinal solubility) may be granted biowaivers under WHO guidance, with certain caveats (see [y]).	[y]
Lamuvidine	BCS Class III. Lamuvidine is also classified as BCS III using the BDDCS approach [q]. No cases of bioinequivalent lamivudine formulations have been reported in the literature. A biowaiver is judged acceptable for any new lamivudine multisource product with certain caveats (see [z]).	[z]
Levetiracetam	BCS Class I. Levetiracetam is a suitable candidate for biowaiver.	[aa]
Levofloxacin	BCS Class I. Levofloxacin is a suitable candidate for biowaiver.	[bb]
Mefloquine HCl	Ambiguous, no BCS classification possible. Mefloquine hydrochloride is not a highly soluble API and no data on permeability are available, as such it cannot be classified. Mefloquine HCl is not a suitable candidate for biowaiver.	[cc]
Metoclopramide HCl	BCS Class III. Metoclopramide HCl is a suitable candidate for biowaiver, with appropriate caveats (see [dd]).	[dd]
Metronidazole	BCS Class I. Metronidazole is a suitable candidate for biowaiver.	[ee]
Nifedipine	BCS Class II. Nifedipine is not a suitable candidate for biowaiver.	[ff]

Table 11A.1 (continued)

Drug	BCS Assessment	References
Piroxicam	BCS Class II. BCS class II drugs are typically not suitable candidates for biowaivers. However, these so called BCS IIa compounds (weak anions with good intestinal solubility) may be granted biowaivers under WHO guidance, with certain caveats (see [gg]).	[gg]
Prednisolone	Ambiguous, no BCS classification possible. Ambiguous, no BCS classification possible. Available data on solubility, on oral absorption and permeability are not sufficiently conclusive for classification (borderline BCS Class I). Available evidence indicates that a biowaiver for IR solid oral dosage forms would be unlikely to expose patients to undue risks. Prednisolone is a suitable candidate for biowaiver with certain caveats (see [hh]).	[hh]
Prednisone	Ambiguous, no BCS classification possible. Available data on solubility, on oral absorption and permeability are not sufficiently conclusive for classification (borderline BCS Class I). Available evidence indicates that a biowaiver for IR solid oral dosage forms would be unlikely to expose patients to undue risks. Prednisone is a suitable candidate for biowaiver with certain caveats (see [ii]).	[ii]
Primaquine diphosphate	BCS Class 1. Primaquine diposphate is a suitable candidate for biowaiver.	[jj]
Propranolol HCl	BCS class I Propranolol HCl is a suitable candidate for granting a biowaiver.	[h]
Pyrazinamide	BCS Class III. Linear absorption over a wide dosing range, indicating that permeability is unlikely to affect absorption. Pyrazinamide can be classified as a narrow therapeutic index (NTI) drug, which is usually a caveat for biowaivers. Pyrazinamide is a suitable candidate for biowaiver with certain caveats (see [kk), and importantly only if the SmPC of the test product indicates the need for monitoring of the patient's liver function.	[kk]
Quinidine sulfate	BCS Class I/III. Biowaivers are rarely granted for drugs, such as quinidine sulfate, with narrow therapeutic windows. Quinidine sulfate is not a suitable candidate for biowaiver.	[ll]

Table 11A.1 (continued)

Drug	BCS Assessment	References
Quinine sulfate	BCS Class I/II. A recent pharmacokinetic study did show bio-inequivalence of two products and it does show dose-related and certain irreversible side effects and toxicities at concentrations slightly above the therapeutic concentration range. Quinine sulfate is not a suitable candidate for biowaiver.	[mm]
Ranitidine HCl	BCS Class III. There are no safety considerations that would prevent a biowaiver approach. Ranitidine HCl is a suitable candidate for biowaiver.	[nn]
Ribavarin	BCS Class I/III. Ribavarin is actively transported across the intestinal lumen, but this system is saturatable. However, no excipients have been found to compete for the transporter (hCNT2) and there are no reported issues of bio-inequivalence. Nonetheless, biowaivers are rarely granted for drugs, such as ribavarin, with narrow therapeutic windows. Ribavarin is a not suitable candidate for biowaiver; however, this should be assessed versus the considerable risks associated with studying BE of ribavirin products in healthy subjects.	[oo]
Rifampicin	BCS Class II. There are many literature reports of bio-inequivalence, but the reasons are not well understood. Rifampicin is not a suitable candidate for biowaiver.	[pp]
Stavudine	BCS Class I. Stavudine is a suitable candidate for biowaiver.	[qq]
Verapamil HCl	Ambiguous BCS I, II If the solubility boundaries between pH 1.2 and pH 6.8 are used this is a BCS class I compound. Verapamil HCl is a suitable candidate for granting a biowaiver.	[h]
Zidovudine Azidothymidine	BCS Class I. A high percentage of failed bioequivalence studies have been reported in the literature (mostly in fixed dose combination products). Whether it is the influence of the other actives is still unclear. Nonetheless, in vitro dissolution does appear to be predictive of these failures. Zidovudine (Azidothymidine) is a suitable candidate for biowaiver with certain caveats (see [rr]).	[rr]

Additional References (Appendix 1)

[a] Kalantzi L, Reppas C, Dressman JB, Amidon GL, Junginger HE, Midha KK, Shah VP, Stavchansky SA Barends DM.Biowaiver Monographs for Immediate Release Solid Oral Dosage Forms: Acetaminophen (Paracetamol). J. Pharm. Sci. 2006, 95(1), 4-14.

[b] Granero GE, Longhi MR, Becker C, Junginger HE, Kopp S, Midha KK, Shah VP, Stavchansky S, Dressman JB, Barends DM. Biowaiver Monographs for Immediate Release Solid Oral Dosage Forms: Acetazolamide. J. Pharm. Sci. 2008, 97(9), 3691-3699.

[c] Dressman JB, Nair A, Abrahamsson B, Barends DM, Groot DW, Kopp S, Langguth P, Polli JE, Shah VP, Zimmer M. Biowaiver Monograph for Immediate-Release Solid Oral Dosage Forms: Acetylsalicylic Acid. J. Pharm. Sci. 2012, 101(8), 2653-2667.

[d] Arnal J, Gonzalez-Alvarez I, Bermejo M, Amidon GL, Junginger HE, Kopp S, Midha KK, Shah VP, Stavchansky S, Dressman JB, Barends DM. Biowaiver Monographs for Immediate Release Solid Oral Dosage Forms: Aciclovir. J. Pharm. Sci. 2008, 97(12), 5061-5073.

[e] Manzo RH, Olivera ME, Amidon GL, Shah VP, Dressman JB, Barends DM. Biowaiver Monographs for Immediate Release Solid Oral Dosage Forms: Amitriptyline Hydrochloride. J Pharm Sci 2006, 95(5), 966-973.

[f] Nair A, Abrahamsson B, Barends DM, Groot DW, Kopp S, Polli JE, Shah VP, Dressman JB. Biowaiver Monographs for Immediate Release Solid Oral Dosage Forms: Amodiaquine Hydrochloride. J. Pharm. Sci. 2012, 101(12), 4390-4401.

[g] Thambavita D, Galappatthy P, Mannapperuma U, Jayakody L, Cristofoletti R, Abrahamsson B, Groot DW, Langguth P, Mehta M, Parr A, Polli JE, Shah VP, Dressman JB. Biowaiver Monograph for Immediate-Release Solid Oral Dosage Forms: Amoxicillin Trihydrate. J. Pharm. Sci. 2017, 106(10), 2930-2945.

[h] Vogelpoel H, Welink J, Amidon GL, Junginger HE, Midha KK, Möller H, Olling M, Shah VP, Barends DM. Biowaiver Monographs for Immediate Release Solid Oral Dosage Forms Based Biopharmaceutics Classification System (BCS) Literature Data: Verapamil Hydrochloride, Propranolol Hydrochloride, and Atenolol. J. Pharm. Sci. 2004, 93(8), 1945-1956.

[i] Charoo NA, Shamsher AA, Lian LY, Abrahamsson B, Cristofoletti R, Groot DW, Kopp S, Langguth P, Polli J, Shah VP, Dressman J. Biowaiver Monograph for Immediate-Release Solid Oral Dosage Forms: Bisoprolol Fumarate. J Pharm Sci 2014, 103(2), 378-391.

[j] Verbeeck RK, Junginger HE, Midha KK, Shah VP, Barends DM. Biowaiver Monographs for Immediate Release Solid Oral Dosage Forms Based on Biopharmaceutics Classification System (BCS) Literature Data: Chloroquine Phosphate, Chloroquine Sulfate, and Chloroquine Hydrochloride. J. Pharm. Sci. 2005, 94(7), 1389-1395.

[k] Jantratid E, Prakongpan S, Dressman JB, Amidon GL, Junginger HE, Midha KK, Barends DM. Biowaiver Monographs for Immediate Release Solid Oral Dosage Forms: Cimetidine. J. Pharm. Sci. 2006, 95(5), 974-984.

[l] Olivera ME, Manzo RH, Junginger HE, Midha KK, Shah VP, Stavchansky S, Dressman JB, Barends DM. Biowaiver Monographs for Immediate Release Solid Oral Dosage Forms: Ciprofloxacin Hydrochloride. J. Pharm. Sci. 2010, 100(1), 22-33.

[m] Dahan A, Wolk O, Zur M, Amidon GL, Abrahamsson B, Cristofoletti R, Groot DW, Kopp S, Langguth P, Polli JE, Shah VP, Dressman JB. Biowaiver Monographs for Immediate-Release Solid Oral Dosage Forms: Codeine Phosphate. J. Pharm. Sci. 2014, 103(6), 1592-1600.

[n] Chuasuwan B, Binjesoh V, Polli JE, Zhang H, Amidon GL, Junginger HE, Midha KK, Shah VP, Stavchansky S, Dressman JB, Barends DM. Biowaiver Monographs for Immediate Release Solid Oral Dosage Forms: Diclofenac Sodium and Diclofenac Potassium. J. Pharm. Sci. 2009, 98(4), 1206-1219.

[o] Jantratid E, Strauch S, Becker C, Dressman JB, Amidon GL, Junginger HE, Kopp S, Midha KK, Shah VP, Stavchansky S, Barends DM. Biowaiver Monographs for Immediate Release Solid Oral Dosage Forms: Doxycycline Hyclate. J. Pharm. Sci. 2009, 99(4), 1639-1653.
[p] Cristofoletti R, Nair A, Abrahamsson B, Groot DW, Kopp S, Langguth P, Polli JE, Shah VP, Dressman JB. Biowaiver Monographs for Immediate Release Solid Oral Dosage Forms: Efavirenz. J. Pharm. Sci. 2013, 102(2), 318-329.
[q] Benet LZ, Broccatelli F. Oprea TI BDDCS applied to over 900 drugs. AAPS J. 20111, 13(4), 519–547.
[r] Verbeeck RK, Kanfer I, Löbenberg R, Abrahamsson B, Cristofoletti R, D.W. Groot, Langguth P, Polli JE, Parr A Shah VP, Mehta M. Biowaiver Monographs for Immediate-Release Solid Oral Dosage Forms: Enalapril. J. Pharm. Sci. 2017, 106(8), 1933-1943.
[s] Becker C, Dressman JB, Amidon GL, Junginger HE, Kopp S, Midha KK, Shah VP, Stavchansky S, Barends DM. Biowaiver Monographs for Immediate Release Solid Oral Dosage Forms: Ethambutol Dihydrochloride. J. Pharm. Sci. 2008, 97(4), 1350-1360.
[t] Charoo N, Cristofoletti R, Graham A, Lartey P, Abrahamsson B, Groot DW, Kopp S, Langguth P, Polli JE, Shah VP, Dressman JB. Biowaiver Monographs for Immediate-Release Solid Oral Dosage Forms: Fluconazole. J. Pharm. Sci. 2014, 103(12), 3843-3858.
[u] Hofsäss MA, e de Souza J, Silva-Barcellos NM, Bellavinha KR, Abrahamsson B, Cristofoletti R, Groot DW, Parr A, Langguth P, Polli JE, Shah VP, Tajiri T, Mehta MU, Dressman JB. Biowaiver Monographs for Immediate-Release Solid Oral Dosage Forms: Folic Acid. J. Pharm. Sci. 2017, 106(12), 3421-3430.
[v] Granero GE, Longhi MR, Mora MJ, Junginger HE, Midha KK, Shah VP, Stavchansky S, Dressman JB, Barends DM. Biowaiver Monographs for Immediate Release Solid Oral Dosage Forms: Furosemide. J. Pharm. Sci. 2010, 99(6), 2544-2556.
[w] Potthast H, Dressman JB, Junginger HE, Midha KK, Oeser H, Shah VP, Vogelpoel H, Barends DM. Biowaiver Monographs for Immediate Release Solid Oral Dosage Forms: Ibuprofen. J. Pharm. Sci. 2005, 94(10), 2121-2131.
[x] Becker C, Dressman JB, Amidon GL, Junginger HE, Kopp S, Midha KK, Shah VP, Stavchansky S, Barends DM. Biowaiver Monographs for Immediate Release Solid Oral Dosage Forms: Isoniazid. J. Pharm. Sci. 2007, 96(3), 522-531.
[y] Shohin IE, Kulinich JI, Ramenskaya GV, Abrahamsson B, Kopp S, Langguth P, Polli JE, Shah VP, Groot DW, Barends DM, Dressman JB. Biowaiver Monographs for Immediate-Release Solid Oral Dosage Forms: Ketoprofen. J. Pharm. Sci. 2012, 101(10), 593–3603.
[z] Strauch S, Jantratid E, Dressman JB, Junginger HE, Kopp S, Midha KK, Shah VP, Stavchansky S, Barends DM. Biowaiver Monographs for Immediate Release Solid Oral Dosage Forms: Lamivudine. J. Pharm. Sci. 2011, 100(6), 2054–2063.
[aa] Petruševska M, Berglez S, Krisch I, Legen I, Megušar K, Peternel L, Abrahamsson B, Cristofoletti R, Groot DW, Kopp S, Langguth P, Mehta M, Polli JE, Shah VP, Dressman JB. Biowaiver Monographs for Immediate Release Solid Oral Dosage Forms: Levetiracetam. J. Pharm. Sci. 2015, 104(9), 2676-2687.
[bb] Koeppe MO, Cristofoletti R, Fernandes EF, Storpirtis S, Junginger HE, Kopp S, Midha KK, Shah VP, Stavchansky S, Dressman JB, Barends DM. Biowaiver Monographs for Immediate Release Solid Oral Dosage Forms: Levofloxacin. J. Pharm. Sci. 2011, 100(5), 1628–1636.
[cc] Strauch S, Jantratid E, Dressman JB, Junginger HE, Kopp S, Midha KK, Shah VP, Stavchansky S, Barends DM. Biowaiver Monographs for Immediate Release Solid Oral Dosage Forms: Mefloquine Hydrochloride. J. Pharm. Sci. 2011, 100(1), 11-21.
[dd] Stosik AG, Junginger HE, Kopp S, Midha KK, Shah VP, Stavchansky S, Dressman JB, Barends DM. Biowaiver Monographs for Immediate Release Solid Oral Dosage Forms: Metoclopramide Hydrochloride. J. Pharm. Sci. 2008, 97(9), 3700–3708.

[ee] Rediguieri CF, Porta V, G Nunes DS, Nunes TM, Junginger HE, Kopp S, Midha KK, Shah VP, Stavchansky S, Dressman JB, Barends DM. Biowaiver monographs for immediate release solid oral dosage forms: metronidazole. J. Pharm. Sci. 2011, 100(5), 1618-1627.
[ff] Gajendran J, Krämer J, Shah VP, Langguth P, Polli J, Mehta M, Groot DW, Cristofoletti R, Abrahamsson B, Dressman JB. Biowaiver Monographs for Immediate-Release Solid Oral Dosage Forms: Nifedipine. J. Pharm. Sci. 2015, 104(10), 3289–3298.
[gg] Shohin IE, Kulinich JI, Ramenskaya GV, Abrahamsson B, Kopp S, Langguth P, Polli JE, Shah VP, Groot DW, Barends DM, Dressman JB. Biowaiver Monographs for Immediate Release Solid Oral Dosage Forms: Piroxicam. J. Pharm. Sci. 2014, 103(2), 367–377.
[hh] Vogt M, Derendorf H, Krämer J, Junginger HE, Midha KK, Shah VP, Stavchansky S, Dressman JB, Barends DM. Biowaiver Monographs for Immediate Release Solid Oral Dosage Forms: Prednisolone 2007, 96(1), 27-37.
[ii] Vogt M, Derendorf H, Krämer J, Junginger HE, Midha KK, Shah VP, Stavchansky S, Dressman JB, Barends DM. Biowaiver Monographs for Immediate Release Solid Oral Dosage Forms: Prednisone. J. Pharm. Sci. 2007, 96(6), 1480-1489.
[jj] Nair A, Abrahamsson B, Barends DM, Groot DW, Kopp S, Polli JE, Shah VP, Dressman JB. Biowaiver Monographs for Immediate-Release Solid Oral Dosage Forms: Primaquine Phosphate. J. Pharm. Sci. 2011, 101(3).
[kk] Becker C, Dressman JB, Amidon GL, Junginger HE, Kopp S, Midha KK, Shah VP, Stavchansky S, Barends DM. Biowaiver Monographs for Immediate Release Solid Oral Dosage Forms: Pyrazinamide. J. Pharm. Sci. 2008, 97(9), 3709–3720.
[ll] Strauch S, Dressman JB, Shah VP, Kopp S, Polli JE, Barends DM. Biowaiver Monographs for Immediate Release Solid Oral Dosage Forms: Quinidine Sulfate. J. Pharm. Sci. 2012, 101(2), 499-508.
[mm] Strauch S, Dressman JB, Shah VP, Kopp S, Polli JE, Barends DM. Biowaiver Monographs for Immediate-Release Solid Oral Dosage Forms: Quinine Sulfate. J. Pharm. Sci. 2012, 101(2), 499–508.
[nn] Kortejärvi H, Yliperttula M, Dressman JB, Junginger HE, Midha KK, Shah VP, Barends DM. Biowaiver Monographs for Immediate Release Solid Oral Dosage Forms: Ranitidine Hydrochloride. J. Pharm. Sci. 2005, 94(8), 1617-1625.
[oo] Goodarzi N, Barazesh Morgani A, Abrahamsson B, Cristofoletti R, Groot DW, Langguth P, Mehta MU, Polli JE, Shah VP, Dressman JB. Biowaiver Monographs for Immediate Release Solid Oral Dosage Forms: Ribavirin. J. Pharm. Sci. 2016, 105(4), 1362-1369.
[pp] Becker C, Dressman JB, Junginger HE, Kopp S, Midha KK, Shah VP, Stavchansky S, Barends DM. Biowaiver Monographs for Immediate Release Solid Oral Dosage Forms: Rifampicin. J. Pharm. Sci. 2009, 98(7), 2252-2267.
[qq] Silva ALL, CristoFoletti R, Storpirtis S, Sousa VD, Junginger HE,Shah VP, Stavchansky S, Dressman JB, Barends DM. Biowaiver Monographs for Immediate-Release Solid Oral Dosage Forms: Stavudine. J. Pharm. Sci. 2012, 101(1), 10-16.
[rr] Soares KC, Rediguieri CF, Souza J, Serra CH, Abrahamsson B, Groot DW, Kopp S, Langguth P, Polli JE, Shah VP, Dressman J. Biowaiver Monographs for Immediate-Release Solid Oral Dosage Forms: Zidovudine (Azidothymidine). J. Pharm. Sci. 2013, 102(8), 2409-2423.

List of Abbreviations

ADAM	Advanced Dissolution, Absorption, and Metabolism
ADME	Absorption, Distribution, Metabolism, and Excretion
ADMET	Absorption, Distribution, Metabolism, Excretion and Toxicity
AIDS	Acquired Immunodeficiency Syndrome
A_n	Absorption Number
ANN	Artificial Neural Network
ANDA	Abbreviated New Drug Application
API	Active Pharmaceutical Ingredient
ASD	Amorphous Solid Dispersion
AUC	Area under the Curve
BCS	Biopharmaceutical Classification System
BDDCS	Biopharmaceutics Drug Disposition Classification System
CD	Cyclodextrin
CheqSol	Chasing equilibrium Solubility
CLND	Chemiluminescence Nitrogen Detector
clogP	Calculated logP
C_{max}	Maximum concentration (in pharmacokinetic experiment)
CMC	Chemistry, Manufacturing and Control
COSMO-RS	Conductor like Screening Model for Real Solvents
CV	Coefficient of Variation
D/S	Dose/Solubility
D_0	Dose Number
DCS	Development Classification System
DDBCS	Dose-dependent BCS
DIPE	Diisopropylether
DMSO	Dimethylsulfoxide
D_n	Dissolution Number
DSC	Differential Scanning Calorimetry
EMA	European Medicines Agency
EtOAc	Ethylacetate
F_a	Fraction of Dose absorbed
FaSSCoF	Fasted State Simulated Colonic Fluid
FaSSGF	Fasted State Simulating Gastric Fluid
FaSSIF	Fasted State Simulated Intestinal Fluid
FDA	US Food and Drug Administration
FDM	Facilitated Dissolution Method
FeSSCoF	Fed State Simulated Colonic Fluid
FeSSIF	Fed State Simulated Intestinal Fluid
FIH	First in Human
FIP	International Pharmaceutical Federation
FL	Flux
FMEA	Failure Mode and Effects Analysis
GI	Gastrointestinal
GI-tract	Gastrointestinal tract
GSE	General Solubility Equation
HCF	Human Colonic Fluid
HGF	Human Gastric Fluid

https://doi.org/10.1515/9783110559835-012

HIF	Human Intestinal Fluid
HPβCD	Hydroxypropyl-β-Cyclodextrin
HPLC	High-Performance Liquid Chromatography
HTS	High-Throughput Screening
ICH	International Council for Harmonization
IMI	Innovative Medicines Initiative
IMI OrBiTo	Innovative Medicines Initiative Oral Biopharmaceutic Tools
iPrOH	Isopropanol
IR	Immediate Release
IUPAC	International Union of Pure and Applied Chemistry
IVIVC	in vitro– in vivo Correlations
IVIVR	in vitro– in vivo Relationships
K_a	Absorption Rate Constant
LBFs	Lipid-based Formulations
LFCS	Lipid Formulation Classification Scheme
LLE	Liquid–Liquid Equilibrium
LLP	Liquid–Liquid Phase
MAD	Maximum absorbable Dose
MDCK	Madin-Darby Canine Kidney
MIBK	Methyl isobutyl ketone
MLP	Multi-Layer Perceptron
MPt	Melting Point
MS	Mass Spectrometry
MW	Molecular Weight
NCE	New Chemical Entity
NDA	New Drug Application
NLT	Not less than
NRTL	Non-random two-Liquid
NRTL-SAC	Non-random two-Liquid Segment Activity Coefficient
NSAID	Non-steroidal anti-inflammatory Drug
OrBiTo	Oral Biopharmaceutics Tool
PBPK	Physiologically based Pharmacokinetic
PC-SAFT	Perturbed Chain-Statistical Associating Fluid Theory
PCB	Polychlorinated Biphenyl
PD	Pre-clinical Dose
PD_0	Pre-clinical Dose Number
Ph. Eur.	European Pharmacopoeia
PK	Pharmacokinetic
PLE	Product Line Extension
PTFE	Polytetrafluoroethylene
PXRD	Powder X-Ray Diffraction
QbD	Quality by Design
SAC-NRTL	Non-random two-Liquid Segment Activity Coefficient
SAD	Single ascending Dose
SBEβCD	Sulfobutylether-β-Cyclodextrin
SEDDS	Self-emulsifying Drug Delivery System
SFI	Solubility Forecast Index
SGF	Simulated Gastric Fluid
SGFsp	Simulated Gastric Fluid without Pepsin

SIF	Simulated Intestinal Fluid
SLAD	Solubility limited absorbable Dose
SLE	Solid–liquid Equilibrium
SLS	Sodium lauryl sulfate
sRNA	Small Ribonucleic Acid
SSF	Saturation Shake Flask
SUPAC	Scale-up and post-approval Change
TRIS	Tris(hydroxymethyl)aminomethane
UNIFAC	Universal quasi-chemical Functional Group Activity Coefficients
UNIQUAC	Universal Quasichemical
UPLC	Ultra-high-Pressure Liquid Chromatography
USP	United States Pharmacopoeia
UV	Ultraviolet
VLE	Vapour–Liquid Equilibrium
WHO	World Health Organization

Index

https://doi.org/10.1515/9783110559835-013